Praise for *The Human Swarm*

"This fine work should have broad appeal to anyone curious about human societies, which is basically everyone."

—Publishers Weekly, Starred Review

"Our times are filled with garage start-ups that become Silicon Valley behemoths overnight. Such scaling-up pales in comparison to humans going from hunter-gatherer bands to our globalized world in the blink of an evolutionary eye—and thus now, a stranger a continent away can be killed when we press a button operating a drone, or rescued when we press a button marked 'Donate now.' In *The Human Swarm*, Mark Moffett charts the science of this scaling up of human societies, and its unlikely evolutionary consequences. This highly readable book is ambitious in its interdisciplinary breadth, rigorous in its science, and deeply thought-provoking in its implications."

—Robert Sapolsky, author of Behave

"A tour-de-force."

—Donald Johanson, discoverer of the famous missing-link fossil
"Lucy" and founder of the Institute of Human Origins

"*The Human Swarm* is a book by a biologist that should fascinate any thoughtful reader and deserves to be taken seriously by psychologists and social scientists alike."

—Roy Baumeister, author of Willpower

"A truer statement hasn't been uttered: 'Our *groupiness* shapes human history.' Moffett's book is a wide-ranging, deeply interesting analysis of how large numbers of individual agents become a society. His travels in the world and through vast intellectual landscapes give him a unique view of why we are the way we are, both in our similarity to other living beings and our differences from them—our ability to include once alien groups into our own, for example. There is no other book I've read recently that made my neurons pop at the rate this book did."

—Mahzarin Banaji, author of Blindspot

"This is a book of amazing ideas, many of them counterintuitive. Mark Moffett's astounding stories of animal societies persuaded me that the future of human cities have been foretold by the ants. Read this manifesto if you like to have your mind changed."

—Kevin Kelly, founder of *Wired Magazine* and author of *The Inevitable*

"In the past quarter century, there has emerged a genre of Big History that includes such epic books as Jared Diamond's *Guns, Germs, and Steel*, Steven Pinker's *The Better Angels of Our Nature*, and Yuval Noah Harari's *Sapiens*. Mark Moffett's *The Human Swarm* is destined to be included in future lists of such books that not only add to our understanding of who we are, how we got here, and where we're going, but change our perspective of how we fit in the larger picture of life on Earth. A magisterial work of monumental importance."

—Michael Shermer, publisher of *Skeptic* magazine
and author of *The Moral Arc*

"Homo sapiens is a small-group social animal that physically seems to be limited to personal relationships with a few individuals. Nonetheless humanity is struggling to deal with societies of billions as human technologies now pose existential threats tied to those numbers. In *The Human Swarm*, Mark Moffett presents an intriguing overview of the biological roots and cultural evolution of this now-critical situation."

—Paul R. Ehrlich, author of *Human Natures*

"*The Human Swarm* is surely the most accurate, most comprehensive, most original explanation of our social existence that we're ever likely to see, one jaw-dropping revelation after another, most of them astonishing, all of them fascinating. It's true without question, which seems obvious as you're reading, and it's very well written—a joy to read."

—Elizabeth Marshall Thomas, author of *The Harmless People*

THE
HUMAN SWARM

THE
HUMAN SWARM

How Our Societies Arise,
Thrive, and Fall

Mark W. Moffett

BASIC BOOKS
New York

Basic Books
Hachette Book Group
1290 Avenue of the Americas, New York, NY 10104
www.basicbooks.com

Printed in the United States of America

First Edition: April 2019

Published by Basic Books, an imprint of Perseus Books, LLC, a subsidiary of Hachette Book Group, Inc. The Basic Books name and logo is a trademark of the Hachette Book Group.

The Hachette Speakers Bureau provides a wide range of authors for speaking events. To find out more, go to www.hachettespeakersbureau.com or call (866) 376-6591.

The publisher is not responsible for websites (or their content) that are not owned by the publisher.

Print book interior design by Jeff Williams.

The Library of Congress has cataloged the hardcover edition as follows:

Names: Moffett, Mark W., author.
Title: The human swarm : how our societies arise, thrive, and fall / Mark W. Moffett.
Description: New York : Basic Books, [2019] | Includes bibliographical references and index.
Identifiers: LCCN 2018038761 (print) | LCCN 2019006348 (ebook) | ISBN 9781541617292 (ebook) | ISBN 9780465055685 (hardcover)
Subjects: LCSH: Sociology. | Communities.
Classification: LCC HM585 (ebook) | LCC HM585 .M64 2019 (print) | DDC 301—dc23
LC record available at https://lccn.loc.gov/2018038761

ISBNs: 978-0-465-05568-5 (hardcover); 978-1-5416-1729-2 (ebook)

LSC-C

10 9 8 7 6 5 4 3 2 1

I dedicate this book to three remarkable individuals.

Firstly, my mentor Edward O. Wilson, out of awe for his poetic spirit, his decades of building connections across the sciences, and his indefatigable support of so many careers,

my off-kilter one included;

To the late and great Irv DeVore, who fostered critical thinking across generations of anthropologists and put up with hours of conversation with this biologist;

And to Melissa Wells, my splendid wife and partner,

who believes in me.

And what, I said, will be the best limit for our rulers to fix when they are considering the size of the State and the amount of territory which they are to include, and beyond which they will not go?

What limit would you propose?

I would allow the State to increase so far as is consistent with unity; that, I think, is the proper limit.

Very good, he said.

—PLATO, *The Republic*

Contents

Introduction

For as long as human societies have existed, people have found themselves changed, the members of those societies transformed in their mind's eye into exalted Human Beings. Yet as powerful as belonging to a society can be in raising the citizens' collective self-image, it's not their fellow members that they see most differently—the foreign undergo the more radical, at times dreadful, transformation. In each person's mind, entire groups of outsiders can turn into something less than human, a kind of vermin, even.

That foreigners might be considered contemptible enough to crush underfoot, like insects, is the stuff of history. Think back to the year 1854, Washington Territory. Seattle, chief of the Suquamish Tribe and namesake of the newly founded city, had just heard Isaac Stevens, the territory's recently appointed governor, speak before the tribal elders. Stevens had explained that the Suquamish were to be relocated to a reservation. Standing to respond, Seattle towered over the slight governor. Speaking in his native Duwamish, he bemoaned the abyss between their societies and recognized that the days of the Suquamish were numbered. Yet he was stoical about the news: "Tribe follows tribe, and nation follows nation, like the waves of the sea. It is the order of nature, and regret is useless."[1]

As a field biologist I make a living thinking about the order of nature. I've spent years contemplating the concept we call "society" while exploring human tribes and nations. I find myself endlessly captivated by the phenomenon of foreignness: the way it turns what objectively are minor differences into gulfs between people that have ramifications reaching into every corner of life, from ecology to politics. The goal of *The Human Swarm* is to take in as much of this broad sweep as possible by investigating the nature of the societies of *Homo sapiens* as well as those of other animals. A principle thesis of the book is that, uncomfortable as it may sound, human societies and the societies of insects are more similar than we might like to believe.

For humans, any little thing can signal foreignness, as I have experienced many times. Mortified souls looked on in India when I picked up my food with the incorrect hand. In Iran I tried to nod yes when to the locals a nod meant no. Sitting on moss in the New Guinea highlands, I watched *The Muppet Show* with an entire village on an ancient television run off a car battery. Knowing I came from America, and that *The Muppet Show* was American, every man and woman looked at me quizzically when a pig, a species they revere, waltzed on the screen in a dress and high heels. I talked myself past machine guns during the Tamil ethnic uprising in Sri Lanka and sweated as wary Bolivian bureaucrats sorted out who this very strange person was and what I was doing—or should be allowed to do—in their country. At home I have seen my fellow Americans behave with equal discomfort, bewilderment, and, at times, anger toward outsiders. In a primordial reaction, both sides think *how strange that person is* despite their profound similarities as human beings with two arms, two legs, and a desire for love, home, and family.

In *The Human Swarm* I examine society membership as a particular component of our sense of self, which should be considered (as I do in the final chapters especially) in tandem with race and ethnicity—identifications that can exert the same primacy and emotional draw. The elevated significance of our societies—and ethnicities and races—compared with other aspects of our identity can seem preposterous. Nobel-winning economist and philosopher Amartya Sen, for one, struggles to make sense of why people collapse their identities into groups that override everything else. Using the deadly conflicts in Rwanda as a case in point, Sen rues the fact that "a Hutu laborer from [the capital city of] Kigali may be pressured to see himself only as a Hutu and incited to kill Tutsis, and yet he is not only a Hutu, but also a Kigalian, a Rwandan, an African, a laborer and a human being."[2] These and other sorts of breakdowns are one theme of the chapters ahead. When convictions about what a society stands for or who belongs come into conflict, suspicion mounts and bonding fails.

The word "tribalism" comes to mind, denoting people drawn together by anything from a love of car racing to the denial of global warming.[3] The idea of tribe, used loosely in this way, is a common subject for bestseller lists. However, when we speak of the tribe of a

New Guinea highlander, or of tribalism with respect to our own connections to a society, we have in mind how a lifelong sense of belonging elicits love and loyalty—yet also, expressed in relation to outsiders, how it can promote hatred, devastation, and despair.

Before we get to these subjects we will attend to the most basic of questions, namely: what is a society? As we shall see, there's a major difference between being social—connecting positively with others—and the situation, far less common in nature, wherein a species supports the separate groups we call societies that endure over generations. Being part of a society is not open to choice; the individuals who count among its members are normally clear to all. Outsiders, with their foreignness unmistakable from their appearance and accents, gestures, and attitudes about everything from pigs to whether tipping is seen as an insult, are admitted with difficulty. And then, in many cases, they are fully embraced only after the passage of time—decades or centuries, even.

Aside from our families, our societies are the affiliations we most often pledge allegiance to, fight for, and die for.[4] But in day-to-day life the primacy of societies is seldom obvious, forming only a portion of our sense of self and our recognition of how others are different. As part of our daily experience, we join up with political parties, book clubs, poker groups, teen cliques. Those of us on the same tour bus may even bond together, thinking more highly of our travel companions than those on other buses for a time, and as a result perhaps working fruitfully as a group through the problems of the day.[5] A predisposition to join groups shapes us as individuals and has been a subject of extensive research. Meanwhile our society hums along, as easily overlooked as our heartbeat and breath. It comes to the fore, of course, in times of group hardship or pride. A war, terrorist attack, or the death of a leader can shape a generation. Yet even in uneventful times our society sets the tone of our days, influences our beliefs, and informs our experiences.

Pondering the sometimes insurmountable differences between societies—be they the swarming populations of continent-wide nations like the United States or local tribes in New Guinea—raises questions of utmost importance. Are societies, and the labeling of others as *foreign*, part of the "order of nature" and therefore unavoidable? Bound by a sense of superiority and vulnerable to the enmity of other groups, is each society doomed to flounder and fall as Seattle supposed, as a consequence of either skirmishes with other societies or a sense of alienation that percolates up among the members of the society itself?

The Human Swarm is my attempt to answer these questions. The argument moves from natural history to prehistory and the fickle course of civilizations—from the mud walls of Sumer to the electronic vastness of Facebook. Behavioral scientists tease apart human interactions in narrow contextual frames, for example by using games of strategy to illuminate how we treat each other. But I endeavor to take a broader approach. Understanding the origin, maintenance, and dissolution of societies—how necessary they are, how they come about, and why they matter—will take us through recent findings in biology, anthropology, and psychology, with some philosophy thrown in for good measure.

History plays into the narrative too, albeit more for the patterns it reveals than for its specifics. Each society has its own saga, yet I propose that there can be common underlying forces that both keep societies together and cause them to crash and burn. The fact is, whether through conquest, transmutation, assimilation, division, or death, all societies—animal or human, low-key hunter-gatherer or industrial powerhouse—reach an end. This impermanence is easily overlooked given their longevity is measured in human lifespans. Obsolescence is assured not by hostile neighbors or environmental ruin (though such factors have played a memorable role in the decline of some societies), nor by the fleeting lives of the people themselves, but rather by the transience of the identities the members present to each other and the world. Differences between people carry great weight, with changes slowly turning what was once the familiar into the foreign.

The human connection with societies has deep origins that extend into our animal past. However, the idea of describing societies in terms of membership, with its implications of ingroup and outgroup, which I adopt from psychology, is unconventional in biology. My colleagues routinely have an aversion, rarely mentioned explicitly, to talking about societies. As an example, even though the English vernacular offers words for the societies of many species (take, for instance, "troop" for monkeys and gorillas; "pack" for wolves and African wild, or painted, dogs; "clan" for spotted hyenas and meerkats; or "band" for horses), researchers often shun those terms and simply assign the term "group," with a consequent loss in clarity and understanding. Imagine sitting through a lecture, as I once did, in which an ecologist spoke about a group of monkeys that "separated into two groups" and later "one of the groups clashed with yet another group." It took feverish concentration to decipher those sentences: what he meant was that the members of a single monkey troop had headed in two directions, with half that troop coming upon and vigorously

defending themselves against another troop. While a troop is undeniably a group, it is a group of a very special sort, being set off from
all other monkeys by a closed and stable membership that makes it
not just worth fighting over, but worthy of being labeled by a term of
its own.

Once a group—a pack, clan, troop, pride, and so on—forges this
sort of singular identity that extends beyond the workaday ties of parents rearing offspring, being part of that society can offer a lot. What
characteristics do we share with those animals? How do we differ—and
more importantly, does it matter?

While animal models are helpful in illuminating the value of societies, they are insufficient to explain how humans got to where we
are today. As natural as our great nations seem to most of the world's
population, they are not necessary. Before the flowering of civilizations (by which I mean societies with cities and monumental architecture), we humans ranged across the habitable surface of Earth in
societies as well, though far smaller ones: in tribes reliant on simple
gardens and domesticated animals, or as hunter-gatherers taking every meal from the wild. Those societies were the nations of the day.
The forebears of every living person were once affiliated with them,
going back eons to a time when humans were all hunter-gatherers.
Indeed, many of the peoples of New Guinea, Borneo, South American
rainforests, sub-Saharan Africa, and other parts of the world retain
their primary connections to the few hundred or thousand individuals in a tribe that carries on for the most part independently of their
national government.

To characterize the earliest societies we can draw on evidence from
hunter-gatherers of recent centuries and the archaeological record.
The vast countries that now cause hearts to swell with pride would
have been unfathomable to our hunter-gatherer ancestors. We will
explore what made this transformation possible, leading to societies
that continue to discriminate against outsiders even though they have
become so numerous that most members are unknown to each other.
The casual anonymity that characterizes contemporary human societies may seem unremarkable, but it is a big deal. The seemingly trivial
act of entering a café full of strangers without a care in the world is
one of our species' most underappreciated accomplishments, and it
separates humankind from most other vertebrates with societies. The
fact that the animals of those species must be able to recognize each
individual in their society is a constraint most scientists have overlooked, but it explains why no lions or prairie dogs will ever erect

cross-continental kingdoms. Being comfortable around unfamiliar members of our society gave humans advantages from the get-go and made nations possible.

The united multitudes that compose modern human societies are unique in the chronicle of life for species larger than a fingernail. However, my training has been with creatures smaller than that: the social insects, exemplified (from a personal bias) by the ants. The idea behind this book entered my head when I encountered a battlefield kilometers long in a town near San Diego, where two supercolonies of the Argentine ant, each billions strong, defended their turfs. These Lilliputians initially led me, back in 2007, to the question of how a vast number of individuals, ant or human, can truly be a society. This book will address how, much like humans, ants respond to each other in such a way that their societies can be anonymous: we—and they—have no need to be familiar with one another as individuals to keep our societies distinct. This ability opened up the possibility for humans to transcend the size limitations of most other mammal societies, as was first seen in hunter-gatherer societies that grew into the many hundreds and in the end paved the way to history's grand republics.

How are anonymous societies realized? Our ant-like approach to identifying each other depends on shared characteristics that mark individuals as fellow members. But these markers—which in ants are simple chemicals and in humans can range from clothing to gestures and language—are insufficient to completely explain what holds civilizations together. The conditions favorable to expanding human societies, compared to those of small-brained ants, have been stringent and fragile. People draw from other epoch-tested skills in their mental toolset to make life tolerable in response to growing numbers of fellow members. The enhancement of differences between individuals by way of jobs and other distinctions (our groupiness) is part of the package. Perhaps more surprisingly, the emergence of inequalities, which I shall discuss especially as manifested in the rise of leaders, is likewise important to building a society's population. We take such phenomena for granted, but they varied remarkably among hunter-gatherers, some of whom lived in roving societies of equals.

The coexistence of various races and ethnicities within a society has largely come about since agriculture, an extension of the willingness, previously mentioned, to accept distinctions between individuals, including the authority of others. Such a union of formerly independent groups was unheard of among hunter-gatherer bands, and indeed is found in no other species. Nations would not have taken hold without people repurposing their cognitive survival tools to take in, and adjust to, different ethnic groups. This allowance for diversity comes with stressors that can ultimately strengthen a society, but may also tear it apart. So while the success of the melting pot is good news, only so much melting goes on, with the issue of we-ness at the root of riots, ethnic cleansings, and holocausts.

Throughout, my aim is to intrigue you with mysteries, some of consequence, others odd but enlightening. For a preview, elephants of the African savannas have societies while Asian elephants lack them. We will tackle the curious issue of why, given that humans are so closely related to two other species of ape, the chimpanzee and the bonobo, ants should do all manner of "human" things like build roads, create traffic rules, have public hygiene workers, and work on assembly lines while those apes don't. We will consider whether a primal-sounding scream called the pant-hoot could have been our remote ancestors' first small step toward patriotic flag waving, and in a sense the foundation of our continent-wide kingdoms. How can I, a foreigner, manage to look past human differences and carry on in other societies when most animals, ants decidedly among them, are unable to do so? Or, a question for the history buff: might the outcome of the US Civil War have been influenced by the fact that most Southerners of the time still thought of themselves as American?

George Bernard Shaw wrote, "Patriotism is, fundamentally, a conviction that a particular country is the best in the world because you were born in it."[6] So what does it mean if an inherent component of the human condition is to hold tight to a society, idolizing it while so often slighting, distrusting, demeaning, or even hating foreigners? That fact is a wonder of our species, and one of my reasons for writing this book. While we have gone from small societies to vast ones, we have retained an uncanny awareness of who fits in and who doesn't. Yes, we build friendships with foreigners, yet they remain foreign. For good or bad, the distinctions remain—along with equally pronounced and often disruptive distinctions mounting within the societies themselves, for reasons I hope to make clear. How we approach the similarities and differences determines the nature and future of societies.

THE JOURNEY AHEAD

Our inquiries will lead us down not one long road but along many interconnected paths. From time to time we will circle back to look at subjects like biology and psychology from new angles. Our course won't always be chronological as we draw not just from human history but from our own evolution in order to come to grips with what we do and how we think. Given what can seem like a journey that meanders through multiple and varied disembarkation points, a preview of what lies ahead is in order.

I have divided the book into nine sections. Section I, entitled "Affiliation and Recognition," takes in the wide range of vertebrate societies. Its opening chapter considers the role of cooperation in societies, which I endeavor to show is less essential than the matter of identity; societies consist of a distinct set of members in a rich tapestry of relationships, not all of which are harmonious. Chapter 2 covers other vertebrate species, especially the mammals, to illuminate how societies, despite whatever imperfections in the system of partnership that exists within them, benefit the members by providing for their needs and protecting them. The third chapter probes into how the movements of animals within and between societies are important to the success of the various groups. One versatile pattern of activity, fission-fusion, creates a dynamic that helps explain the evolution of intelligence in certain species, humans most obviously among them, and the subject will come up repeatedly in this book. Chapter 4 investigates how much the members of most mammal societies must know about each other for their societies to stay together. Here, I reveal a limiting factor in the societies of many species: all their members are obliged to know each other as individuals, whether they like each other or not, restricting the societies to, at most, a few dozen individuals. This sets up a puzzle about how the human species broke free of such a constraint.

The second section, "Anonymous Societies," addresses a group of organisms that readily crash through this population limit: social insects. One of my objectives is to break down any aversions you, the reader, may have about likening insects to "higher species," especially humans, by making clear the value of these comparisons. Chapter 5 reports on how social complexity generally climbs with an increase in the size of insect societies with features like infrastructure and division of labor becoming more complex, a trend paralleled in humans. Chapter 6 looks at how most social insects, and a few vertebrates such as the sperm whale, demonstrate affiliation with a society by using something that marks their identity: chemistry (a scent) in ants, and a sound in whales. These simple techniques are not constrained by the limitations of memory, and thus permit the societies of certain species to reach immense sizes, in a few cases without an upper bound. The chapter after that, "Anonymous Humans," spells out how humans employ the same approach: our species is attuned to markers that reflect what each society finds acceptable, including behaviors so subtle they may only be noticed subliminally. By this means people can connect with strangers in what I call an anonymous society, thereby breaking the glass ceiling in the size societies can achieve.

The three chapters included in Section III, "Hunter-Gatherers Until Recent Times," ask what the societies of our species were like before the advent of agriculture. I cover people who existed as hunter-gatherers up to recent times, ranging from those who lived nomadically in small, spread-out groups, called bands, and others who settled down for much or all of the year. Although the nomads have gotten most of the attention and are treated as the gold standard for our ancestral condition, a readily defensible conclusion is that both options have been within the reach of human beings likely going back to the origins of our species. We can also conclude that hunter-gatherers were not archaic people living an archaic mode of existence. Their people must be recognized as essentially no different from us: humans, as it were, "in the present tense." Despite traces of ongoing, even rapid human evolution in the past 10,000 years, the human brain clearly hasn't been restructured in any fundamental way since the appearance of the first *Homo sapiens*.[7] This implies that notwithstanding any human adjustments to modern life, we can look to the lifestyles of hunter-gatherers in recorded history and consider the nature of early human societies as the bedrock that underlies our own.

What concerns us most are the extraordinary differences between the nomadic hunter-gatherers—equality-minded jacks-of-all-trades, who solved issues by discussion—and settled hunter-gatherers, whose societies often became open to leaders, division of labor, and disparities in wealth. The former social structure points to a psychological versatility we still possess, even if most people today behave more like settled hunter-gatherers. Two conclusions of Section III are that hunter-gatherers had distinct societies and that those societies were distinguished, just as societies are now, by markers of identity.

What that means is that at some point in the distant past, our ancestors must have taken the crucial but heretofore overlooked evolutionary step of making use of badges of membership that would, in time, permit our societies to grow large. For clues about how this happened, Section IV, with one chapter, transports us into the past and also scrutinizes the behavior of modern chimps and bonobos. I put forward the hypothesis that a simple shift in how the apes use one of their vocalizations, the pant-hoot, could make that sound essential for identifying each other as society members. Such a transformation, or something like it, could have easily occurred in our distant ancestors. Ever more markers would have been added to this initial "password," many of them connected to our bodies, transforming them into flesh-and-blood bulletin boards for displaying human identity.

Having looked at how markers of identity originated, we are in a position to explore the psychology underlying those markers and society membership. The five chapters of Section V, "Functioning (or Not) in Societies," review a fascinating range of recent findings about the human mind. Most of the research has focused on ethnicity and race, but should apply to societies as well. Among the topics are the following: how people see others as possessing an underlying essence that make societies (and ethnicities and races) so fundamental that they think of these groups as if they were separate biological species; how infants learn to recognize such groups; the role stereotypes play in streamlining our interactions with others, and how those stereo-types can become tied to prejudices; and how the prejudices are expressed automatically, and unavoidably, often leading us to perceive an outsider more as a member of his or her ethnicity or society than as a unique individual.

Our psychological assessments of others are many and varied, including our penchant for ranking outsiders as "below" our own people or in some cases as subhuman altogether. The fourth chapter of Section V elucidates how we apply these assessments of others to societies as a whole. People believe that the members of foreign groups (and their own people as well) can act as a united entity, with emotional responses and goals of its own. The final chapter steps back to draw from what we have discovered about the psychology of societies and the underlying biology to pose more sweeping questions about how family life fits in the picture—whether, for example, societies can be understood as a kind of extended family.

Section VI, entitled "Peace and Conflict," takes on the issue of the relationships among societies. In its first chapter I document the evidence from nature, which shows that while animal societies need not be in conflict, peace between them is relatively rare, present in just a few species and supported by situations of minimal competition. The second chapter then highlights hunter-gatherers to examine how not merely peace but active collaborations between societies became options for our species.

Section VII, "The Life and Death of Societies," examines how societies come together and fall apart. Before writing about people, I survey the animal kingdom, concluding that all societies go through some sort of lifecycle. Although, as we shall see, other mechanisms for starting new societies exist, the pivotal event in most species is the division of an existing society. The evidence from chimpanzees and bonobos, bolstered by data on other primates, is that a division is preceded by the emergence, over months or years, of factions in the society, which increases discord and ultimately causes a split. The same formation of factions, usually over the passage of centuries, takes place with humans also, except for a key difference: the primary pressure that severed human factions was when the original uniting markers keeping a society together were no longer shared, leading people to see themselves as incompatible. This section lays plain how people's perceptions of their own identities change over time in a way that could not be stopped in prehistory, mainly the result of poor communication across hunter-gatherer bands. For this reason, hunter-gatherer societies split apart when they were tiny by today's standards.

The expansion of societies into states (nations) was made possible by the social changes I lay out in section VIII, "Tribes to Nations." Some hunter-gatherer settlements and tribal villages with simple agriculture took the first tentative steps in this direction as leaders extended their power to take control of neighboring societies. I begin by describing how tribes were organized into multiple villages, each of which acted independently much of the time. What leaders these loosely connected villages had were not very proficient at sustaining social unity and curtailing social breakdowns, in part because they lacked the means of keeping their people on the same page with regards to identifying with the society—things such as roadways and ships that connected people with what their compatriots were doing elsewhere. Growth also required societies to expand their dominion over the territories of their neighbors. This didn't occur peacefully: across the animal kingdom I find little evidence of societies freely merging. Human societies came to conquer each other, thereby bringing outsiders into their fold. Occasional transfers of membership take place in other species too, but in humans such exchange was taken to a new level with the advent of slavery, and finally, the subjugation of entire groups.

Now that we understand the forces that can cause small societies to scale up to large ones, including the nations of today, the final chapter of Section VIII evaluates how these societies tend to meet their end. What's typical of societies put together by conquest isn't division between factions, as we saw earlier for hunter-gatherers, nor utter collapse, though it can happen, but rather a fracturing that almost always occurs roughly along the ancient territorial lines of the peoples that have come to make up the society. Large societies may be no more durable than small ones, fragmenting on average once every few centuries.

The final section carries us along the circuitous route that led to the rise of ethnicities and races and the at times muddy waters of current national identities. To become an interlocking whole, a conquering society had to make the shift from controlling what had been independent groups to accepting them as members. This requires an adjustment in people's identities, in which ethnic minority groups adjust to the majority people—the dominant group that most often founded the society and controls not only its identity, but also most of the resources and power. This assimilation would be accomplished only to a degree, for this reason: ethnicities and races—as demonstrated earlier in the book for individual persons and for societies as well—will be most comfortable together if they share some commonalities and yet differ enough to feel distinct. Status differences emerge among the various minorities too, and may change over the course of generations—though the majority almost always stays firmly in control. Bringing the minorities into the fold as society members entails allowing them to intermix with the majority people, a geographical integration of populations that not all past societies have permitted.

The second chapter of Section IX tackles how modern societies have made the friendlier incorporation of large numbers of outsiders possible through immigration. Such movements have seldom occurred easily, and, as in the past, have assigned lower power and status to the immigrants, who may face the least resistance when they take on social roles that minimize competition with other members while giving them a sense of value and esteem. The identity immigrants had once treasured in their ethnic homeland is often recast into broader racial groups. The shift in perception may initially be pushed on the newcomers, but they can accept the changes because of the advantages of having a more extensive base of social support in the adopted society. The chapter closes by describing how criteria for citizenship have come to deviate from the psychology of how people register who has a rightful place in a society. The latter is heavily influenced by people's attitudes about how important a society should be in providing for different individuals or groups versus protecting themselves—attitudes relating to patriotism and nationalism, respectively. Variation among the members in these points of view may well be required for a healthy society, even though it also compounds the social conflicts that make headlines today. Given these stresses, a final chapter, "The Inevitability of Societies," raises the issue of whether societies are necessary.

In making what inferences I can in this book, I admit up front that a unified field of study of societies is a distant dream. All too often, academic disciplines foster a habitual concentration on certain modes of thought and a disdain for the unfamiliar by dividing the intellectual world into mutually alien societies known as biology, philosophy, sociology, anthropology, and history, thus leaving much room in the nooks and crannies between for debate. For instance, "modernist" scholars of history view nations as a purely recent phenomenon. My contention will be that the national pedigree has ancient roots. Some anthropologists and sociologists go a step further and see societies as entirely optional, with people forming such unions when it serves their interests. My goal is to show that membership in a society is as essential for our well-being as finding a mate or loving a child. I will frustrate some in my own discipline of biology, too. I have listened to biologists adamantly oppose the idea that societies ought to be examined as groups of distinct identity and membership when their study species don't quite match with this criterion—a passionate reaction that more than anything makes plain the cachet of the word "society."

Disputes among the specialists aside, readers of every political persuasion will find both good and bad news in the current science. Whatever your social views, I urge you to consider insights from fields beyond your usual interests to become aware of how your own, often subliminal, biases and those of people around you—writ large, across multitudes—might affect both the actions of your country and your daily conduct with others.

SECTION I

Affiliation and Recognition

CHAPTER 1

What a Society Isn't
(and What It Is)

S een from the top of the stairs in the main concourse of New York's Grand Central Terminal, swarms of people whirl and eddy under the famous four-faced clock. The staccato clatter of shoes on Tennessee marble and the clamor of voices, swelling and subsiding like an ocean in a seashell, reverberate across the cavernous acoustical marvel. The vaulted ceiling, depicting 2,500 stars frozen in place along their regimented paths one October New York night, makes a perfect counterpoint to the tumult of humanity below.

The sheer number and diversity of people rushing past each other, or clustered here and there in conversation, make this scene a microcosm for human society in its entirety: not a society as a voluntary association of people, but a society as an enduring group, the kind that occupies a territory and inspires patriotism. When we think of such societies, we may think of the United States, ancient Egypt, the Aztecs, the Hopi Indians—groups that are central to human existence and are the cornerstones of our collective history.

What are the characteristics of a population that make it a society? Whether you have in mind Canada, the ancient Han dynasty, an Amazon tribe, or even a lion pride, a society is a discrete group of individuals amounting to more than a simple family—more than one or both parents with a single brood of helpless offspring—whose shared identity sets them apart from other such groups and is sustained continuously across the generations. Indeed it may eventually come to produce other such societies, as when the United States split off from Great Britain or one lion pride becomes two. Most importantly, the membership of a society changes rarely and arduously; such a group is closed, or "bounded." Though its constituents can vary in the intensity of the passion they attach to their national consciousness, most prize it well above any other sort of membership—their ties to those nuclear families aside. This significance is conveyed among humans by commitments to struggle, even die, for the sake of the society, should the situation demand it.[1]

Some social scientists see societies primarily as constructs of political convenience, an arrangement that emerged in recent centuries. One scholar with such a view, the late historian and political scientist Benedict Anderson, conceived of nations as "imagined communities," since their populations are too numerous to allow for members to meet face to face.[2] Indeed, I concur with his basic idea. By serving to distinguish *us*, those who belong, from *them*, the outsiders, shared imaginings are all we need to create societies that are true and tidy entities. Anderson also proposed that these made-up identities are artificial products of modernity and mass media, and this is where he and I diverge. Our shared imaginings bind people with a mental force no less valid and real than the physical force that binds atoms to molecules, turning them both into concrete realities. This has been the case for all time. In fact, the concept of imagined communities holds true not just for modern societies, but for all the societies of our ancestors, likely from their remote, prehuman origins. Hunter-gatherer societies, held together by a sense of common identity, did not depend on their members establishing one-to-one relationships—or knowing each other at all, as we shall examine; among the other animals, too, societies are represented firmly in the minds of their members, and in that regard are also imagined. This is not to diminish the societies of human beings. Societies are rooted in nature and yet have come to flower in elaborate and meaningful ways that obviously are peculiar to our species, a subject this book will pursue.

I think the point of view I lay out captures what most people have in mind when we speak of a "society." Of course, any word must encompass some variation, and no animal society is equivalent to a human society just as no two human societies are equivalent. I offer this to those concerned about where to draw a line: The utility of a definition is best shown by how much we learn from anomalous situations where the word doesn't quite work. Pushed sufficiently hard, any definition outside those for mathematical terms and other abstractions will break down. Show me a car and I may show you a pile of junk that once functioned as a car (and maybe in a mechanic's mind still is one). Show someone a star, and an astronomer points to a mass of convergent superheated dust. The hallmark of a good definition is not just that it tidily delimits a set of *x*'s, but that it also breaks down when things get conceptually intriguing about *x*.[3] Hence countries exist that strain my view of a society as a discrete group with a shared identity in ways that are informative. Iran, for example, counts many ethnic Kurds among its citizens despite the government suppressing their identity as a group, while at the same time those Kurds conceive of themselves as a separate nation and claim rights to their own territory. Situations where groups such as the Kurds have identities that clash with the society throw light on the factors that can serve, over time, to further empower and expand societies or rend them and start new societies.[4] Conflicts over identity can arise in other animal societies, too.

Many experts in my field of biology, and anthropologists also, present a different definition of a society, describing it not in terms of identity but rather as a group organized in a cooperative fashion.[5] Though sociologists recognize collaboration as vital to a society's success, it has become rare in that field to equate a society with a system of cooperation.[6] Still, thinking of a society this way comes easy, and for obvious reasons: we humans have evolved such that cooperation is central to our survival. Humans out-cooperate other animals, having sharpened our skills of communicating our own intentions and deducing those of others in the pursuit of shared goals.[7]

WHAT HOLDS US TOGETHER

In considering cooperation, as contrasted to social identity, as an essential feature of societies and a basis for distinguishing one society from another, one place to start is with a hypothesis developed by anthropologists to explain the origin of intelligence. It postulates that as our brains have gotten bigger, our social relationships have likewise grown, each driving the greater size and complexity of the other.[8] Oxford anthropologist Robin Dunbar has described a correlation between the size of a species' brain—to be precise, the volume of its neocortex—and the number of social relationships individuals of that species can sustain, on average. Our close relation the chimpanzee, per Dunbar's data, manages about 50 coalition partners or allies: call these 50, with whom an individual cooperates most generously, its friends.[9]

By Dunbar's calculation, when it comes to humans, the average Joe can sustain about 150 close relationships, his specific bosom buddies changing over time as friends are made or lost. Dunbar characterizes this figure as "the number of people you would not feel embarrassed about joining uninvited for a drink if you happened to bump into them in a bar."[10] This has become known as Dunbar's number.

The "social brain hypothesis" leaves plenty of room for argument. For one thing, it's reductive: no doubt you find advantages to having lots of gray matter besides keeping track of the Toms, Dicks, and Janes you meet—finding food, making tools, and other skills also require cognitive effort. Context matters, too. At a professional conference, for example, an academic is likely to share common interests with a high proportion of the attendees and might willingly join any number of folks uninvited at that bar. Also, friendship isn't a binary, yes/no category. If Dunbar's number proved to be 50 or 400, it would simply have indicated a greater or lesser degree of basic intimacy and rapport.

No matter how much human brainpower is devoted to relationships, though, our social circles don't come remotely close to matching the size of nation-states. The disparity between your capacity to make room in your life for 150 chums and a chimpanzee's competence at handling 50 is too small to explain human societies today, with their staggering sizes—or the smaller societies of our past, either. In fact there has never been a human society, from the Stone Age to the Internet era, that consisted solely, for more than a passing minute, of a band of brothers: a clique of mutually shared friends and family living in requited admiration. To think otherwise would be to misapprehend the nature of friendship, and hence our private networks of friends. Whether in overpopulated India, or on the Polynesian island nation of Tuvalu with its 12,000 citizens, or among the miniscule El Molo tribe on the shores of Lake Turkana in Kenya, no one befriends or cooperates with everyone in a society: they pick and choose. When Jesus said to love your neighbor as yourself, he didn't mean that you must be a friend of everyone. The El Molo aside, our societies encompass at least some folks—more often a great many—whom we will never meet, let alone befriend. And never mind those we don't select as buddies, or those who reject us—our worst enemy almost certainly carries the passport of our country.

Data on how individuals interact reveal the same discrepancy between Dunbar's number for a species and the size of its societies. A chimpanzee society, called a community, often has well over a hundred members, but even a community of 50, which by Dunbar's calculation could consist entirely of bosom buddies, in truth never does.[11]

The "cognitive constraints on group size" (Dunbar's phase) that puzzle some devotees of the social brain hypothesis do so because of the confusion of social networks (for which social ties vary in strength and depend on each person's perspective—described, for example, by Dunbar's number) and distinct groups (most notably the societies themselves).[12] Both play a part in the lives of humans and other animals. Societies, with their unambiguous boundaries, provide the richest soil within which long-term, though perhaps ever-changing, cooperative networks can grow. While networks may at times encompass everyone, they flourish best among those members who get along with each other well, built on the smarts and cooperative skills at each person's disposal.

Societies, then, staked out as different from each other by the identities of the members, rest on more than personal networks of allies. And unlike other species, humans keep social life functioning and social networks strong with rules administered in a host of formulations from one society to the next. We thrash out practices—and penalties— to enforce fair exchanges and ethical behavior that can operate across the many for the mutual good. The trash collector does his part by picking up the garbage of strangers in return for a paycheck. He buys coffee from the corner shopkeeper he doesn't know and may speak to unfamiliar faces by the hundred at a church or union meeting. But the governance of such interactions has limits. Despite the shared economic and defensive benefits a society confers, disagreements between factions, especially about what counts as doing one's part and mutual good, can be painful. Such melees are the least of it, however. No society is without crime or violence (the antithesis of cooperation) committed by one full-fledged member or group of members against another. Yet societies can endure for centuries even if dysfunction speeds their dissolution—the Roman Empire immediately comes to mind, though there have been countless others.

In general, however, societies will favor cooperation. How much selfishness or divisiveness it takes to rupture a society is likely much greater than intuition would suggest. In *The Mountain People*, the British-born anthropologist Colin Turnbull chronicled moral decay among Uganda's Ik people during a disastrous famine in the 1960s that brought a disregard for social bonds and led to the deaths of children and the elderly. Turnbull's account showed just how far a society could unravel under stress; nevertheless, the Ik carried on.[13] Likewise

Venezuela remains intact despite repeated economic collapses and a murder rate in its capital city of Caracas that in some years exceeds that of a war zone. Whenever I visit an intrepid friend there, he drives us at high speed over harrowing backstreets to avoid shootings by *motorizados* along the highway. Despite it all, he loves the place. The surprising fact is that Venezuelans show every bit as much attachment to and pride in their nation as Americans show in theirs.[14] Societies have survived worse. For example, during the California gold rush the homicide rate was dramatically higher than that of modern-day Venezuela.

While disagreements and disagreeableness can fray the fabric of societies, their positive counterpart, cooperation, does not necessarily knit societies together or separate them from other societies. This is true even when cooperation contributes to the social capital that builds up among the members and enhances the productivity of the whole. The biggest problem with predicating life in a society on cooperation is it simply ignores much of what makes existence in a society a challenge. The nineteenth-century social theorist Georg Simmel interpreted cooperation and conflict as inseparable "forms of sociation," each unimaginable without the other.[15] To focus overmuch on cooperation is to cherry pick.

In the societies of our ape relatives, kindness and cooperation are likewise just a fraction of the picture. Chimpanzees intimidate each other or fight outright over status, with those on the losing end occasionally ostracized or killed. The paltry assistance the apes render one another, outside of mother-infant bonds, usually materializes when several animals work together to unseat competitors, fighting so that one of them can attain the alpha male status. Chimps also join up to hunt red colobus monkeys, by some accounts making the kill by acting in parallel more than by collaborating. Whichever chimp ends up with the meat may give others a taste, but only after being begged.[16] The bonobos, which look like chimps with small heads and pink lips, are more beneficent yet will steal food from society-mates when they can, and are no more inclined to teamwork.[17]

Even social insects, the very symbol of mindless cooperation, can face domestic strife and act selfishly. Though across most social insect species the queen is normally the only individual to reproduce, in honeybees and some ants a few workers subversively lay eggs. Their nests can be veritable police states, with workers on a vigilant quest to destroy any egg that isn't the queen's.[18] Whatever the species, individuals that don't pull their weight also drag others down; how societies, ranging from arthropod to human, deal with cheaters and enforce fairness is a field of study in itself.[19]

Given a society with its clearly defined membership, how much co-operation is required to hold it together, for animals generally? In theory, not much. Expelling foreigners may be the minimum collab-oration needed. Imagine a loner creature that controls an exclusive space, or territory, by throwing rocks at anyone drawing near. Now imagine a few such creatures settle a territory together. Each throws stones at outsiders exactly as before but with one difference: they leave each other alone. This tacit agreement to "do no harm," so to speak— to coexist in peace—amounts to cooperation of a rudimentary sort.

Of course, societies could not have evolved if they did not offer competitive advantages to either the group (evolutionary biologists call this group selection), its individual members, or both.[20] What might be the attraction for such an uncharitable bunch? Such a soci-ety would make sense if, say, ten animals throwing rocks this way could hold on to more than ten times the territorial space that they could have alone, or seize a higher quality territory with less effort or risk to each member. It may also be that purely by keeping others out, they exclude those whose steadfastness at rock throwing is uncertain while sharing exclusive opportunities to mate with each other (even if a lot of infighting remains over who has sex with whom).

Surveys of the animal kingdom show societies can exist with just a little "prosocial" behavior between the members: call it proto-cooperation, doing good turns by accident, perhaps.[21] Madagascar's ringtailed lemurs come close to displaying such minimal expectations, in that troop members help each other little except to join forces in at-tacking outsiders.[22] One expert opines that marmots, a kind of alpine squirrel, don't even *like* each other, yet still find such behavior as hud-dling for warmth sufficient motivation to stay together, while another authority has described a social badger clan as "a tight community of solitary animals."[23] Even people retain commitments to groups whose members don't get along, with our helpfulness often dependent on what our society demands of us.[24]

SOCIETIES THAT GET ALONG

When the Spanish ships the *Santa María*, the *Niña*, and the *Pinta* first arrived in the New World, one society greeted, and the other en-slaved. The local group of Taíno Indians, an Arawak tribe depicted by Christopher Columbus as "stark naked as they were born, men and women," swam or paddled canoes out to greet the newcomers. Without understanding a word the Spaniards said, the Indians showered them with fresh water, food, and gifts. Columbus recorded a more cynical, European reaction to the Indians: "They would make fine servants. . . . With fifty men we could subjugate them all and make them do whatever we want. . . . As soon as I arrived in the Indies, on the first island that I found, I took some of the natives by force in order that they might learn and might give me information of whatever there is in these parts."[25]

The stark contrast between these alternative outlooks, one of open trust and the other of cunning and exploitation, is disturbing, but hardly shocks us. We humans have a flair for identifying who's in our society and who isn't, then drawing a sharp line between what psychologists call the ingroup and the outgroup, even when we are friendly with the latter. We learn from childhood to regard foreigners as a possible threat or—as the Arawak and Columbus both did in differing ways—as an opportunity.

Thus we are presented with another reason why cooperation doesn't always indicate where one society ends and the next begins. Some among the horde in Grand Central concourse are no doubt foreign nationals navigating productive relationships with US citizens. Hence just as it's possible to have enemies as well as allies in one's own society, it's possible as well for members of one society to communicate with members of another toward the goal of friendship and cooperation. Camaraderie occurs between nonhuman societies, too—in rare cases. The bonobo has been called the hippie ape for preferring peace to provocation. Yet I would wager that individual bonobos find the occasional adversary in other communities. Even a peacenik doesn't get along with everyone.

The ease with which people today jet off to other countries has brought our contacts with foreigners to a new level, one that, as we will come to grips with later, is without parallel in nature. Indeed, modern life challenges our tolerance for others by twisting and stretching our identities in novel ways. But societies have remained with us throughout it all.

COOPERATION WITHOUT SOCIETIES

Following Terry Erwin into a Peruvian rainforest means getting up at dawn. While the wind is still low, this entomologist at the Smithsonian's natural history museum loads a machine called a fogger with a bio-degradable insecticide and aims its spout upward so a pale gray mist climbs into the trees. There's the light sound of rain, but no water is falling—instead, the patter is the sound of tiny bodies hitting sheets stretched over the ground. Over the years, Erwin has learned just how rich the tropics are. By his estimation, 30 billion individuals belonging to 100,000 species live in a single hectare of the rainforest.[26]

Wherever I go, I am awestruck by the diversity of living things. The data of Erwin and others push us to look at societies from the broadest possible perspective of global biodiversity. What is striking is that the vast majority of organisms get by just fine as lone individuals. That's true of more than 99 percent of the species perched in the trees, in Peru or elsewhere. Putting aside the obligation to mate and possibly raise young, it's not always clear why one should stay close to others at all. As beings capable of taking pleasure in the company of others, we humans seldom think about this question. But birds of a feather, whether people or actual birds, are potential competitors for the same resources: food and drink, opportunities for sex, and a place to call home and rear offspring. In many species, individuals cluster together only incidentally, to fight or scramble over food like so many squirrels grabbing nuts. Solitary existence is a safe approach for keeping what one has worked hard to get. For life in a crowd to pay off—any crowd, let alone a whole society—something must be gained in dealing with needy and greedy others.

One option is to cooperate with those others when the situation is right, and this possibility suggests a final difficulty with associating co-operation with societies: while cooperating individuals can be said to be social, that doesn't mean they compose a society. In his important book *The Social Conquest of Earth*, my hero and mentor, the ecologist Edward O. Wilson, observes that animals that are social—that come together at some point in their lives to pull off something mutually advantageous—are ubiquitous.[27]

Even so, few species have taken the step of evolving societies. Con-template the two most basic social units: a mated pair, and a mother with offspring. Not all animals display even these sorts of social pair-ings. Salmon cast off eggs to be fertilized in the water column, and turtles abandon their clutches after hiding their eggs in the sand. Yet newborns and hatchlings are fragile, so it can be strategic to support them while they are helpless. In all birds and mammals, and in some species of other animal classes, mothers are the caretakers during this critical time. In a few cases, as in the American robin, dads lend a hand. Still, that's usually as extensive, and as lasting, as the group gets: most such little families operate solo, not as part of an enduring society.

Nor do networks of allies or intimate friendships need a society to blossom. For example, orangutans don't have societies and are loners much of the time, but primatologist Cheryl Knott tells me females who meet in adolescence will hang out now and then later in life. Or consider that two or more cheetahs, often though not always brothers, will collaborate to hold a territory.[28] Yet in my estimation, friendships, as distinct from sexual partnerships, prosper most often in societies—a pattern that suggests societies' dependable memberships provide a stable foundation for the sort of close relationships that intrigue adherents of the social brain hypothesis.

Nevertheless, being with others—even if in fleeting relationships not specifically established for childrearing or friendship—may be beneficial. Think of chorusing birds that come and go like rowdy teens at a soiree. Such flocks are social affairs that can attract whoever's nearby, protecting the birds from predators, connecting them to mates, or stirring up bugs to eat.[29] Some birds burn less energy migrating in a V formation than flying alone. Schools of sardines and herds of antelope offer similar payoffs even when the participants make no commitment to a particular ensemble.[30] In addition to such joint benefits, there can also be cases of altruism where an animal helps others at some cost to itself. A few fish in a school of minnows will swim forward to examine a predator, with the school apparently learning how dangerous the situation is from how aggressively the predator reacts.[31] When relatives are involved, such generosity has a special evolutionary logic because individuals can promote their own genes by helping those kin, as shown in an elementary way by a pair of robins rearing their progeny: that's kin selection.

Proximity can offer a safety in numbers that's outright self-serving, as I observed in my first tropical expedition while an undergrad. In Costa Rica I joined lepidopterist Allen Young, who asked me to record the behavior of the caterpillars of the orange-spotted tiger clearwing butterfly. The lumpy larvae ate the leaves of a weedy plant, rested, and moved in tight groups. Spiders and wasps were the bane of their existence; the caterpillars positioned on the outside of a cluster were the ones these predators picked off and ate first. I concluded that their survival instinct was to pack together, each pushing between the others, leaving the weak ones to their demise. Writing up the results I discovered that W. D. Hamilton, a famed biologist, had already proposed this centripetal behavior for fish schools, mammal herds, and the like, and given it a name: they are selfish herds.[32] Despite their selfishness, my caterpillars helped each other, if only by accident. Alone they had a hard time cutting past the tough hairy leaf cuticles to begin feasting; as a group they fared better since the first caterpillar to succeed opened up the leaf for everyone to feed.[33]

The important point is that my caterpillars—like schools of minnows, nesting robins, and flocking geese—cooperate but don't have societies. As long as their companions were of similar size, the larvae seemed to get along fine with each other when I combined the broods of their species. The same is true whenever individuals can come together willy-nilly, for example when another species, the tent caterpillars, join up to weave a silk tent that's bigger and better sheltered from cool weather.[34] Similarly, the African bird known as the sociable weaver inserts its nest among many others to produce a large communal structure that provides air-conditioning for all residents. The birds move into and leave from these colonies as they choose, albeit over months rather than constantly like birds in many flocks. While some birds get to know each other, a colony, like a flock, isn't closed to outsiders. Any new arrivals are tolerated so long as they find a spot to nest.[35]

In short, those who equate societies with cooperation have it backward. A typical society encompasses all manner of relationship, positive and negative, amicable and embattled. Given that cooperation can flourish both within and between societies and where there are no societies at all, a society is better conceived not as an assembly of cooperators, but as a certain kind of group in which everyone has a clear sense of membership brought about by a lasting shared identity. Membership in societies of humans and other species is a yes/no matter, with ambiguity rare. The prospects for alliances, whether from friendship, family ties, or social obligations, may rank among the paramount adaptive gains of having societies in many species, yet aren't necessary to the equation. A misanthrope with no family, filled with contempt for all humanity, can still claim your nationality. That's true whether he lives as a hermit outside the system or as a parasite of others within it.[36] The members of a society are united by their identity, whether or not they are in regular contact or willing to help each other out—though the membership they have in common can be a solid first step in making such relationships a reality.

So what came first, chicken or egg—membership or cooperation? Whether more than a bare minimum of collaboration had to exist for societies to evolve, or memberships had to be worked out before long-term cooperation was at all likely, remains an open question. Whichever was originally the case, the next chapter introduces us to the many advantages that societies offer our vertebrate cousins in nature.

CHAPTER 2

What Vertebrates Get out of Being in a Society

Animals in a society struggle just as solitary species do. They face conflicts over who gets what, as well as the right to mate, make a home, and raise young. Not all succeed. What membership provides is a certain security in facing the broader world. This is true even in societies whose members do little for each other beyond driving off outsiders. The gamble is that being part of a successful, or dominant, society will allow each member to ultimately get a more substantial slice of the overall pie than the portion it would have earned alone or in a weaker society. While loose and temporary groups have advantages, once animals adapt to life in permanent societies, it can be problematic for them to go back to surviving on their own. Anyone not in a society, or in a failing society, is at risk.

Vertebrates, and to be more specific the mammals, provide a good starting point as we consider the benefits of societies, especially since we are mammals ourselves and our evolution as such is a core theme of this book. That's not to say no other vertebrates have societies. In some bird species, such as the Florida scrub jay, fledglings help their parents raise younger siblings; given this "overlap of generations," as biologists describe it, their groups represent a simple sort of society. Or consider the shell-dwelling cichlid, *Neolamprologus multifasciatus*, a fish native to Lake Tanganyika, Africa.[1] Societies of up to 20 cichlids guard a cluster of shells they dig out of the sediments. Each fish has a personal shell abode in what one biologist described as a "complex of

apartments that would do modern public housing proud."[2] An alpha
male does the breeding, with outsiders of either sex inserting themselves into the colony membership only at rare intervals.

The mammals that live in societies are far more widely known, and
talked about, than society-dwelling fish or birds.[3] Even so, revisiting
them with the issues of membership and identity in mind yields fresh
perspectives. Consider two beloved examples, the prairie dogs of the
North American plains and the African savanna elephants. Both have
societies, but those turn out not to be the groups of either that garner
the most attention. People traditionally think of prairie dogs as living
in colonies or towns, and elephants in herds; yet a colony or herd is
rarely one society but rather a multiplicity of societies, antagonistic to
each other in prairie dogs and often convivial in the elephant.

No individual prairie dog identifies with, or fights for, a colony;
rather, its loyalty is to one of the small land-holding groups within
it, sometimes called a coterie (the word is appropriate, meaning an
exclusive group). In perhaps the best-studied of the five prairie dog
species, the Gunnison's prairie dog, each coterie contains up to 15
breeding adults, including one or more of each sex, occupying a vigorously defended area up to a hectare in extent.[4]

Savanna elephants are, in contrast, widely sociable across their
populations, yet one grouping merits being called a society.[5] The core
groups, or simply cores, have a membership of up to 20 adult females,
accompanied by their calves. The societies are female affairs. Each
male goes his own way when he matures and never becomes part of
a core. Cores often can be picked out by eye from their responses
to each other, even at times when hundreds of elephants, and many
cores, socialize. To retain their distinct memberships, the cores ordinarily keep outsiders, even animals the members are fond of, from
lingering with them in a long-lasting way. The relationships between
cores are complex. The cores form connections, called bond groups,
but these social networks are inconsistent, with differences in opinion
as to who belongs—core A can bond to B and C at the same time that
C avoids B. Only the elephant cores themselves retain a distinct membership over the long haul.

The savanna elephant's life in a core differs from the two other
elephant species, the forest elephant of Africa and the Asian elephant,
which, though social, lack distinct societies.[6] Whether the absence of
societies makes those species less sophisticated depends on what is
meant by "sophisticated." The answer may well be that they are more
so, since Asian elephants have heavier brains relative to their body

weight than the savanna elephant; maybe the savanna elephant's focused dependence on its core simplifies its life by reducing its everyday social obligations to just those few core-group companions. Loner species such as weasels and bears need to be totally, and constantly, self-reliant, which may explain why they can be smarter than many society-dwelling species, as measured by their adroitness at solving puzzles.[7] Asian elephants, socially gregarious as they are, could well face constant cognitive challenges because they live with few clear social boundaries compared to Africa's savanna species, with its cores.

THE ADVANTAGES OF SOCIETIES FOR MAMMALS

Looked at in sweeping terms, societies give mammals—from savanna elephants and prairie dogs to lions and baboons—multiple avenues for providing their members with both security and opportunity, shielding them from the perils of the external world while offering them entrées to the resources they jointly hold. Loosely speaking, these safety nets fall into two, loosely overlapping categories: societies provide, and they protect.

Among their functions in providing resources is access to dependable long-term helpers, which can be an asset for mothers feeding and housing the young. Gray wolves and birds such as the Florida scrub jay, among others, are cooperative breeders, with their societies based around extended families in which offspring assist the parents or close relatives in raising siblings. Babysitting while mother goes off to feed herself is a common task across many species, but helper meerkats also tidy the burrow and supply the young with insect prey.[8] In some monkeys a female helper contributes little of value; however, if she has never had young of her own before, she profits from having the practice of handling a baby—with the nervous mother keeping a close eye on her.[9]

Other benefits of societies evince how lifelong members who know each other well can obtain meals by efficient group efforts. Big game predators dispatch prey that amount to a windfall for everyone. However, cooperation is more obligatory in some species, such as painted dogs, than in others—lions can be lazy about joining a group hunt and often don't obtain more meat for themselves than they would by stalking prey alone. Some behaviors common across mammal societies are important even though little improved over what we see for temporary aggregations like bird flocks. Meerkats and ringtailed lemurs, for instance, rely on their numbers to locate patches of food, and by staying clustered together stir up lots of insects for easy catching. Baboons are known to hang tight with the most successful forager in their bunch, occasionally even stealing his or her morsels.

Among the bottlenose dolphins of western Florida, societies perhaps serve primarily for another purpose: enabling the animals to adapt to local conditions. Raising a dolphin is a communal responsibility, with the pups learning traditions passed down through the generations. The elders of some communities, for example, teach the young to team up to circle fish until their prey clump up and can be driven onto the beaches. There the dolphins snatch their flopping prey before wriggling back down into the water again. Such social learning likewise plays a role in species such as the chimpanzee.[10]

The protection a society affords its members can be just as crucial as the resources they can access. Of course the two are linked. Because chimpanzee females can defect to another society if they don't get what they need for their offspring, the males commit to the brunt of the toil and the danger of guarding the members and keeping their resources from bellicose outsiders. There's no better immediate motivation for the males, it would seem, than sex.[11] Beyond food and drink, many species secure a den for the young or hold on to valued topographic features: for gray wolves, a lookout; for horses, a windbreak; and for many primates, safe spots to sleep. Spaced out across their "towns" like suburban homes, the mounds controlled by prairie dogs, much like the shells of the cichlid fish, are important as living spaces for individual members—or two or three members per mound in the case of the prairie dogs. All must be defended from outsiders.

With regard to the control of resources, the costs incurred by coping with rivals among the members themselves can be compensated for by the effectiveness of having more eyes and ears to detect competitors and other threats from outside, more voices to give warning should enemies appear, and more teeth and claws to fight back. Safeguarding young is critical, as when elephants in a core group shield everyone's calves against lions or when horses in a band encircle the foals to kick at wolves. At times all concerned take part in defense. An entire baboon troop, right down to mothers carrying their infants, will commit to mobbing a leopard together, with some of the monkeys receiving massive slashing wounds in their attempt to corner and, occasionally, kill it.[12]

A society's most intense competition is typically with other societies of its own species. The best defense can be a good offense. A common way to monopolize commodities from conspecifics isn't to protect the assets directly but to lay claim to the space where they exist. Territoriality is an option if an area can be exclusively controlled or at least heavily dominated. So it is that species take a wall-building approach to international relations: in the shell-dwelling cichlid fish, there are literal walls of sand thrown up between the apartment complexes of different groups, whereas mammals like gray wolves and painted dogs mark their spaces with a scent. Prairie dogs take advantage of visual landmarks such as rocks or brush; however, even borders cutting across open ground remain stable from parent to offspring. The rodents know exactly where they are and protect their space by driving out or killing trespassers. When a species is territorial its societies can be identified simply by mapping out the movements of the animals to show that separate groups occupy different regions.

The societies of a few species, such as the horse, the savanna elephant, and the savanna baboons, aren't territorial. Instead these animals cohabit the same terrain as other societies of their kind. Even so, they rarely wander haphazardly, but stay in a general region each group knows best. These societies don't quarrel over access to the land itself, but rather over the resources it contains, which are generally too scattered to make defending a whole stretch of ground practical. A strong society can seize either the territory or its resources from a less populous society or lone animal. Not all mammals struggle with their closest neighbors, however. Both bonobo and coastal Florida bottlenose dolphin societies stick to what are essentially private plots of ground but fight little when outsiders visit, a border-without-walls accommodation that suggests they face less competition with the neighbors.[13]

Sometimes being in a society serves in part to cut down on the harassment animals experience with their own species merely by reducing their interactions to a merciful few. A band of horses usually contains adults of both sexes, for example, but some bands do fine with no stallion. "Better the devil you know than the devil you don't" would be a spot-on slogan for such all-female bands when they acquiesce to a male joining them. No matter how pushy that stallion may be toward the mares, he can repel the endless stream of other pestering bachelors that come along.[14] In other species, members of a society will mate with outsiders, liaisons that the animals of the opposite sex try to stop. A passerby ringtailed lemur male finds eager sex partners if he can slip into a troop undetected by its resident males. Similarly, a female prairie dog can cross into another territory for a dalliance, although she will be attacked if discovered in the act.

A final bonus of societies is their internal diversity. Numbers are good for more than the mere repetition of eyes and ears, teeth and claws: the members' varied strengths make up for their individual deficits. A monkey with bad eyesight or an injured leg, or which is simply poor at finding food, can tag along and benefit from those with eagle eyes and healthy limbs, even when the strong don't intentionally set out to aid the weak. And it may be the weak can also fill a social role, perhaps by tending the babies.

RELATIONSHIPS WITHIN A SOCIETY

In all these matters—from dodging or fighting predators or enemies to avoiding harassment, finding resources or mates, harvesting food, doing chores, and teaching and learning—opportunities arise for cooperation or altruism between animals. Though identifying as members, rather than cooperation, is the feature that unites a society, cooperation can obviously be a benefit of society life, even if the members have conflicting stakes in what goes on or when helpfulness doesn't extend uniformly to everyone. Selfishness, as seen in the caterpillars that hide behind each other to escape being someone's lunch, is seldom a prime motivator for associating with others. But that members commit such perilous acts as baboons ganging up on a leopard points to an alignment of interests—though such alliances are better developed, or expressed more widely, in some species than in others.

Many instances of affiliation *within* a society, such as those personal social networks designated by Dunbar's number, are customized to the individual. Often one's closest ally and best friend is a family member or potential sex partner, but not always. Both gray wolves and horses turn to special companions for solace and support.[15] Similarly lionesses that happen to rear cubs at the same time form a tight partnership called a crèche. Bonds between male bottlenose dolphins, often childhood buddies of different moms, last a lifetime and serve the pair as they court females together and drive away adversarial males.

Tight friendships and networks of allies don't guarantee that life in a society is a cakewalk, as the previous chapter made clear. Not every person can afford a penthouse suite, and not every wolf is top dog. Societies can become veritable battlegrounds of a social, and at times physical, sort, with each individual's portion of group resources at stake. Animals often stick to a society in spite of the power plays and upsets, pain and persecution, holding out for the opportunities that come along, each struggling for its version of the American dream.

In this, some have it harder than others. Power differences can exist not only between societies, but among the individuals within them. Control by dominant individuals, often at the top of a social hierarchy, presents a tough bottleneck to advancement in many primate societies, resulting in intense physiological stress.[16] The same is true of spotted hyenas—the only hyena species to form societies. In particular, the most alpha male spotted hyena still has a smaller penis and less testosterone than his savage female counterparts (which have a sexual appendage called a pseudopenis) and a status so low that any cub can drive him away from food.[17] By comparison, savanna elephants, bottlenose dolphins, and bonobos lead idyllic lives, yet they still face discord: unpopular elephants are mistreated by their sisters, dolphins spar over mates, and a mother bonobo intervenes to frighten off any males that pester her son while he's trying to have sex.

Dominance has its advantages, even for those who fail to achieve it. Certainly once a hierarchy is set in place, influenced by each individual's physical and mental endowments—and in some species, based on the status of its mother—conflict may decline, a bonus for all. With its position worked out, a monkey of low standing can stop wasting time confronting much higher-ranking individuals and focus on improving its station among those at its own, perhaps miserable, level. Without reconciling in this way to their situation, members would wear themselves out. For people as well as other animals, societies would fall to pieces under the general, unrelenting struggle to get ahead.[18]

It seems, then, that both the affable bonobo dependent on his mother's protection and a baboon stressed over its social standing live by the rule Michael Corleone pronounced in *The Godfather, Part II*: "Keep your friends close, and your enemies closer." Within the confines of a society's predictable membership, animals can keep tabs on both friends and adversaries, refining how they handle their relationships, positive and negative, and sometimes even working with competitors to their advantage.

Perhaps this explains the bonobo, a species for which the advantages of societies are not obvious given that these apes, in striking contrast to the chimpanzee, compete little, cultivate foreign friends and sex partners, face few predators, and seldom need help finding meals or catching large prey.[19] Indeed, bonobos can be outgoing, even generous to outsiders.[20] It seems they could behave the same way if they lived with no distinct communities to speak of, yet the community borders remain in place. Why, then, have societies? Possibly their communities exist for the most minimal of reasons: so bonobos can structure their days around a clear and manageable set of individuals. This hypothesis is reasonable, albeit not altogether satisfying. I am inclined to think the societies offer them more than is clear to scientists at the moment. Be that as it may, one thing is certain: bonobos, like the chimps, and like savanna elephants and prairie dogs, spotted hyenas, and bottlenose dolphins—and other vertebrates such as Florida scrub jays and shell-dwelling cichlids—have lives firmly rooted in their society.

One psychologist has declared of humans that, as a consequence of our evolutionary past, our "sense of personal security and certainty" is strongest when we belong to a group perceived as different, and distinct, from outsiders.[21] This statement could well be true of any number of animals that depend on being provided for and protected by their fellow society-mates. In fact, much of society life bears on how those members interact, yielding dynamic movements within and among societies that have been critical in shaping the social evolution of species, humans among them, a subject that will be addressed forthwith.

CHAPTER 3

On the Move

O ur Land Rover lurched to a stop in view of something wonderful—a painted dog, ears held up to form satellite dishes. I couldn't contain my excitement—these group hunters are rare here in Botswana and throughout sub-Saharan Africa. Painted dogs live in packs, and so others had to be nearby. Yet this one was alone and nervous about it. She paced, then paused, made a loud hoot, listened, hooted again. Within a few seconds of her third try we heard the response of her pack companions and instantly the female bounded away in their direction. One bumpy, banging minute later our vehicle landed among a crowd of the wild dogs, dozing, leaping, growling, sniffing, and playing in a tight bunch all around us.

Societies offer benefits that a solo existence cannot—that much is clear. But the shape those benefits take has everything to do with movement. The patterns by which the individuals of a society move and occupy space establish the interplay of individuals and groups. For a painted dog or monkey, being separated from society-mates is an emergency situation on the order of a three-alarm fire. Should a meerkat, for example, be too engrossed in eating a scorpion to notice the departure of its clan, the laggard gives a lost-call until it hears a response and reunites with the others, its distress reflecting the risk that a predator or enemy takes advantage of its plight to go on the attack. For horses, too, there can be a panic for a stallion on a walkabout or a mare that falls behind with her foal. The wanderer may climb a hill to catch sight of its band and promptly run to catch up.

That is not the case in all animal societies: in other species, being spread out is the normal state of affairs. Prairie dogs are an obvious case, distributed as they are among the semiprivate mounds from which each of them forages every day. Their burrows stay put for generations. However, the most intriguing of these habits of dispersal is a dynamic kind of nomadic life. In these fission-fusion species, society members temporarily cluster here and there in social groups that form, dissolve, and reform elsewhere. Since most animals seldom have to be in immediate proximity, a bit of fissioning and fusioning can be detected in the societies of just about any species. Still, fission-fusion is the workaday behavior of a handful of mammals.[1] Spotted hyena, lion, bottlenose dolphin, bonobo, and chimpanzee societies rarely converge on one place. For other animals, fission-fusion is more context-specific: packs of gray wolves and cores of savanna elephants split up for times when it's beneficial for finding food. Fission-fusion sounds esoteric, but there's an abundance of reasons to look at this lifestyle. The fission-fusion species include virtually all the brainy mammals that anthropologists studying the social brain hypothesis salivate over—most significantly, us, *Homo sapiens*.

The animals of a fission-fusion society travel as best serves their social success. Put simply, this is where social smarts come into play. Moving with few constraints, these animals have the luxury of picking their companions, spending quality time with a friend or mate while finding relief from adversaries. For spotted hyenas such relief takes effort—the species is competitive from birth, at times to lethal effect. Observing the animals with hyena expert Kay Holekamp in Kenya, I delighted in the sight of cubs frolicking at the den until I realized the object of their tug-of-war was a playmate's corpse. As soon as the young are big enough to leave the den, they scatter across the community, seeking out allies while approaching any other clan member with caution.

Freedom of movement allows social interactions to grow more complex, and it also pays off whenever personal relationships become hard to manage. Options open up that are impossible for animals living "face to face"—monkeys, for example, that have no choice but to stay constantly together in a troop. Object to the local scene? A cunning chimpanzee of low social rank may find an opportunity to slip off to a quiet spot in his community's territory for a private liaison with a female. Even more surreptitiously, a gray wolf or male lion can sneak away to visit another pack or pride as a first step to defecting—transitioning to become a member of another society can require this sort of painstaking duplicity. No surprise that many fission-fusion species are so bright.

Fission-fusion offers other advantages. Spreading out allows for a territory to sustain a greater population, as everyone in the society isn't clambering on top of each other to gain access to the same resources. Imagine what would occur if a whole community of 100 chimpanzees stayed in one tight troop: the land they passed through would be picked clean, forcing them to move relentlessly and scramble for each tidbit like shoppers on Black Friday. Instead, the chimpanzees keep apart, meeting in temporary groups, called parties, of a large size only where they hit a resource jackpot, such as a tree weighed down with fruit.

Fission-fusion has its downsides. When members are spread thinly across their territory, enemies can enter and strike small groups or singletons with little danger to themselves. The attackers are also less likely to become the target of a massive counterattack before they make an escape. Such an assault on at most a few individuals would be suicidal if directed at 100 or so chimpanzees in one party.[2]

Perils of assault aside, the scattering of individuals means fission-fusion animals can better monitor a big space for thieving trespassers, since the society has eyes and ears almost everywhere. Maybe that's why chimpanzees never attempt incursions to feed from the neighbor's fruit trees. By comparison, animal societies that stick together rarely know when an intruder enters the far stretches of their home range. For instance, baboons and painted dogs, which spend their days very near others of their troop or pack, can't do much about a rogue foreign animal visiting the far corner of their territory. In this way, spreading out can help the animal society defend more ground.

TOGETHER OR APART

It can be easy to claim that a species lives by fission-fusion based on our human perceptions. However, it's important to consider that animals perceive space differently depending on their sensory acuity and how they stay in touch. In short, whether the members of a society are close together or far apart isn't for us to decide. Writing of the hardships baboons face, the South African naturalist Eugène Marais dubbed their lives "a continual nightmare of anxiety"[3]—but it's not hard to imagine that anxiety as being aggravated by their constant presence together. GPS data from a troop of baboons confirms that, at one particular area where they stay quite packed, the animals never venture more than a few meters apart.[4] In reality, though, while a baboon is almost always aware of the presence of others, it is unlikely to register the dozens in its troop from moment to moment. Most of its fellows will often be out of sight, vanishing behind its back or looping around a bush. Since baboons' sight and hearing are not much keener than those of a human, the best they can do is keep tabs on whichever comrades are nearest and catch up when they move. When practiced by all, these behaviors should be enough to keep a troop in formation.

Species in which society members stray farther apart are often associated with superb sensory abilities. Elephants are aware of the sounds of their companions across several kilometers. While a person might presume a couple of elephants resting under a tree are isolated from those few eating leaves out of view, the animals, emitting a continual low-frequency call known as rumbling, are connected. This ability to communicate over long ranges permits a core's scattered members to coordinate their activities.[5] Spread out they may be, but until they wander very far indeed, the elephants can be better informed about each other's movements than the baboons in a compact troop.

Togetherness, then, is in the eye (or ear or nose) of the beholder, and even fission-fusion animals can panic when truly isolated. All of which is to say that for society members to act in concert, or at least respond to each other effectually, what makes a difference isn't how close to each other they are in our eyes, but their knowledge of the location of others. This knowledge can be limited to a local area, as in baboons, or extend far and wide, as in elephants. The members of a society are *together* as long as they are aware of each other.

Contrast elephants (which scatter, yet can still often act as one since they remain connected by their acute senses) with similarly spread-out chimpanzees. With vision and hearing like our own, a chimp may ordinarily just perceive those near it. In fact, the sensory limitations of chimps, for all intents and purposes, render their communities much more fragmented than the cores of the savanna elephant. If a danger presents itself—a hostile outsider comes by—only those chimps that happen to be together at that place have any chance in organizing a defense. So while a community is always dispersed, the chimps spend most of their time among others, in one of a community's scattered parties.

Yet in even the most dispersed fission-fusion societies, animals usually have some means to check in at intervals with those who aren't too remote. Spotted hyenas whoop and holler at any provocation; chimpanzees stay in touch by loud pant-hoots; lions roar; wolves howl. As long as lines of communication exist, individuals aren't completely cut off.[6]

We are still learning what information the cries convey. The animals may use them simply to update others on their whereabouts and space themselves accordingly, though some vocalizations can be calls to action. Lions can summon assistance to battle or to chase down dinner. As many as 60 hyenas have converged in an attack on an enemy clan, a nasty and deafening event, while fewer come together to stage a hunt. Communiqués can be hacked. Hyenas make so much ruckus they often draw in enemies and end up in a brawl.

Among the chimpanzee calls, the pant-hoot is the noisiest, reverberating for a couple of kilometers. The name is onomatopoeic—no human makes any noise like it in polite society (unless you are Jane Goodall impressing an audience). The cry expresses excitement among apes on an outing together, a *yay team* cheer that strengthens the bonds between males and helps parties within earshot keep tabs on each other. Discovering fruit sends chimps into a pant-hoot frenzy, attracting others to any spot where there's plenty to eat.[7] But it's the pant-hoots from a foreign community that really agitate the apes. They embrace, stare toward the uproar, and sometimes pant-hoot back in a hollering match.[8]

Most chimpanzee territories are expansive enough, though, that the parties separated by the greatest distances are likely to be out of earshot; hence in practice the entire community is never alerted to a situation, and certainly never acts as one. By comparison, bonobos form similar parties, but they are larger, longer lasting, and are often all in listening distance. There can be an active exchange of information across an entire bonobo community. That skill is on display at sunset, when the bonobos voice a special "nest-hoot" that guides everyone to the same general area for bedtime. Primatologist Zanna Clay tells me that should a bonobo party wander beyond the hearing range of the others, they forgo giving the nest-hoot, apparently aware they are on their own—and fine with it.[9]

Someday we may learn that animals impart more information than we have given them credit for. Gunnison's prairie dogs communicate in a complex range of tones; it seems they are able to formulate a slightly different squeak of alarm when, say, a tall fellow in a red shirt runs into view compared to a short, slow-moving guy wearing yellow, or when a coyote comes along versus a dog. Whether, and how, the little fellows use this information isn't known.[10] Other species may similarly communicate more than we know now. Could a howling wolf be giving its pack-mates instructions over a great distance? *We over here smell the blood of the enemy—come prepared for battle!*

What of humans? I will have more to say later on the topic of dispersal in our species, and its significance to the origins of our intelligence and our societies—but here's a preview. Before agriculture tied people down, hunter-gatherers had the option to spread out. As it does for other species, this fission-fusion reduced competition within a human society and enlarged the population the land could support. At the same time it permitted each individual to sort out how, and how much, to interact with special others. Chimpanzees might be incapable of taking full advantage of the fusion aspect of fission-fusion to draw together faraway members, whereas for humans, keeping abreast of each other at a distance has been a key to success. Most people in early societies were usually too far off to yell, making bits of ingenuity such as smoke signals and drums indispensable. Such techniques had limitations, as did all methods of long distance communication before Morse code transmitted news in real time. Many prehistoric signals might have conveyed about as much information as a text message reading "Hello."[11]

Nevertheless, early records are full of clues that hunter-gatherers communicated quite well, especially in an emergency. Messengers making the rounds perhaps represented the pony express of the day—running was no problem; the human body is built for endurance.[12] By such means a thirsty populace gathered at the last puddle, and anyone stumbling on game or an enemy drew in others for feast or fight. In one of the first European contacts with native Australians, the crew of a Dutch ship kidnapped a man on April 18, 1623. The next day they were faced with 200 Aborigines brandishing spears: apparently word had spread fast.[13]

SHIFTING ALLEGIANCES

Societies are closed, yet not impregnable. Movement between societies is essential to the health of each group. Without opportunities for individuals to transfer from society to society, the rare sneaky infiltrator aside, populations would become inbred, especially when the societies are small. Savanna elephants circumvent this predicament with males that range free. Bulls spend much of their time with a mix of other males, sticking closest to a male chum or two. Unaffiliated with a core, they mate with whomever they choose (but rarely without a fight from another bull). In most cases, though, societies comprise

both adult males and females. In such species, transfer between societies can be a mandatory step in reaching adulthood. Youngsters
may strike out from their birth society after its members, who knew
them growing up, avoid them as mates. Primates represent a diversity
of sexual dispersal patterns. In the highland gorilla and a few other
primates, either sex can make the move. In other species, just one sex
relocates. In the hyena and many monkeys it's the male that leaves;
in the chimp and bonobo the female does—although a few female
chimps stay where they were born. An animal may relocate again later
in life if its status is so miserable that it's pushed out of a society or escapes voluntarily. Not only are hyena males beaten up by the females,
but the females quickly lose interest in newcomers as sex partners,
forcing many males to try their luck in another clan.

Sometimes an animal is on its own for a while before it finds a new
home. Actually, acceptance by a new society can be hard to come by—
to keep their boundaries distinct, barriers to entry are high. A forceful
individual may initially bully its way into the group by fighting those
of its own sex, who see it as a contender for the society's private supply
of mates. But attaining permanent group membership can require
being found desirable by the opposite sex, too. Hence a freshly arrived
male baboon insinuates himself into a new troop by beating off the local males while befriending the ladies. A wannabe female member of
a chimpanzee community typically makes her first appearance when
she is sexually receptive. This strategy attracts an entourage of male
enthusiasts that shield her from females who would rather not have
her stay.

Even so, the female chimp must contend with those females after
she is taken in, just as a newcomer baboon is stuck dealing with the
current male hierarchy; with chutzpah and female support he rises to
its top. This aggravation doesn't exist in species where newbies force
the previous male occupants out of a society. Male lions and horses
hang with other bachelors until they earn a place in a society, typically
by driving off its current male residents. They may not do it alone:
two or more male painted dogs, lions, and at times horses can team
up to stage a hostile takeover and are likely to remain longtime allies
thereafter. Male ringtailed lemurs and Ethiopia's baboon-like geladas
cultivate similar bromances. While the geladas maintain their ties,
lemurs often forget their partnership once they have inserted themselves among the other males of a troop.

In other cases, joining a new society is a matter of perseverance
rather than force. From time to time a female elephant will defect

to another core, sometimes after its members repeatedly nudge her
away. A male hyena or female monkey hovers on the fringes of a clan
or troop, putting up with an onslaught of abuse until its presence becomes routine and tolerated. The ease of acceptance can depend on
the straits of the destination society. A wolf pack will normally admit
a lone male only after days or weeks of attempts, if ever; but under
the right circumstances a lucky dog can be welcomed almost immediately. That was demonstrated when, in 1997, a Yellowstone pack,
whose alpha male had been killed by humans, was observed taking
in a roaming wolf. Keeping their distance for hours, the loner and
the pack howled back and forth until they came face to face. Eventually a juvenile broke the ice, running forward to greet the male. Only
six hours later, the whole pack mobbed him enthusiastically. Any dog
owner would recognize their behavior: tail-wagging, sniffing, and play.
Even though his adopted pack had killed two of his siblings the year
before, he instantly became the new alpha animal.[14]

In species where only one sex leaves, it's the less adventuresome
sex that has the easier life. Like a town resident who inherits the family
farm, such an animal retains childhood friends and relatives as well
as familiarity with its home turf, an area it knows like the back of its
hand. The grown male chimpanzee, a lifelong homebody, goes back
to his favorite childhood haunts, whereas a sister who moves to a new
society must start from scratch, severing not just her personal relationships but also all bonds to her birthplace.

In all of the foregoing, we have seen that the ways animals travel
and keep track of each other inside their societies can influence the
benefits societies have to offer; we have also looked at how movement
between societies, accomplished at rare intervals and with difficulty,
allows for changes of allegiance that turn a former foreigner into an
established part of the social scene. Next we consider what, and how
much, mammals must actually know about each other to reduce the
flow of individuals in and out of their packs, prides, troops, and cote-
ries, so each society can function as a clear, independent, and enduring unit.

CHAPTER 4

Individual Recognition

S ocieties, as we have seen, are not impregnable. An elephant can seek her fortune in a new core group, male hyenas and female chimpanzees can jump ship to other clans and troops, and a lone male wolf can be crowned leader of a foreign pack. This movement of mammals between societies brings to the fore the conundrum of how membership is achieved and how it stays crystal clear.

It's possible for groups other than societies to stay cleanly apart indefinitely, even when the animals involved know nothing about each other. For instance, certain social spiders come together by the hundreds to weave communal webs to catch prey, the average person's bad dream. Because the webs are spaced and the spiders stay put, the colonies normally never mix. Yet put one colony next to another, and the spiders merge without making any distinction between them.[1] You wouldn't know it without doing this experiment, but the colonies prove to be completely permeable—open to outsiders coming and going. Thus it would overstate the matter to say the spiders are part of a *society*: they display no affiliation with the colony that one could construe as membership. Too much permeability can sometimes be an issue even for species with clear societies. Where domestic honeybees evolved in western Asia and Africa there was little opportunity for confusion between hives. These days, beekeepers put the nest boxes so close together that some workers drift—fly astray and go about their business in blissful ignorance, working for the wrong colony, resulting in a loss of beepower for the home hive.

Among most vertebrates, permeability is restricted to rare and arduous transfers between societies of individuals seeking mates. These species keep outsiders out by forming individual recognition societies: each animal must recognize every other member as an individual, regardless of whether that peer was born in the group or admitted from outside. In the mind of a savanna elephant or bottlenose dolphin, then, every Tom, Dick, and Jane in the society must be identified as Tom, Dick, or Jane. Not that they use personal names—except perhaps the dolphins. These cetaceans can catch the attention of a friend with custom "signature whistles" that some researchers take to mean "over here, Jane!"[2]

Know everyone by heart and of course you know who belongs—piece of cake. Animals can also learn to identify individuals who are not part of their society. And those individuals are often, but need not be, enemies. Primatologist Isabel Behncke tells me that certain bonobos groom each other when communities come together. While not exactly international diplomacy, grooming appears to be a sign of extrasocietal friendships. This openness doesn't lead to misunderstandings as to who goes where. After some frivolity, the apes go home, memberships intact. Chimpanzees fail to have such foreign friends except in the Taï Forest of Africa, where they have to be circumspect about it: females from different communities, who presumably knew each other in their youth before they emigrated elsewhere, will get together. The two meet surreptitiously, as if aware they would be killed if anyone else found them grooming each other.[3] A society, then, seldom includes *everyone* animals know. Rather, an animal sees its society as a particular set of individuals: an "us" versus "them" distinction.

Such capacities in an animal may seem surprising, but many vertebrates can store information about others of their kind, tagging each with a category that we (a language-using species) might call "citizen," and then further classifying their peers within that group.[4] Consider baboons: they recognize rank, family, and coalitions within their troop and use these categories to predict how others will behave. Because, for example, the females largely inherit their social ranks from their mothers, forming support networks known as matrilines, baboons expect a low-status female, no matter how assertive her personality, to back down in a face-off with a female of high rank, no matter how infirm or shy. Indeed, "in the mind of a baboon . . . social categories exist independent of their members," biologists Robert Seyfarth and Dorothy Cheney have said.[5] Doubtless that applies not only to their matrilines but to their societies—their troops—as well.

Africa's vervet monkeys don't just recognize which monkeys are foreign, they often know which troops those others are linked to. Cheney and Seyfarth found that vervets recognize each other by voice and paid more heed when the cries of a member of a neighboring troop came from the wrong place—from the territory of yet another troop. The vervets bounced around manically like spectators anticipating the blows in a boxing match. Their expectations fit what one would guess to happen in this circumstance. The monkeys' reaction indicates that they realized the neighbor whose voice they recognized must have entered another troop's land, behavior that generally occurs during a brawl between troops. So beyond differentiating their own society from the catchall category of "everyone else," vervets grasp that foreign monkeys are split up among multiple troops as well.

A membership system based on individual recognition works because the members agree on who belongs. At times there can be differences of opinion, but across all society-forming species these are rare and temporary, limited to the transitional moments when an individual is being expelled from, or joining, a society. A stallion, excited at the prospect of another mate, may encourage a mare to join his band while the band's mares, still seeing her as an outsider, simultaneously try to drive her off. The endgame for the mare, as she works to wear down this female opposition, is not merely to be recognized, but to be recognized as one of the group.

RECOGNIZABLY DIFFERENT

Recognizing one's group-mates, of course, requires that each member be distinguishable in some way. Across the animal kingdom, recognition can come in many sensory modalities. The vocalizations of most social mammals, from the calls of vervets to the roars of lions, vary from one individual to the next. Vision can also be important: capuchin monkeys can quickly distinguish between photos of their own troop members and those of outsiders.[6] Depending on context, the distinctive spots of spotted hyenas serve variably as camouflage and as means of identification at a distance on the open savanna. Lacking markings as distinct as a hyena's spots, chimpanzees are drawn, as people are, to faces, notably the eyes, as a key distinguishing feature. Their voices, too, are as distinctive as those of humans, and they can also tell each other apart perfectly well by their rear ends, a gift of discernment yet to be tested in humans.[7] Horses give way at a water hole if they see members of a dominant band approach within fifty meters.[8] Monkeys grow so familiar with one another that in Uganda the red colobus and other species—blues, mangabeys, and redtails—play together with everybody zeroing in on certain interspecies friends.[9] In studying how well lions know each other, George Schaller, a senior conservationist at the Wildlife Conservation Society, has thoughtfully observed their behavior both within and across prides:

> No matter how widely females are scattered or how frequently they meet each of the other members, they still constitute a closed social unit which strange lionesses are not permitted to join. . . . A pride member joins others unhesitatingly, often running toward them, whereas a stranger typically crouches, advances a few steps, then turns as if to flee, and in general behaves as if uncertain of its reception.[10]

The cats pounce on a lion that seems strange until it is identified as one of their pride. Mind you, no one has set out to methodically prove that lions or any other animals recognize each and every member of their society, but the assumption seems safe based on this sort of observation.

Recollecting others must have been a necessary precursor to the evolution of societies in vertebrates.[11] Indeed, I wouldn't be surprised if individual recognition is universal among mammals and birds, given that fish, frogs, lizards, crabs, lobsters, and shrimp all have the capacity.[12] We should expect this. Even for animals that don't live in societies, telling one individual from another is still important, whether in fighting over a territory, dominating others, finding a mate, or distinguishing one's infant from somebody else's. And so hamsters, asocial as they are, integrate the smells from different parts of the fuzzy bodies of others of their species into a representation of each individual, much as our minds process facial lineaments to create a comprehensive representation of a person.[13] Emperor penguins and their chicks experience long separations when the parents leave for days to bring home a fish. How do they find each other among the thousands in a colony? By listening.[14] In the same way that we filter the din of a cocktail party, the birds selectively hear the calls of their kin: *Did I hear Tom on the other side of the iceberg?*

However crowded their conditions, neither hamsters nor penguins have societies. And however widespread such feats of memory, life in a society presents a whole different ball of wax when it requires a blanket knowledge of *everyone*—or at least a knowledge of all other members to some minimal degree. But how minimal?

Biologists studying the issue of recognition and relationships have focused on the strongest connections in a society: the individuals that know each other best, and how they interact. Yet as a consequence of this seemingly sensible choice, an equally compelling area of investigation has been neglected—the question of how much the individuals that interact *least* in a society actually know about each other. It is conceivable that two members who happen never to approach each other may be ignorant of the other's existence. Their lack of contact could also be a sign of indifference, disdain, or an inability to start a relationship because they move in different social circles. The two may overlook or avoid one another as a strategic choice, or it could be that, as with the guy you have seen sitting in the coffee shop a hundred times, they never get around to introducing themselves because there are just not enough hours in a day.

It may be even simpler. You might be completely unaware of the fellow drinking his coffee, but still register him in the back of your mind. Does the following experience sound familiar? One day I felt that something was different about my favorite café. After a moment I realized a regular patron was gone, a person who, for the life of me, I couldn't recount in detail even if I had to. When we take a moment to perceive someone, like a coffee shop patron, as a person, we are said to individuate him—render him as a Tom, Dick, or Jane. People, and presumably many animals, don't have the endurance to individuate everyone. We store our knowledge of many others at a schematic, often subliminal, level.

Here's a thought experiment: Suppose humans related to everyone entirely in this schematic way. A scientist would pronounce that we are *habituating* to the distinctive quirks of each individual, which is to say that our subconscious registers his or her personal attributes at the same time we tune out those attributes from our everyday awareness, much as we do with ambient noise we don't notice until it goes quiet. A species might still form societies by individual recognition at this subliminal level, without being mindful of a single individual around them—what a weird and impersonal world that would make!

Despite that possibility, I suspect most vertebrate societies are small enough that animals are cognizant of the other members, not just in the minimalist way in which I was vaguely aware of that café patron, but very well indeed, whether they choose to take notice of them constantly or interact with them often.[15]

DEMANDS OF MEMORY

The members of a troop of baboons or vervet monkeys stay close enough together to encounter each other daily, if not every few minutes. Constant exposure to the same familiar faces should make individual recognition a breeze. However, because out of sight can mean out of mind, and some animals in a fission-fusion society can be absent from their companions for a while, recollection can sometimes be a challenge. For example, biologists spend months between sightings of certain shy chimpanzees, often beaten-down females who are members of the community but keep to themselves in a private corner of its territory. Almost certainly other chimps see such loners only occasionally and after long intervals, too.[16] Clearly chimps need good memories.

Anthropologists have designated our species as "released from proximity" because we not only remember others but also keep track of relationships with people we don't see for lengthy periods of time (even maintaining them through intermediaries—friends of friends and the like). For us, trust is easily regained or suspicions rekindled.[17] This "release" pertains to other mammals as well, with society-mates reuniting peaceably after a protracted absence. There are many records of animals that, recognizing others despite changes wrought by age, take up their relationship without skipping a beat. Biologist Bob Ingersoll recalled visiting a chimp that hadn't seen him in more than 30 years. "At first she didn't really recognize me, and I didn't recognize her. Then I said, 'Mona, is that you?' And she immediately signed to me, 'Bob. Hug hug hug.'"[18]

Memories aren't always so long. Horses don't recollect their own offspring after separations of more than 18 months. By that time the colt or filly will have bonded to a society—a band—of its own.[19] Perhaps, in species where animals permanently depart, good recall would be a waste of mental energy and a burden.

The demands of memory include inserting oneself into the memory banks of others. In an individual recognition society, being correctly identified, let alone being identified as one of the group, ultimately takes becoming known to all. An outsider joining a society faces risks due to its initial status as a stranger: anyone who doesn't recognize the newcomer may attack. To sidestep this difficulty, as primatologist Richard Wrangham pointed out to me, the new arrival will likely stick near those who are acquainted with it already, so that those animals it still hasn't met will see it as belonging there—*a friend of our friend must be a friend*—or at minimum, a compatriot.

Youngsters born in a society are not exempt from this quandary. The process of becoming a familiar presence begins early, with a mother picking out her infant. Birds distinguish their young by an age when there are likely to be mix-ups, for example when the chicks leave the nest and might intermingle with other chicks: species that pack into colonies, for example, can tell one chick from another almost from birth.[20] For a youngster to become part of a society, though, not just mom but everybody else has to learn to identify it eventually. Fortunately, a baby is harmless at first, reasonably ignored by everyone other than the mother. Problems come when it reaches an age at which others may mistake it as a possible foreign threat. All the adults in a troop of dwarf mongoose jointly anoint the juvenile members with an excretion from their anal glands, perhaps a sign of acceptance.[21] Until then, like the immigrant, youngsters of every species stick by those who know them best, safely presenting themselves as *the friend of a friend who must be a friend.*

INDIVIDUAL RECOGNITION AND THE SIZE OF SOCIETIES

All the societies of the mammals I've mentioned share one conspicuous feature in common: they are small, with often a dozen or two and seldom over 50 members. Lions don't sweep over the Serengeti in prides of a thousand to fell a herd of wildebeest. Prairie dogs never take dominion over a territory in the style of a human nation, their burrows spreading across the landscape with a population at ready to repel all foreign prairie dogs. And other apes don't rise up in armies like those in *Planet of the Apes.*

In some instances the ultimate explanation is ecological. One thousand lions would have trouble feeding themselves day after day, and such an enormous pride would starve. But it's safe to predict that the societies of most of our vertebrate cousins are consistently small for a more prosaic reason: as Dunbar recognized for friendships, keeping track of many individuals, in this case sometimes even very minimally, is hard.

With the exception of humans, and our astounding capacity to live together, ape societies peak at about 200 members in the chimpanzee. Most vertebrate groups much larger than this are aggregations—schools, flocks, colonies, herds, and the like, from which individuals come and go with impunity—not societies, which have rigid memberships. A school of herring located not far from Manhattan (and rivaling the size of that island) once held 20 million fish. The sky's the limit for the red-billed quelea, which turn and wheel over Africa in flocks of a million, while equal-sized herds of wildebeest thunder below. As for nesting birds, the largest colonies may be the 4 million sooty shearwaters that breed on Isla Guafo, Chile.

Such groups consist of opportunists. Penguins huddle yet, bonds between parents and chicks aside, are indifferent about whom they huddle with. During migrations, "the assembled wildebeest are a crowd of strangers," as one wildlife biologist put it.[22] The same can be said for caribou and American bison, which may stick by their calves and a couple of friends but join herds made up of varying collections of unknown others.[23]

A mammal society reliant on individual recognition faces a limit to the number of others, whether those members be friend or foe, that each animal will be able to track and recall. As societies approach that maximum, memories fail, putting a hard cap on their size.[24]

The grass-browsing gelada, looking like a baboon in a Halloween mask, is a telling example of the constraints of memory. Since many hundreds of the animals forage in a herd, we might imagine a gelada's recall of others would be legendary. Yet when primatologist Thore Bergman watched how the animals responded to recordings of other geladas, he concluded that they are almost as ignorant of those in the crowd around them as are the bison. Almost, but not quite: their recognition of others was capped at twenty to thirty individuals. Bergman deduced that groups within the herds, each composed of one to a few males and several adult females, are the true social entity of the gelada: in short, its societies. Herds come to be when dozens of these troops, technically called units, forage across the same ground.[25] An alpha male gelada doesn't seem to know anyone outside of his troop. This incomprehension leads him to treat the many outside males moving around him all day in the herd as interchangeable potential philanderers of his females and usurpers of his position.

This may make geladas seem dim-witted, yet an important lesson emerges. Animals reliant on mutual recognition are compelled to be at least minimally aware of everyone in their society, and that renders sociality at a massive scope prohibitive.[26] Our own kind, *Homo sapiens*, has obliterated the glass ceiling of those other species. No matter what our intelligence, humans would never have been as successful as we are today without going large.

Before turning to humans, however, we continue our exploration of nature with another high point in social evolution, the insect societies. Not only do these arthropods include the vast majority of the society-dwelling organisms, but among their colonies are some of a scale and complexity that are indeed immense, and which I believe shed light our own societies. Here's a cryptic preview of the conclusions ahead:

> *Chimpanzees need to know everybody.*
> *Ants need to know nobody.*
> *Humans only need to know somebody.*

And that has made all the difference.

Anonymous Societies

CHAPTER 5

Ants and Humans,
Apples and Oranges

When I first made my way into a tropical rainforest as a young man, what enthralled me most wasn't the monkeys, parrots, or orchids, alluring as they were, but sliced leaves the size of coins, hoisted in a parade of ants a foot wide and the length of a football field.

You don't have to like entomology to be a fan of ants. We modern humans may be genetically close to the chimp and bonobo, yet ants are the animals we resemble most. The similarities between our species and theirs tell us a lot about complex societies and how they emerge. In fact there are many modes of life among the over 14,000 or so ant species, and even more for social insects in general, a category that includes termites, social bees, and certain wasps.

The leafcutter ants of the American tropics exemplify the ant's potential for social complexity. Within their nest, the green leaf banners are broken down to a substrate on which the ants raise their food crop, a domesticated fungus they tend in globular gardens ranging from baseball to soccer-ball size. There are ants with a nomadic, even fission-fusion lifestyle, but species that settle down, such as the leafcutters, can do so in spades. Basing foraging operations out of a nest or central lair is uncommon in vertebrate societies except for species that stash their young in a den or primate troops that keep returning to a tree to sleep, yet is commonplace in ants. The size of leafcutter settlements can be gargantuan: in the French Guiana jungles I came upon a nest the square footage of a tennis court. A drawback of such a metropolis is the same faced by a human city: pulling in enough resources means a lot of commuting. From the far corners of that large nest spring a half-dozen speedways along which the workers doubtless hauled hundreds of pounds of fresh foliage over the course of each year. To unearth just part of another colony I once hired six men with pickaxes and shovels near São Paulo. The bloody bites I sustained that week didn't stop me from feeling like an archaeologist exhuming a citadel. Hundreds of gardens grew in chambers arrayed along meters upon meters of superbly arranged tunnels, some at least six meters below the surface. Scaled to human dimensions, their subway system would be kilometers deep.

The superabundance of activity in the nest of any ant species makes plain to the most casual observer that, among the social insects, the payoffs of group living are many and varied. Ant workers stake out territories, boldly gathering food even from our dinner plates, and rear their progeny in elaborate safe havens. Communicative, persistent, hard-working, battle-ready, risk-taking, and highly organized, whether they are agronomists, herders, or hunter-gatherers, ants form elaborate labor forces of superb military operatives and diligent homemakers, masters at both protecting and providing for their colonies. The leaf-cutters, for one, have societies decidedly more complicated than any other nonhuman animal, and carry out mass-scale agriculture to boot.[1]

Likening people to ants can raise hackles. Comparing ourselves to other mammals comes easier, because we are mammals ourselves, as our hair, warm-bloodedness, and ability to lactate confirm. For all that, while watching a nature documentary about mammal societies, you probably won't find yourself exclaiming, *Eureka, they are so like us!* What similarities exist can be subtle. More often we are struck by the differences: quirks such as the fact that male elephants are outcasts—not members of any society at all, properly speaking. As for those relatives of ours, the chimpanzee and bonobo: how like them are we? Physically we resemble both apes, owing to the genetic closeness of our species. But what of our mode of social life? Most of the similarities that have been brought up, often in the context of evolutionary psychology or anthropology, bear more on broad facets of cognition than on the details of the apes' social organization we would otherwise think of as specific to people.[2]

Those similarities are seldom as great or exclusive as they might seem. Insomuch as chimpanzees and bonobos think as we do, the parallels often extend to other animals, too. Both are like us in recognizing themselves in a mirror; then again so do dolphins, elephants, and magpies, and there's a claim, widely doubted, that ants do, too.[3] At one time chimps were thought to be unique among nonhumans in making tools—employing twigs to snare termites is one example. Yet now we know of other toolmakers, such as woodpecker finches that poke twigs to nab bugs.[4]

Chimpanzees do resemble us in how they handle conflicts. For example, some individuals achieve influence through brawn, while others rely on brains. To mention one situation: a female chimp can stop irate males from using a rock in a fight by walking over to take it away—repeatedly, if necessary.[5] Such political maneuverings, at least, might seem unique to chimps, bonobos, and us, but in fact they likely aren't. The discovery of such habits is a direct result of our opinion that these apes, which we recognize as human relatives, are worth close study—as one scientist put it, the experts have been "chimpocentric."[6] The assassination of "40F," the alpha female of the Druid Peak wolf pack at Yellowstone, is a case in point. The evidence suggests the pack rose up to kill her after she ferociously attacked two of them. The researchers who observed these acts wrote, "Her life and death could be summarized with the old phrase that often applies to human tyrants: if you live by the sword, you may die by the sword."[7] The politics of wolfdom can be complicated indeed. Once this kind of scrutiny is applied elsewhere, doubtless more species will be identified that rival apes and wolves in their mastery of social machinations.

The fact is that though we share 98.7 percent of our genes with chimpanzees and bonobos, it's our differences that are most striking.[8] Indeed we are as different from them as apples are from oranges. In both apes, relationships are dictated by strict hierarchies of power, which are tyrannical in the chimp—especially in the males. On maturing, females of both species abandon childhood kith and kin for another community, never to return. The females are sexually receptive only on occasion, this condition made obvious by their swollen rear ends. The female chimp can pretty much either be beaten up or ignored by males except on the rare days she's in heat, at which point she often has sex forced on her. No wonder neither chimps nor bonobos have pair bonds or an extended family life, and mothers get little child support from dad—or anyone.[9] Females aren't especially skilled at befriending each other, either; in fact a harried chimp mom must give birth at a private site to avoid having her baby killed.

So while I will point out engrossing similarities between the societies of humans and other vertebrates in the pages ahead—oftentimes in regard to the benefits reaped from being in a society, and how those societies interact—for the most part the social lives of other mammals, those of our ape cousins included, can seem downright weird, if not altogether inhuman. And next to their weirdness is ours as mammals: No chimp has to grapple with the rules of a speedway or the upkeep of a homestead. Nor does it contend with traffic congestion, public health issues, assembly lines, complex teamwork, labor allocation, market economies, resource management, mass warfare, or slavery. As alien as insects seem to us in appearance and intelligence, only certain ant and human societies do such things, along with a few other social insects such as honeybees and some termites.[10]

PUTTING TOGETHER A LARGE SOCIETY: LESSONS FROM THE ANT

The parallels between certain social insects and modern humans are largely an outcome of one underlying commonality between us: our societies are populous. Scientists investigating animal behavior have too often been narrowly devoted to the evolutionary relationships between species, when many features of societies have more to do with scaling—sheer numbers—rather than pedigree.[11] The societies we have looked at among the nonhuman vertebrates, including the non-human apes, contain at most a few dozen individuals. A large leafcutter nest harbors a workforce in the low millions.

Once a population grows that large, all types of complexity can emerge—in fact often have to emerge to get jobs done. The coordination seen among hunting groups in a chimp or painted dog society is mercurial enough to amount to wishful thinking compared to the elaborate way some predatory ants organize hunts, with some workers stopping the quarry in its tracks, others applying the death blow, and others dismantling the corpse into slabs of flesh carried off in coordinated teams. Most vertebrates lack the labor force to take on such specific roles; nor must they operate in this way for the members to get the food they need.

The same is true for housing and infrastructure. The burrows of prairie dogs can be intricate; they connect underground with hibernation chambers and dead ends that thwart predators. Yet, contrasted with the monumental architecture of a leafcutter nest or a honeybee hive, the dwellings of these rodents seem a relic of the Animal Stone Age.

But—I imagine your protest—no human hunters have ever caught prey in the way ants do, and no human residence resembles one of their nests. Any two things, apples and oranges included, have innumerable characteristics in common and an equally immeasurable number of dissimilarities. What interests someone, similarities or differences, depends on that person's point of view. Identical twins aren't identical in the eyes of their mother; and, as will become important when we look at psychology, the members of one race, though outsiders may fail to differentiate among them, don't look alike to each other. If nothing else, remember this: *comparing identical things is deadly boring.* Making comparisons is most fruitful when parallels are noticed between ideas or things or actions ordinarily treated as distinct. Hence slavery in ants, where individuals work against their interests immersed in another society, is different from how Americans carried out slavery, and which in turn differed from the treatment of those defeated in war by ancient Greeks.

People and ants reach different solutions to the same general problems, sometimes by using completely different approaches; but then again so can different human societies or different ant societies. In some parts of the world, we drive on the left side of the road, in others on the right. On busy routes in colonies of Asia's marauder ants, incoming traffic streams down the highway center while outbound ants take to its flanks, a three-lane approach no human district has tried. Both patterns bespeak the importance of getting goods and services to the right places safely and efficiently when the populace depending on them is enormous and doesn't, and possibly can't, all go out foraging.

Think about the allocation of goods and services. Humans show a lot of variety here: Marxist societies don't handle problems the way capitalist ones do. Ants have solutions of their own. In the red imported fire ant, for example, the flow of commodities is regulated according to what is available and what's needed, a supply-and-demand market strategy. The workers monitor the nutritional cravings of other adult ants as well as the brood, and change their activities as the situation requires. In a nest laden with food, scouts and their recruits hawk merchandise by regurgitating samples to "buyers" in the nest chambers, who in turn roam through the nest to distribute meals to whoever wants them. If these middlemen find their customers sated on meat (dead insects, perhaps), they peruse the marketplace for other commodities, until they find, possibly, a seller offering something sugary. When a market becomes glutted and sellers can no longer peddle their wares, both buyers and sellers engage in other jobs, or take a nap.[12]

How does division of labor play out in the ants? Certain jobs can be farmed out to specialists that execute them exclusively, or with greater frequency and precision than others. Age has a role in determining that frequency: because young ants happen to still be near the brood in the nest chambers where they themselves were reared, they start as nurses, grooming and feeding larvae on behalf of their mother (reversing the chronology for humans, in which the elderly help tend the grandkids). But things can get complicated. Just as aspects of people's appearance offer clues about their career—the person with a suit and briefcase may well be a lawyer; those with hardhats and lunch pails are probably construction workers—division of labor in ants can also be associated with appearance. The fairness of stereotypes of spindly human office workers aside, in many ant species different workers do come in body sizes and proportions to suit their role.

For ants and most other social insects, the most fundamental specialization is unlike anything known in people, and in fact is closer to the situation for gray wolves or meerkats: usually only one female (the queen in social insects) breeds. This aspect of the life history of ants, whereby several generations of young raise siblings, makes the group to which they have membership more than an everyday family and hence worthy of the word "society."

Furthermore, ant societies are sisterhoods (which is why I often describe workers using the pronoun *she*). That's not unheard of in vertebrates. For the savanna elephant, too, all the adult members of a society, or core, while not always sisters, are female. However, for many ants, sex matters little in practice because workers can't reproduce; the ovaries of leafcutter workers don't function. Meanwhile ant males, like honeybee drones, are socially useless: their sole contribution is to have sex and die. Termites, on the other hand, show sexual parity: there's a king in the nest, as well as a queen, and the workers are of both sexes.

DIVIDING UP THE WORK IN AN ANT EMPIRE

Division of labor in the most extreme leafcutter species is like nothing else. The growth spurt of some larvae yields a soldier caste that guards the other nest denizens. The large soldiers additionally function as a crew for heavy-duty roadwork, clearing a highway so that food, materiel, and personnel flow smoothly. The hefty leafcutters that carved slices into my skin as I dug up the colony in São Paulo belonged to this soldier caste.

To tend their gardens, ants form assembly lines involving all but the soldiers.[13] Medium-sized workers cut the foliage and pass it to slightly smaller ants to be conveyed to the nest in the long parades. Once the greenery is dropped off in a garden chamber, yet smaller ants mince it into pieces that still littler ones crush into a pulp. Then even more diminutive workers integrate the resultant mulch into the garden with their forelegs, where tinier ants "plant" tufts of fungus and then prune them over time. The tiniest ants of all meticulously weed the gardens of inedible fungi and contagions. The ants also apply their version of pesticides, produced by their bodies, to the gardens to keep the yields high. The necessity of all this labor, from planting to tending and harvesting the crop, should resonate with any farmhand. No vertebrate other than humans, regardless of how smart or populous, has taken even a rudimentary step toward domesticating its food, something leafcutters, and a few other insects, have done many times.[14]

One problem with mass production is handling the waste. The matter wouldn't occur to a chimpanzee, I assure you. Nor should it: as in the sparse human communities of Tibet, where toilets have never been required and going off in the woods remains the fashion, chimp excrement vanishes in the soil before it can build to a health catastrophe. But leafcutter nests need full-time disposal squads.[15] Moreover, their nests are built to keep fresh air circulating. After over 150 million years of colonial evolution, ants have come to invest much more of their GDP in public safety and recycling efforts than we do.

COMPLEXITY IN SOCIETIES, LARGE AND SMALL

A wonderful thing about ants is that it's possible to compare complicated societies like those of leafcutters with the colonies of species that live very simply. Some colonies are small indeed, like those of *Acanthognathus teledectus*, or trap-jaw ants, containing a couple dozen workers.[16] Each colony takes over the hollow core of a twig on the floor of a tropical American rainforest—the ant analog of a cave. Such a colony doesn't need a highway or assembly line or complicated teams any more than a similar-sized hyena clan, gray wolf pack, or small human tribe, would, beyond perhaps coordinating a few individuals to take down prey. The trap-jaw ants face no crisis distributing supplies or managing waste, and when there is danger, they just run away. Other than the queen, they are equal in size and do the same things. To get all the colony's tasks done, each has what amounts to a Swiss army knife built into her face. The "trap jaws" are lengthy mandibles that end in spikes; every ant is, in effect, a bear trap, configured to carry out the killing and retrieval of prey entirely on her own. Specialization is minimal: not only are there too few individuals to fill different roles, but overspecializing is risky when a group is small. A military squadron that loses its sole radio operator may be doomed. A trap-jaw colony is simply so small that its workers have to be generalists. Populous societies offer a redundancy that allows a more tailored labor force, as a comparison of the range of job listings in New York City to those in a village will show.

What makes the leafcutter story all the more remarkable is that the species evolved agricultural practices by taking much the same path as humankind. The leafcutter's progenitors made their first move to gardening 60 million years ago, as analysis of the genetics of the ants and their fungi reveals.[17] Ants that live as those ancestors did still exist, and their social complexity is nearly as modest as that of the trap-jaw ant. Such societies consist of a few dozen to a few hundred workers raising tiny gardens of wild fungi; in many ways, in fact, they resemble the small human tribes that often cultivated wild species they took from nature, growing only what they needed in small plots of ground near simple abodes. Twenty million years ago, long before *Homo sapiens* appeared, some of these easily overlooked ants domesticated their fungus—it came to depend on the ants' care.[18] Thus modified, the fungus no longer propagated in the wild yet now could be farmed at a mind-boggling scale; the colonies exploded in population much the way human societies did after agriculture got its start in places like the Nile Valley, supported by domesticated grain.

Many commonplace features of such large societies, such as infrastructure and systems of waste management, are virtual requirements when many settle in one place. Yet complexity and numbers don't invariably go together, as an invasive species, the Argentine ant, confirms. Most Argentine ant "supercolonies" encompass far greater multitudes than any leafcutter nest, but the workers are one size, show no specialization, lack aptitudes for assembly lines or elaborate teamwork, and erect no monumental central abode. Relying on roaming habits akin to those of the spotted hyena or bonobo, Argentines ramp up fission-fusion to the *n*th degree. Spreading every which way, they turn any suitable corner of the territory at once into part of the nest and a place to forage. The simplicity of a supercolony reminds us that while we are sure not to see an Empire State Building in a tribal village—there's only so much complexity a few souls can create and manage—there won't be skyscrapers or good plumbing in every major metropolis.

Despite their societal simplicity, however, the effectiveness with which each Argentine ant supercolony remains identified as one unit can help us answer the very big question of how humans put together truly immense societies, as we shall discover next.

CHAPTER 6

The Ultimate Nationalists

Though it's debatable whether insects erected the first society, without question today's social insects are the master society builders. Insects have had this success despite their tiny nervous systems. Many aspects of cognition don't require much of a brain, however. Chances are good that insects have the mental capacity for subjective experiences—a unified perspective on the world that yields a sense of "self," even if that self is simple from our point of view. Unable to get inside their heads, we can't say for sure.[1] Yet a part of their triumph can be chalked up to something straightforward indeed: an efficient approach to distinguishing societies. No species illustrates this better than the Argentine ant.[2]

Birders casually call any nondescript bird a "little brown job." If there ever were a little brown job among the ants, it would be this species, originally from northern Argentina but globally invasive. Argentine ants the color of weak tea routinely overran my pantry when I lived near San Francisco, as they do in millions of homes in the Bay Area. Feeble of jaw, those nuisances didn't look like they amount to anything, lacking even stingers. Yet the species represents a pinnacle of social evolution. I could have plucked an Argentine ant from my house in Berkeley, driven 800 kilometers to the Mexico border, dropped her off, and she'd have been just fine. In a very real sense she would still be home. Improbable as it might seem, the ants swarming around the feet of the customs officers inspecting my passport would have been members of the same ant nation that ran roughshod through my kitchen up north.

If, on the other hand, I had taken the same ant about 60 kilometers north of Mexico, to the outskirts of San Diego, and deposited her a centimeter or two beyond a border that means nothing to us but that ants mark off with their lives, things would have gone very differently for her. There, too minute and hidden in the grasses to be noticed by human suburbanites, she would have confronted the ants' own border patrol and likely joined the tiny corpses heaped along a narrow line extending block after block beneath the grass of manicured lawns, where over a million ants die each month in what is arguably the largest battlefield of all time.

To the west of that border lies the holdings of the Lake Hodges Colony, a kingdom of the same Argentine ant species spread over 50 square kilometers. The dominion to the east is claimed by what the experts dub the Large Colony, a single social entity whose territory stretches from the Mexican border to California's Central Valley past San Francisco. Given that a backyard in Southern California can sustain a million Argentine ants—a footstep on the tidiest of lawns stirs up a host of them—Large Colony clearly contains billions of the workers. Little wonder, then, that entomologists call these ant republics supercolonies.

Four supercolonies are known in California: the abovementioned two, and two others. Given the right level of moisture, nothing seems to stop their continued growth—except combat, along contested contact zones that extend for kilometers, bringing to mind the trenches of the Western Front in World War I. *Homo sapiens*, it turns out, isn't the only imperialist organism. What stupefies is the length of time the supercolonies have been at war. The species, immediately recorded in the newspapers, arrived in California in 1907. Each supercolony started off as a few ants, possibly lodged in the potting soil in separate shipments of houseplants. Their domains expanded incrementally over the ensuing decades, wiping out other ant species until they came to abut each other. Then the fights began. The front lines shift glacially month after month, a few meters one way, then the other.

Internally, however, the supercolonies run so smoothly and at such an immense scale that people in nations—with their meddling, sharp differences of opinion, cheating, selfishness, outright aggression, and homicides—look positively dysfunctional by comparison.[3]

Before they stumbled on the war zones near San Diego, everywhere scientists went they found Argentine ants living in bliss, an observation that led them to conclude the ants all belonged to one happy family. That is, until 2004, when researchers collected samples of ants from different neighborhoods that, by pure happenstance, fell within the territories of two different supercolonies. To the scientists' shock, a fight raged the instant those ants were combined, killing many of them. The abrupt about-face this caused in the experts' understanding of Argentine ants hints at how toilsome it can be to interpret societies in nature.

THE IGNORANT ANT

If big-brained vertebrates can typically manage only societies of a few dozen individuals, what allows the ant, with its minimal nob of neural tissue, to do infinitely better? As hard as it is to fathom a leafcutter ant nest of a million, the Argentine empires boggle the mind.

Clearly an ant can't be acquainted with each fellow member of its society as wolves and chimps are with theirs. This isn't because insects inevitably fail to recognize each other as individuals. For instance, the North American paper wasp *Polistes fuscatus* excels at face recognition, much as humans do. Recognizing each other is important when they first join up to construct a nest because of the wrestling matches that determine who breeds.[4] Later in the season, a colony can grow to 200—a lot of faces to recall. However, by then there's little to fight over and in all probability the wasps no longer keep track of everyone.

Nevertheless, the paper wasp is an exception: most social insects don't have individual recognition. Ant and honeybee workers don't know anyone as an individual.[5] Associations between ants—the team pinning down an enemy combatant, for example—are impersonal. The best a worker can do is discriminate among types of individuals—tell soldiers from worker ants or a larva from a pupa, for instance, and most essentially a queen from everybody else.[6] The ants do manifest personality differences, with certain workers putting in more effort than others, for example, but those do-gooders go unrecognized for their trouble. That means ants avoid ever having to confront a rival or build an alliance within their societies the way we vertebrates do—the queen aside, no ant picks favorites, and every other worker in the colony is equally on the same side. To the ant, only the society, not the individual, matters.

The fact that ants don't need to know each other explains why you can move an Argentine ant anywhere in its supercolony and she will go right into interacting as she had before, but now with whatever random colony-mates happen to be there. In fact, as far as experts know, an Argentine ant, lacking a central nest, wanders indiscriminately across her supercolony's territory until the day she dies. If ants are forever strangers, moving through a crowd of unknown others, how can they tell where their society ends and the next one begins?

ANONYMITY

In 1997 two chemists were rearing Argentine ants and cockroaches in a lab for pest control studies. An assistant decided the roach supply would make good Argentine ant chow. Pure scientific serendipity ensued. One day the ants began killing each other rather than carving up the roaches. The cause was quickly discovered: That morning a technician had tried feeding the ants a different roach, a pest from Africa. Any ant that so much as touched one of these brown-banded cockroaches was summarily slaughtered by her fellows.[7]

The problem was the smell. Most insect communication is based on chemicals known as pheromones, with specialized glands releasing compounds to signal emergencies or signpost a route to food. Colony membership is marked by chemistry, as well. Although ants don't tell individuals apart by their personal aromas the way hamsters do, they do recognize each other as nest-mates—or as foreign—using an odor as a shared sign of identity. As long as an ant displays the correct emblem—as long as she smells right, which requires that she have the right combination of molecules known as hydrocarbons on her body—her colony-mates admit her as one of their own. The scent (or taste, if you wish to call it that, since ants detect the emblem by touch) is like a flag pin, one that every ant must wear. An ant that shouldn't be there is quickly detected by her alien scent. Since ants have no white flag of surrender, more often than not the outsider is killed, like the unfortunate Argentine ant that snoops across a territorial border. Brown-banded roaches, perhaps by coincidence, have critical components of the scent the ants use to cue in on colony mates, and on those from foreign colonies. When an ant touched the roach provided as a food source, she transferred these hydrocarbon molecules to herself, in effect putting on the uniform of the opponent, and was then misidentified as an enemy.

Those identifying markers are what enable social insects to transcend the widespread vertebrate requirement of having firsthand acquaintance with each other. Whether we are talking about several trap-jaw ants intimately clustered inside a twig or a billion Argentine ants spread far and wide, the members need never have met, nor even come close to each other, let alone recall one another. Species that mark their aggregate identity have what I call anonymous societies.[8]

Markers can have a few origins. The roach example is proof that the environment can influence colony identity, but all the members of a supercolony are not likely to share a single environment. That the abutting borders of supercolonies are measured to the centimeter for territories that extend across kilometers of varied habitat indicates that genetics must play the overriding role. Indeed, the hydrocarbon scent is coded in the genes, with diet generally having little effect.[9]

It's easy to assume that the actions of ants, including their response to a colony marker, will be genetically prescribed in a fixed, simplistic way. Actually, most aspects of behavior have built-in, or innate, components regardless of species—even human activities like learning a language, which our infants do as a matter of course. Social psychologist Jon Haidt has described such innate features as organized in advance of experience.[10] When a living thing is likely to encounter a range of situations, the blueprint of its nervous system—its organization in advance of experience—must allow for flexibility. For social insects, just as in other animals, that means less flexibility than in humans. Yet don't give them short shrift. Tiny they may be, but insects are flexible nevertheless.

In ants this flexibility can extend to how they come to know their national scent. This is demonstrated by the slavemaking ants, which exploit the way ants learn their identity to take slaves that do all the domestic chores of the slavemaking colonies. The slavemakers aggressively raid other nests, most often those of another ant species. Mature individuals are no good to them: Nationalists to the core, adult ants will die before accepting a foreign flag. Instead the slavemakers seize the colony's young, favoring the pupae, the quiescent stage before the insects turn into what we recognize as an ant. Pupae don't have the colony scent yet. Normally an ant emerges from the pupal state in the nest of her mother queen and quickly learns the scent of her own society, finding it agreeable for the rest of her life. But a kidnapped ant is duped, imprinting on her captors in much the way a hatchling chicken will imprint on you if you are the first thing it sees rather than its mother. Emerging in a slavemaker nest—and ignorant that anything is amiss—a stolen ant adopts the local colony scent as her "nationality" and dutifully sets to work unaffected by differences between slave and slavemaker in size or color that seem more important in our eyes. Even so, occasionally the slaves figure out something's wrong and make a run for it, but their captors usually force them back.[11]

The favorable reception of slavemaker by slave is the least of it. Here's where an ant's adaptable brain comes in: To steer clear of a social meltdown, each slave and each slavemaker must welcome all the other slaves in the nest as well, no matter how many colonies they have been pilfered from. Yet despite the differences in scent produced by each individual, neither slavemaker nor slave has a problem identifying all the others as members of "their" society. Underlying this adaptability is grooming.[12] Among primates grooming serves to bond friendly individuals, but the experts suspect that grooming in ants hardens attachments at the society level by blending the odor of the nest-mates to ensure they achieve a standard scent—hence some of the slavemakers' scent rubs off on young slaves, marking them as part of the colony, and the slaves modify everyone's odor as well. Commingling the aromas has unexpected results. Should a slave blunder onto the colony of her birth—the one to which she truly belongs, with her actual colony-mates and sisters—she is attacked as the enemy. It's a story with the trappings of a Greek tragedy.[13]

Acquiring a national marker can be like receiving the key to the city: all is possible. In an orchard in Australia, I tore apart a weaver ant nest to find an orange arachnid five millimeters long. Many painful ant bites later I identified the species, *Myrmaplata plataleoides*, a spider that doesn't spin webs but joins the weaver ants as if it were a citizen of the colony itself. This identity theft is achieved when the spider absorbs the colony's aroma by stealing its brood. Cloaked in the stench of nationalism it enters the nest unimpeded to eat more larvae snatched from the nursing workers. Once it takes on the identity of the colony, the eight-legged interloper has it made. Still, there's also a risk: should it amble away from that nest to another weaver ant colony, it will be attacked, not as a spider, but as an invading ant.[14]

ANONYMITY WITH A BACKBONE

At this point, I must admit I've been holding back. There's at least one vertebrate that lives in anonymous societies, that in fact marks its societies with a scent much as ants do: the naked mole rat.[15] Neither mole nor rat, these wrinkly, furless, pink rodents, residents of African savannas, "violate even the most liberal standards of animal beauty," admit two leading experts on the species.[16] Their markers may explain how the colonies edge past the population maximum of roughly 200 that is characteristic of most other mammal societies, with the largest ever recorded having 295 members.

Unlike worker ants, naked mole rats recognize each other as individuals, as well. Whether the average mole rat gets much out of this—whether it has a best buddy, maybe—isn't clear. Also unknown is whether a mole rat in a large colony can recollect each of its fellow members or, as I predict, must rely on scent when it comes across those wrinkly blobs less familiar to it.

The decidedly ant-like nature of the naked mole rat astonishes. The only cold-blooded mammals, they shiver together on cold nights, as honeybees do. Their snuggle sessions may give rise to the colony-wide scent, similar to the function of grooming in ants. Naked mole rats, like Damaraland mole rats, another African species, employ a division of labor with a bulky reproductive queen. In this they resemble termite more than ant: The queen selects two or three males to serve as kings, her exclusive mates.

These mole rats are termite-like in their subterranean proclivities, too. By necessity they are all about infrastructure, digging tunnels that meander beneath thousands of square meters to gain access to bulbs and tubers, their sole nutrition. Instead of the colony moving about in a troop-like unit, workers come and go from a central, insect-like nest, a collection of chambers packed in an area a half meter to a meter in diameter. The rodents live nomadically, shifting their nest location in the burrow labyrinth every few weeks. Among the workers, the biggest individuals tend to act as soldiers, defending the colony from snakes and any foreign naked mole rats detected by their alien aroma.

I wouldn't be surprised if markers signaling society membership are to be found among other mammals, especially if they involve smells or sounds people can't detect. Spotted hyenas, for instance, rub their rears on clumps of plants, an activity called pasting. Each hyena has a scent of its own, and the exchange of smells at posts is thought to be a way for the animals to be in tune with other individuals of their clan. However, the blending of the pastes of the different clan members generates an odor unique to the group. The hyenas could, in theory, tell clans apart by the smell. Even if they did, though, each clan has only a few dozen individuals, few enough that the animals are already well versed about who belongs and who doesn't. The clan scent would serve at best as a kind of backup, while during the course of a hyena's daily life individual recognition would be key.[17]

More persuasive evidence of anonymous societies exists in birds. The pinyon jay of the southwestern United States flocks in the hundreds, a normal enough behavior for birds. Yet here's a clue that something fabulous is going on: when one flock meets another, the jays combine into a cloud that later cleanly separates back into the original flocks without the least uncertainty. This reminds me of how savanna elephants peacefully intermix, at certain times in great numbers, but always go back to their original cores. What's remarkable is that while a core consists of just a few elephants, as many as 500 jays make up a flock. So what superficially amounts to just a modest throng by bird standards is on closer inspection actually a populous and well-formed society, each "flock society" retaining a membership of multiple extended families throughout the year.[18] A flock concentrates on a tract of land averaging 23 square kilometers, though it isn't territorial: the jays don't defend the area or the food in it, and they often enter the airspace of neighbors. For much of the year the flock flies as a tight mass to hunt seeds and insects. In the breeding season, pairs of birds, devoted to their mates for life (the pinyon jay is more monogamous than humans), spread out over one stretch of the flock's land to rear their young in nests—and yet even then they identify with the same group of birds.

No one can say for sure how the flock societies retain their clean separation, how the members identify their society-mates both when the birds are in tight flocks and when they disperse among nests. Certain of the jay's vocalizations—including the "near" call, a soft nasal sound—are distinct for each bird. Even so, it's hard to credit the jays with employing individual recognition, because that would seem to require them to keep track of hundreds of others while on the wing. Indeed, studies have shown that no jay can recognize every flock member, just its mate and young and a select few others in its personal support network that it can draw on as a respite from hassles with dominant flock members over food and nest material. Hypothetically, knowledge of some, but not all, of the members might seem sufficient to keep the full contingent of 500 threaded together into a flock, perhaps if each bird were comfortable so long as it heard the near call of at least one familiar jay. In practice, however, such alliances within the society wouldn't keep a flock intact. Over time it would fragment or combine permanently with other flocks.

With so many birds involved, something more than individual recognition must be operating to keep birds of a feather flocking together. Almost certainly the pinyon jay forms anonymous societies, and indeed its vocal repertoire includes a couple of likely markers for flock membership. The jays repeat a *rack* call in response to predators, a sound that varies enough to identify both the specific caller and the flock to which it belongs. Even more significant is the *kaw* the jays utter when in flight. This sound, which birders listen for to identify the species, likewise differs between flocks, as we would expect for a marker. I have no doubt the *kaw*, the *rack*, or both express the bird's identification to a flock, explaining how the jays keep their memberships straight.

Pinyon jays receive definite inducements to live in permanent societies rather than in temporary flocks. First, the safety bonus: Each jay buries a private stash of seeds for hard times (though much like human societies, pinyon jays have their share of thievery). Sentinels stand watch in the trees overhead on the lookout for predators like foxes that can kill the digging birds. The members also coordinate their activities. When the babies are guarded in the nest by mom, all the fathers go off to hunt together, scaring up more bugs as a group than they could as individuals, and returning all at once every hour to bring meals to their respective nests. After the fledglings leave those nests, they socialize in a nursery group under the watchful eye of a couple of adults who pull guard duty for the day while the other parents are off finding meals for their offspring. A year later the youngsters separate from the fledglings, joining what amounts to a boisterous teenage gang, until they are ready to commit to a mate, often after being admitted to a new flock. Pinyon jay lives are organized to a tee.

Finally, evidence is accumulating that yet another group of vertebrates, the cetaceans, create anonymous societies, and that these can sometimes reach stupendous populations. The case for such societies is especially compelling for the sperm whales, the squid-chomping toothed species of *Moby Dick* fame.[19] Indeed, this behemoth has what I'd grant are societies operating independently at two separate levels.

At first blush, the sperm whales appear to have rather middling societies for a vertebrate, with units consisting of 6 to 24 adult females plus their offspring, all of which stick together. (Adult male sperm whales, like the male elephants, roam as they wish, mate with whoever they can, and don't participate in the female-only societies.) Unit societies persist for decades; in fact, most females stay with theirs for life, though a few transfer for unknown reasons, so that many units come to contain unrelated individuals.

Young sperm whales learn short sets of clicks called codas; think of them as a letter or two extracted from a Morse Code message. Certain codas differ slightly from one unit society to the next. The whales emit these when their unit draws near another, apparently permitting the units to recognize each other and coordinate their movements.

Remarkably, the units are part of clans, an arrangement that neatly defines the whales in a dual way. Five clans range across the Pacific, each marked by a specific set of codas. The clans are made up of hundreds of units spread over thousands of square kilometers. While their daily lives are spent in their units, the whales also value clan membership. Only units of the same clan come close to each other, and indeed will hunt together for a time—think of it as a sort of fission-fusion. It's unlikely whales from different clans fight, given how easily such large animals could hurt themselves. Moreover they aren't territorial (albeit in the Atlantic some clans live far apart). They simply avoid each other.

The units provide sperm whales many of the payoffs of societies in other animals: protection against predators, joint care of the young, and opportunities for sharing their accumulated knowledge; whereas the advantage of belonging to a clan appears to reside in how the whales forage—whether the units of a clan dive or travel together, or stay near islands or in open water. These details matter, as they result in each clan catching different species of squid. One clan does really well during warm, El Niño years, when squid are hard to come by. The hypothesis is that the whales capture prey proficiently only when they buddy up with other units of their own clan, making use of the same hunting technique.

These foraging differences are not genetically determined. With males open to mating with females in any clan, all the clans have the same genetic makeup. The strategies have to be cultural: the whales learn the clan ways from their elders, much as dolphins learn fishing tactics.[20] The simplicity of the codas must make errors about clan membership exceedingly rare.

BREEDING AN ANT KINGDOM

The pinyon jay, the insect-like naked mole rats, and the sperm whale represent evolutionary sidebars among the societies of the vertebrates, most of which operate by individual recognition. Yet the anonymous societies of ants, and most definitely the tiny minority of ant species with vast supercolonies, still stand out for their intricacy, their efficiency, and their size. How does the identity of an ant colony initially fall into place, and how is it that Argentine ants scale those identities up to supercolony status?

An ant society typically germinates from a preexisting one. The process begins when a mature colony rears winged queens that flutter skyward to mate in midair, often with a few winged males from other colonies, before each queen drops to the earth and digs a little "starter" nest by herself to raise the first of her many broods of workers. The workers become the sinew of a society. Genetic and environmental factors cause them to produce a scent and identity independent of all others, including the queen's birth colony. The entourage expands in population through the generations until the colony approaches a mature size, specific to its species, at which point part of its brood production shifts over to queens and males that depart to birth yet another generation of societies, one year after the next. The original mother queen stays with her colony, which survives as long as she does. That can be long indeed: a quarter century in leafcutters. With her demise the workers go into a funk and expire soon after. Give a colony all the food and space in the world and it won't last longer than its queen.

Argentine ant supercolonies owe their ever-burgeoning populations to a twist in the story: a big supercolony is home to not one queen, but millions, because the queens never fly away. Moving on foot between nest chambers spread throughout the territory, they stay in their native society to lay more eggs, which hatch into ants that stay as well. Year after year the supercolony expands its range to occupy any suitable nook or cranny.

So long as the same scent is produced throughout the supercolony's often vast domain, the society remains intact. This consistency may seem impossible to achieve, yet we can imagine a method for self-correction built into the system. Suppose a gene affecting the colony marker mutated in one queen out of the many. Any other genetic change in her behavior or morphology wouldn't matter to her acceptance. However, if her smell no longer matched the identity of the ants around her, the workers would kill her before she laid eggs. The mutation would vanish with no trace. The outcome of this incessant purging is the ants' adherence to a common identity not just in one nest, as in most ant species, but over distances of hundreds of kilometers. With a uniform identity from one end of their range to the other, supercolonies achieve a kind of immortality. Think of California's four colonies as the very same societies that invaded the state a century ago. Occasional rumors to the contrary, they show no sign of slowing down.[21]

This reality is hard to grasp when you stand in California with the ants of Large Colony swarming to every horizon. Some biologists question whether a supercolony can really be one society. A few go through contortions to propose that, because a supercolony's population is never a continuous throng of ants across the ground, it isn't really a society at all, but rather a constellation of many societies. This patchy distribution of ants, however, has more to do with the suitability of the habitat than with the inhabitants' social behavior or identity. The ants avoid overly dry areas, for example. Yet turn on lawn sprinklers on a parched day and what had been two patches of ants expand and combine, trouble free, into one.

When pressed on the question of whether these ants in their billions can represent a society, the same experts are likely to respond cautiously that they *act* as if they are one society, even with their patchiness and the existence of genetic variation from place to place within a supercolony. My response is to say, but of course! What criterion makes sense in determining what qualifies as a society other than the choices of its own members as to who should be there and who shouldn't? As long as the ants accept one another and reject outsiders, the span of earth they occupy, and how varied the members are, doesn't matter any more than it matters in a nation such as the United States, with all its ethnicities and political squabbling.

For the ants, simple markers are up to the task. When entomologist Jerome Howard took a leafcutter worker from one side of her colony to the other—a distance of several meters—the ants at the new spot sometimes paused to inspect the newcomer. Maybe the population of such a metropolis, with its many highways and byways, doesn't completely intermingle, so that subtle distinctions in scent accumulate from place to place—a trifling difference in the national flag. Nevertheless, after a second of mild interest the new arrival was permitted to continue about her business without a hitch—she is still treated as a member of the colony.

Given the scale of their colonies, Argentine ants seem staggeringly united. The species highlights how individuals can remain members of a society regardless of how little they cooperate or interact. The species does well on ocean voyages; that's how the four supercolonies arrived from Argentina to the United States in the first place. Leapfrogging about as well on our planes, trains, and automobiles, the supercolonies have made their way across the globe while still holding on to their identity, much the way Hawaiians retain a common nationality with denizens of the US mainland. Large Colony has traveled to, and seized control of, 3,000 kilometers of European coastline and other faraway parts of the planet, Hawaii included. Meanwhile, other supercolonies have toeholds in places like South Africa, Japan, and New Zealand.

The combined weight of the ants in invasive supercolonies can reach and exceed, literally, the mass of a sperm whale, which raises the question of how they got that way. Perhaps the most remarkable thing about the transcontinental societies is how much larger they are than those of the Argentine ants in their original homeland. The societies of this species in Argentina are positively petite: at most a kilometer wide, certainly jaw-dropping for an ant but not much to look at by California standards. This difference appears so radical that one might conjecture that a major evolutionary change must underlie it. I imagine that aliens might have assumed much the same for humans if they had landed on Earth 20,000 years ago to find societies of a few hunter-gatherers, then returned centuries later to discover China with its billion souls. A far simpler explanation for the superlarge societies of modern humans and Argentine ants alike is that no dramatic transformation was required: in both species the expansion of societies became certain when the conditions were right. This extravagant capacity for indefinite growth, not size as such, is what sets supercolonies apart from the societies of other species. Even several dozen Argentine ants in a shipment of plants are, then, a supercolony (or at least a fragment of one). The capacity for societies to grow indefinitely is a true rarity, characteristic of just a few ant species, possibly the clans of sperm whales, and human beings.

Aside from their rapacity for expansion, Argentine ant societies are not much different from those of other ant species. They direct their aggression at outsiders and show little animosity toward colony-mates, like all ants do. Furthermore, the societies in Argentina operate no differently from the immense colonies overseas, except that their growth is choked by an overabundance of nasty neighbor ant colonies. The condition that triggers the explosive expansion of a supercolony abroad is the lack of competition. The supercolonies that got to California had nothing to stop them from conquering the state—until their growth came to a halt when they met and started fighting each other.

In the upcoming chapters I will argue that humans likewise did not need to change in any essential way for the small unions of prehistory to grow, given the opportunity. All the elements needed for the success of empires were already built into the Paleolithic mind, right down to a very human fixation with markers of identity.

CHAPTER 7

Anonymous Humans

There is little in the history of life more remarkable than a human strolling through a coffee shop. The patrons can be utterly unknown to us, and—nothing happens. We do fine, stay calm encountering those we have never met. This tells us something unique about our species—opposable thumbs, upright stance, and smarts aside—because most other society-dwelling vertebrates cannot do this. A chimp bumping into an unknown individual, never mind a café full of them, would fight or flee like the damned. Only a young female stands a chance of surviving the confrontation without risking combat—but she had darn well better be receptive to sex. Not even the bonobo would walk by an individual it doesn't know with indifference.[1] Yet humans have a talent for dealing with strangers and moving seamlessly in their company. We enjoy being surrounded by a sea of others at a concert, theater, public park, or fair. We grow accustomed to each other's presence and befriend the few we want to, in a kindergarten, summer camp, or job.

We allow such anonymity by recognizing certain signs in others that fit our expectations—characteristics that act as markers of identity.[2] Among our markers are those indicating every imaginable aspect of identity. A six-carat diamond ring signals wealth and status, and some markers, like a quirky way of making an arrowhead, can be specific to a person. However, in this book the word "marker" and its synonyms will generally specify the attributes people associate with their society.[3] The recognition of markers is a human aptitude that most animals lack, with the exception of a few vertebrates such as the naked mole rats and sperm whales and most social insects. Few social scientists will have anything to do with such comparisons, however; indeed, they may be puzzled by (even unreceptive to) the idea that ants, presumably lacking conscious behavior, have an identity at all. Yet though they lack the capacity for self-reflection, ants and other social insects resemble us in an elementary but striking manner by having anonymous societies.

Most mammals, indeed most vertebrates, lack anything they could reliably use to mark their society. The horses in a band never have the same gait or style of whinnying, for example. In most situations the lack of markers keeps vertebrates focused on managing individual relationships, in contrast to ants, for which such familiarity is non-existent. People fall in the middle, with a selective focus on cultivating key social connections without taking on the obligation of keeping track of everyone in the society. We form variable relationships based on our social histories with others, treating only some people as individuals.[4] Distilling the differences to fewer words, we return, at last, to the formula given at the end of chapter four: to function as a society, chimpanzees need to know everybody; ants need to know nobody; humans only need to know somebody.

The "somebodies" to which each of us are connected fall into expanding circles of social relationship, from most intimate to most abstract: our spouse, our nuclear family, our extended family, our 150 or so friends, and the many hundreds we know less closely. Beyond that is everyone who identifies with our society as a whole, whether it's a tribe or nation, which in all but the most diminished societies like Kenya's El Molo encompasses many who are strangers to us. Most of these connections, aside from our overarching allegiance to our society, are social networks that differ from person to person.[5] Overlying this list are the people with whom we share special affiliations; some proudly display markers of their own, like a Chicago Bears fan wearing a team cap. And of course our knowledge of others isn't restricted to our society. We resemble bonobos and savanna elephants in not just knowing members of other societies, but even befriending them.

MARKING OUR SOCIETIES

Flags, patriotic songs, and similar in-your-face signs of nationality are only the most obvious of the varied ways people indicate and perceive their connections to a society. Any characteristic of a society can serve as a marker, consciously created or not, so long as when aberrations in it become great enough, the members detect that something is amiss. Whereas ants and naked mole rats rely on an odor to identify society-mates, and sperm whales exclusively on vocalizations, for *Homo sapiens* just about anything will do.

Certain markers are displayed continuously, and often conspicuously, such as socially prescribed forms of dress. Others come into play on occasion and have to do with values, customs, and ideas. Some markers require conscious intent—carrying a passport, for instance—while others are cues over which the members have no control, such as, in certain societies, skin color. A marker need not be connected with a person at all; the identity of a group does not just permeate the people but may extend to anything the society claims, perhaps a location such as Bunker Hill or an object like the Liberty Bell.[6] Historical events enter the national consciousness, too. A few may be as old as the societies themselves, yet novel ingredients come into being all the time. In America, the September 11 attacks of 2001 were a defining moment. The site of the Twin Towers in New York, the numbers 9/11, and the date they signify all became part of how Americans view themselves and how others perceive them, which demonstrates how quickly new aspects of our identities can gain prominence.

The markers of a society can appear trivial or just plain weird to outsiders. For example, the custom of eating with the hands in India; or, in Thailand, primarily with a spoon, never chopsticks. There are capricious matters of taste, though an open-minded outsider can still appreciate their beauty: think of the tonal system of Indian music or the black-on-black pottery designs of Pueblo American Indians.[7] Silly and arbitrary or not, some of the differences can have life-and-death consequences, such as the practice of driving on either the left or right side of the road. Still other markers are icons that bear a relationship with something out there in the world, such as many Egyptian hieroglyphs that are instantly recognizable for what they stand for. Even the oddest markers can seem logical to those who use them. The bald eagle and the bear, mighty predators, respectively represent the strength of the United States and Russia, and ties to such markers can be a powerful social glue.

Markers can be reinforced, or outright determined, by biology. Certain mutations spread among people with the advent of animal domestication several thousand years ago that allowed adults to digest the lactose found in milk. In Tanzania, the cattle-herding Barabaig, who find milk delicious, live near Hadza hunter-gatherers, who are nauseated by dairy products. Doubtless this gulf in diet further amplifies the physical and cultural separation of their peoples.[8]

While sight and hearing are the primary senses involved in humans perceiving markers, the importance of taste is evident. "What is patriotism but the love of the food one ate as a child?" goes a Chinese proverb, and there's wide latitude in what people salivate over.[9] I've eaten fried centipedes in China, pickled wasps in Japan, pig embryos in Thailand, toasted ants in Columbia, dried mopane worms in South Africa, raw termites in Namibia, beetle grubs in New Guinea, and diced rat in Gabon. To those who crave such delicacies, the disgust of outsiders is a shock. And while we never accord scent the primacy ants do, people sometimes comment, unfavorably, on the breath or body odor of ethnic groups.[10]

Many of the attributes I call markers fall under the rubric of culture. Though this word is often associated with the pantheon of a society's intellectual and artistic achievements, it refers broadly to the full complement of characteristics passed between generations, mainly by active teaching. Among the most studied of these expectations are norms. Those are the understandings citizens share regarding their values and moral codes, including predilections to be generous or helpful and beliefs about what's fair and proper.[11]

Striking cultural norms and attributes like food taboos or national banners get the most attention, but a miscellanea of more subtle markers are both profoundly important and easy to overlook. A scene from the 2009 Quentin Tarantino film *Inglourious Basterds* comes to mind. A British spy masquerading as a Nazi orders three beers at a German bar by raising the three middle fingers of one hand rather than extending his index finger, middle finger and thumb as Germans do—the ensuing firefight is classic, heart-stopping Tarantino.

A study of culturally specific gestures alone could fill volumes. In Italy everyone's hands seem to be constantly in motion. Waving one hand horizontally in front of the body with the first finger and thumb touching to produce an o means perfect; chopping with one hand is a warning to be careful; waving both hands up toward the face says this is boring. Isabella Poggi, a psychologist, has classified more than 250 such gestures, many unique to Italy, centuries old, and often more reliable in their meanings than spoken vocabulary.[12] Gestures are plausibly more primeval than words, coming so automatically that blind people wave their hands when addressing a blind audience.[13]

Easier still to miss than gestures are nonverbal accents that are absorbed subliminally rather than taught. These often fall below our radar even though they vary from one society to the next. In the final decade of his life, Charles Darwin published *The Expression of the Emotions in Man and Animals,* in which he found that the most basic human emotions are universal across our species.[14] Nevertheless, the facial muscles involved in conveying those emotions are brought into play ever so slightly differently across societies. It has been shown, for example, that Americans who fail to distinguish Japanese nationals from Japanese Americans in photographs when both are expressionless can pick out Japanese Americans, their fellow citizens, when the faces in the photographs display anger, disgust, sadness, fear, or surprise.[15] In another experiment, without being able to explain how, Americans often guessed whether a person was American or Australian by watching how he or she waved a hand or walked, which suggests that the entire body can be turned into a nonverbal accent.[16] Such differences, unlike gestures but like many smells, are impossible to express in words.[17]

The ability to register such details comes from repeated exposure. Their comprehension is akin to that of a few individuals who could distinguish the incoming Nazi fighters and Allied airplanes during World War II. The British were desperate for more plane spotters. The trouble was that no one with the skill could explain how they did it. The only successful training method was for students, not having any idea what to look for, to guess. The spotter corrected them until they got it right.[18] Societal markers may often be acquired and recognized this way, without comprehension or calculated thought.

RECOGNIZING MARKERS

Our societal markers are so clear that some can even be noticed by animals that don't use markers for their own species. Elephants detect different human tribes and have expectations about their behavior. In Kenya they fear the Maasai, who spear pachyderms as a rite of passage, but are indifferent to the Kamba, who don't hunt elephants. Elephants hide in high grass when Maasai approach, possibly because they can distinguish the body odor of those tribal members, whose diet is based heavily on cattle, from that of the vegetable-loving Kamba. The elephants seem to be able to distinguish the tribes by attire, as well, and will attack cloth of the red color favored by Maasai.[19]

Humans evolved markers of identity for the same reason: our safety and well-being can depend on them. The psychologist Gordon

Allport, who founded the study of personality, has explained that "the human mind must think with the aid of categories . . . orderly living depends on it."[20] The ability is clearly rooted in our animal legacy: pigeons, for example, not only put things in categories like bird and tree; they can be trained to distinguish Picasso's paintings from Monet's.[21] All animals can categorize. No elephant or human, baboon or hyena is forced to recognize each and every elephant that passes by to know one when it sees one, picking out traits like the trunk and its trumpeting call. Still, an elephant *society*, having no physical or behavioral traits specific to its group, isn't as obvious to the eye as elephants themselves are as a species.

While most vertebrates don't wear their group identity on their sleeve (or in their voice or odor), their knowledge of who belongs yields societies no less valid and concrete than those based on markers. In fact, humans also have groups that are defined purely by individual recognition. Take the sports teams in my neighborhood as a kid. We couldn't afford team jerseys, so no outsider could have picked out who was on which team if we were all mixed together in a lineup. Yet we were never confused. What set our team apart was in our heads. We knew each other as individuals, after all. Much as observers can discern a wolf pack or lion pride by watching the animals interact, onlookers could have figured out our teams by watching us play.

Even when humans do know each other, markers are advantageous. My childhood team would have been jazzed to wear team jerseys if we'd had them. Not only might they have bolstered team pride, but they would also have deemphasized our differences and perhaps have given us a more unified and intimidating appearance (all subjects for later in this book). Jerseys would have helped us play, too: wearing them, we could have identified our team members by color even out of the corners of our eyes when the action unfolded fast. Speedy and accurate recognition is a real plus when mistakes are costly; without it, you can lose the ball to the other team or, in the case of societies, your resources or life to a hostile foreigner. Jerseys might additionally have sped the process of bringing on board another member. Sure, we might not have completely trusted any new kids at first, but given satisfactory appearance (the jersey), and behavior (including how they got that shirt), they would likely have been quickly identified as part of the group even by the players who didn't know them.

As for societies, markers add up to an indelible awareness of who *we* are, yoking people who haven't met to a shared worldview even when those traits do not require our attention. During the average moment, our markers are so familiar and expected that we notice them no more than we see the sky's precise shade of blue. Yet we hunger for them when they are absent. This is why, when starved for others "like us" on a trip abroad, we seek out a bar, restaurant, or hangout of people from our nation, and greet the expatriates there, somehow familiar though unknown to us, as old friends.

Despite their usefulness, adding markers to distinguish our society from outsiders brings no fundamental change: we can be a society with them; we can continue to be a society if we lack them—up to a point. Having markers becomes increasingly valuable, and eventually essential, for keeping societies straight as their populations grow—and, in humans, for distinguishing all sorts of groups within each society, as well. The fez of a Shriner or a Chicago Bears cap are not like anything in nature: the same fellow can identify himself as a Bostonian, a fire fighter, a conservative, with none of the categories invalidating any other—or his identity as an American. Many aspects of human identity come with markers that we may display proudly, but most can be taken on and off as needed. Beyond discriminating worker from queen (and in some species, from soldier), no ant breaks down her identity into such groups. Nor, as it turns out, do vertebrates, which never separate cleanly into lasting cliques within societies, and certainly not aided by markers: even though the wolves in a pack can vary in coat color, the blacks never set themselves apart from the grays. Hence the billions in an Argentine ant supercolony appear to be a homogeneous continuum, while you and I differ in a hundred ways in how we fit into our society (for instance, I suspect you don't study ants).

THE ROLE OF LANGUAGE, AND WHAT REALLY MATTERS

Languages, dialects, and accents are the most studied, and perhaps the most potent, human markers. Most societies, and many ethnicities such as Jews or the Basques, have their own language, or version of a language.[22] The evolutionary linguist Mark Pagel writes of the biblical story of the Tower of Babel, in which God gives people different languages to prevent them from uniting to build a tower tall enough to reach heaven: the irony of this story, Pagel points out, is that "language exists to stop us from communicating."[23] Above and beyond the vast differences between languages, every language offers words that describe its own speakers and how they see themselves—and words that portray those who belong to different societies. Labels matter, including the names of societies themselves, which possess "phenom-enal power," as one African scholar argues;[24] even six-year-olds prefer kids whom they are told are from their country.[25] One of the greatest threats a human society can face is a competing claim from another society for the same name, illustrated today by the impassioned dis-pute over the word "Macedonia," which the country of Macedonia has taken for itself but which millions of Greeks apply to their ethnic group.

Nevertheless, language differences are neither necessary nor suffi-cient to keep societies distinct. "Once a group has lost its language, it will generally lose its separate identity," one linguist claims.[26] Certainly this can be true when one society overpowers another and turns its people to serfs or slaves, circumstances under which many markers, language included, are lost or recast. Yet groups can lose their native tongue and keep their identity and independence, disproving the ab-solute sway of language over other aspects of identity. Africa's Pygmies, part-time hunter-gatherers, abandoned their own languages over the past 3,000 years in favor of the speech of their agricultural neighbors, whom they live with part of the year. Pygmy societies remain culturally intact even though mere traces of their original tongues exist—a few words, mostly describing forest animals and plants.[27]

The crucial point is that the rich combination of markers enables people to unmask those who don't belong without hearing them speak a word. That said, languages stand out in their importance for human identity. A language or dialect is almost impossible to replicate precisely unless you learned it growing up, giving languages a priority in unmasking the foreigner among us.[28] The book of Judges (12:6) in the Bible recounts how the Gilead soldiers rooted out the people of an Israeli tribe that spoke with a faint accent:

> [A Gileadite soldier] would tell him, "Say the word shibboleth." If the fugitive would say sibboleth, because he could not pronounce the word correctly, they would grab him and kill him at the shal-low crossings of the Jordan River. At that time, 42,000 men from Ephraim died.

The value of language as a "spot check" of identity is heightened by the level of nuance it conveys: people glean far more from it than is incumbent on them to understand what's being said.[29] Even a child registers a speaker's native tongue before she or he finishes uttering one word.[30] The same is true of other markers, of course: we are more aware of how our comrades walk or smile than we need to be to walk or smile ourselves. But what makes language especially worthy of attention is that speaking the local tongue is a matter of pride for humans in a way that carrying oneself like a local is not.[31] Perhaps only traditions like fire walking, practiced by some Polynesians, among others, signal a commitment to a group no language can convey. However, while people rarely perform such extreme rites, they talk all the time in voices that carry around corners and through crowds, their accents immediately signaling their native or foreign status. Few other aspects of identity elicit such an instantaneous gut response.

PERMISSIBLE VARIATION, ECCENTRICS, AND OUTLIERS

People treat their differences in markedly varied ways. For example, it's not always the case that "a language is a dialect that has an army and navy," as the Yiddish linguist Max Weinreich once quipped.[32] Societies don't need to be marked off by their way of speaking, as the Pygmy example shows. The ways societies use languages reveal that a facility with foreign practices can be not only tolerated but encouraged. Where countries pack into a small space, such as Europe, people commonly speak languages in addition to their native tongue, oftentimes those of neighbors, trading partners, or erstwhile colonial powers. Multilingualism isn't a recent novelty. In pre-contact Australia, many people spoke multiple tongues, often because they, or a parent, had married into the society. Some nations, such as Switzerland, may have several native languages. Others may share a single language with other societies, as do the United Kingdom, Australia, and the United States, with each having multiple dialects of its own.

The key of ethnography, the science of documenting cultures, isn't so much to figure out the similarities and differences among societies, but to find out what their people consider important, in language or in anything else. What a linguist might determine to be two languages could be interpreted by the speakers as one, or vice versa.[33] What I pointed out for Argentine supercolonies is true as well of the societies of all species: it's the choices the individuals in a society make concerning what signals identity that determines who is a member, and not what you, as an outsider, might assume is important. Tweak the scent of an ant colony, and you will find that some variations make a difference and some don't. As for people, variations, for example in language, similarly don't necessarily jeopardize our identities—it depends what the variations are.[34] I once met a Peruvian who is often mistaken for a foreigner in his homeland because of his failure to produce the rolling Spanish *r*. Whether he's less a Peruvian for this awkwardness is up to other Peruvians. Societies make room for differences of equal significance to speech. Modern-day religions, to give one example, extend across, and coexist within, territorial borders, yet they generally (but admittedly not always) do so without confusing people about who belongs where.

Society members aren't stamped from a cookie cutter—societies allow for similarity *and* diversity. For a person to be welcomed, his or her behavior has to lie within permissible social boundaries, as was true even for hunter-gatherers: "What he does, and how he does this is his, not his neighbor's affair, as long as he remains within the realms of accepted norms of behavior," wrote Bushman expert Hans-Joachim Heinz.[35] Social guidelines, often unspoken, inform these evaluations and constrain our choices; as the Japanese saying goes, the nail that sticks up gets hammered down. Naturally there can be issues of interpretation and taste. Ultraconservatives, for example, have little tolerance for the archliberal and vice versa, even though each grudgingly recognizes that the other belongs to their society. By expecting an appropriate level of conformity—such as the tacit agreement and childhood training not to readily show anger in some Asian cultures—the members of a society detect renegades and aliens.

Societies differ in the degree to which they value conformity. Some promote eccentricities as a sign of individuality and enterprise, and proclaim as a core value the right to be different.[36] Still, all societies, including those taking pride in their commitment to personal freedom, exist by exacting a loss of choice on members for the sake of safety and predictability. The precision with which we must learn from and imitate each other—the narrowness of the boundaries—is more stringent for some behaviors and in some situations than others. The fact is the repulsion felt toward a person who ignores important norms of behavior can be so profound that the social deviant can be treated more harshly for the same offense than a foreigner—an overreaction known to psychologists as the black sheep effect. Outliers poorly matching expectations are ostracized, stigmatized, pressured to change, or treated as foreign, depending on the kind and extent of their aberrance. Such censure reins in what goes on in a society.[37]

Animals also have limited tolerance for odd behavior. Even species with individual recognition societies endure minimal deviation. As a rule, elephants support the sick, but they, and chimpanzees, are also known to ill-treat gimpy or unhealthy individuals.[38]

Social insects are the supreme conformists. Ants allow for very little individuality when it comes to identifying with their colonies.[39] This is reflected in the ants' absolute commitment to their anonymous societies based on a scent alone. Their rigid identities give us good reason to call an ant colony a superorganism. Individual ants are bound to a colony as cells are to an organism in the following manner: the ants identify each other by detecting the hydrocarbon markers on their body surfaces, and in a healthy society invariably evade or destroy foreign ants with different markers. Likewise, the cells in a body identify each other by detecting chemicals on their surfaces, with the immune system killing foreign cells bearing the wrong signals. On this basis your body, with its trillions of member cells, represents a society of a microbial sort.[40] While disgruntled people can leave their society, even defect to another, an ant worker, linked for all time to nest-mates sharing her scent, can be stressed to death and still not abandon her nest-mates. Admire the devotion of the cell to its body or the social insect to its colony, but from the human point of view their total immersion into a greater whole feels dystopian. Human societies aren't superorganisms, nor would we want them to be—we cherish our choices as individuals too much.

As in other animal societies, the young in human societies are outliers as well. Not unlike ant larvae, which don't yet have a colony scent, children have at best a nascent identity. No child earns a respected place in a society merely by being alive, however.[41] In contrast to the young in individual recognition species, which simply must come to know and be known by everyone, human youngsters have to detect and learn the markers that permit them to fit in. Until our progeny master that task, they are weak and safely ignored. At that young age, adults recognize that they belong from their proximity to parents or friends.

STRAIN ON THE BRAIN

As much as humans can still rely on individual recognition for memberships in some groups, might species with societies that depend on individual recognition have at least the potential to use markers? While we know that primates can learn to associate a plastic token with a nonsocial object (such as food),[42] we understand far less about their ability to represent other social agents with markers. So while monkeys don't ordinarily use signs to differentiate themselves from foreigners, I'd love to introduce a marker into a troop to see if it catches on. Any able-bodied primate will tear off a red jersey, but what about paint: can it learn to expect troop-mates to have a dab of red on their foreheads? If so, would it welcome a foreigner wearing the mark of its "team"? Could a supersized troop stay together if its young acquired identical red splotches, enabling them to keep track of each other without knowing one monkey from the next? And would tensions mount if half the troop were dabbed with blue? I anticipate the answer to these questions is no. Monkey minds aren't wired for markers, and would fail to recognize them even should such differences exist.

Now, before we get all superior about humankind's use of markers to recognize and respond to one another, let us consider the cognition required for marker recognition. To suggest that monkeys aren't wired to readily grasp markers of identity isn't to imply that the ability requires a great intellect. The social brain hypothesis posits that handling more social relationships requires a large brain. But for those who point to this theory to predict that populous societies require formidable mental power, the lowly ant presents a serious challenge. Given their social complexity and flexibility, ants do a lot with a mere 250,000 or so neurons. As Charles Darwin proclaimed, "The brain of an ant is one of the most marvellous atoms of matter in the world, perhaps more marvellous than the brain of man."[43] Of course, human markers are vastly more numerous and diverse than those of ants, which don't wear ethnic hats or speak with an accent. Yet the aroma that serves as the ant badge of identity could be more elaborate than one might suppose. Each scent is composed of a cocktail of hydrocarbon molecules that differ in both kind and concentration from one colony to the next.[44] In practice, then, the scent represents a whole collection of markers, not just one, with some molecules probably influencing the ants' interpretation of the scent more than others. Ants are also nimble in responding to foreign colony odors, for example distinguishing members of neighbor societies they know well from members of a colony they have never encountered. The latter, being an unknown threat, are likely to be attacked much more violently.[45]

Many academics may protest that human markers can be supremely complicated, and indeed some are, such as the memorization of an arcane religious text. Humans turn labeling into an art by imbuing many of our markers with significance, often at multiple levels, to create symbols—an impulse that arguably distinguishes us from other animals. For the Irish the shamrock is at once a plant claimed to predict the weather; a Celtic emblem of good luck; and a tool Saint Patrick used to teach the Druids about the holy trinity.

Each person develops beliefs regarding "his personal affiliation with certain symbols, or, more accurately, with what certain symbols stand for," Edward Spicer, an anthropologist who studied American Indian tribes, pointed out.[46] Sociologists balk at the notion that any animal, lacking symbolic capabilities, could have anything resembling a nationality, while anthropologists consider the employment of symbols as critical to the emergence of humanity.[47] Tellingly, however, most people can't tell you much about the meanings of the symbols they prize.[48] Americans belt out *The Star-Spangled Banner* without knowing what it means to spangle—or remembering the words. "It is likely that even people who are expert in the use of symbols—shamans, priests, or sorcerers—cannot state precisely what a particular symbol is all about," social anthropologist Mari Womack reminds us.[49]

In truth, it isn't necessary to understand how or why something is profound, or for that matter to possess a deep meaning at all, for us to be sensitive to it—or to its absence, or appropriateness. People don't have to burden their overloaded cranial wetware with what a symbol signifies. What meaning it has, if any, may reside in the eye of the beholder.

Producing and perceiving markers therefore need not take much effort, and, once learned, the markers can be applied to indefinite numbers of individuals with no additional mental demands and no obligation to maintain a relationship. Among ants, brain size actually *declines* in species that have large societies.[50] Workers in a small colony are mentally challenged, as much as ants ever are, because they are jacks-of-all-trades. In a big colony, a soldier ant may attack enemies but seldom, if ever, nurses the brood, a job for smaller workers. Not only do ants save mental exertion by being ignorant of each other as individuals, but in large colonies this reduction of skills further diminishes the demands on their brainpower.

For people, too, as long as the strangers around us look and act reasonably, ignorance can be bliss. Imagine if you felt obliged to introduce yourself to, and get to know, everyone you came across. The mental demand would be overwhelming. By comparison, markers are simplicity itself. While managing the markers of our society could have contributed to the enlargement of the forebrain of humans compared to other animals, it isn't likely to be the primary cause. Sitting in a public place, you register the physical, cultural, and other traits of those around you without much thought, a minimal investment of effort. A psychologist would say that markers reduce the cognitive load of social surveillance, freeing you up to read this book or converse with friends at the café. Even small tribes like Kenya's El Molo, with its handful of people, take advantage of the low cost of social surveillance made possible by their shared dress, language, and so on—they may know every other El Molo intimately, but the El Molo are still an anonymous society. In fact, since the advent of farming, humans have undergone a reduction in overall brain volume the size of a child's fist, attributable, perhaps, to the extent to which we have grown more dependent on others for tasks from cooking to construction.[51]

Of course, humans adapt their behavior to circumstances, and that can include adjusting to foreign cultures. After months in rural India I subconsciously took on the accent and side-to-side head wobble while speaking. Subsequently in Singapore I shifted to their accent, adding the word "*lah*" to the end of sentences for emphasis. The tweaks in my speech patterns seemed to help the locals understand me, and I once proudly helped an Indian tourist communicate with a Singapore shopkeeper by "translating" between the two English dialects. Even so, I'm sure my foreign manner of speaking was transparent—if my ineptitude at putting on the Indian men's sarong, called a lungi, or a hundred other things didn't give me away.

As for your sizable brain, that has less to do with the population of your society than with your ability to immerse yourself in the lives of those who matter to *you*. Markers not only take the cap off the size of societies; they make social life less complex.[52] Why is it then that so few other vertebrates employ them? Possibly their societies function best with populations so small that individual recognition suffices. Yet even in a small society keeping track of everyone individually requires more mental bandwidth than the simplicity of noticing a reliable marker. Think of the difference between anonymous and individual recognition societies as another take on the apples and oranges idiom: animals know who fits in which category by registering either their similarities (picking out a marker they share, much as one picks out oranges by their orange color) or their dissimilarities (picking out each member from its personal features). The former demands less concentration. Animals reliant on individual recognition do what they can to circumvent this cognitive burden. Primatologist Laurie Santos tells me that when macaque troops face off, both sides bunch up. Clumping serves as a defense shield, but also minimizes confusion. So long as each monkey sees at least one foreign individual on the other side, it can take for granted the whole lot of them is foreign. In the same way, it can assume a monkey grooming its friend is a member even when the groomer's back is turned.

Complicated personal interactions, mostly with the few dozen souls in our intimate social networks, are at once a hallmark of humanity and a holdover from individual recognition societies in other primates. At the same time, humans know not just friends and relations, but many others to varying degrees of familiarity. So while most mammals treat each society member as an individual and from that personal knowledge build a collective identity, people enjoy the ant-like option of ignoring and even being ignorant of each other. Like the ants, we relate to strangers based on whether they share our identity.[53]

This section laid out how the societies of ants show the same trend of increased complexity with population size that human societies do. Yet adding citizens to a nation, like adding ants to a nest, doesn't need to entail any additional strain on the brain. By employing markers of identity, we, as members of anonymous societies, are gifted with the ability to think of a stranger as one of us.[54] Modern human societies, in all their sometimes continent-swallowing grandeur, are underlain by this feat of the imagination, as was true, as well, for the small societies of our ancestors. That includes the societies of the agriculture-lacking humans of the past millennia, who were really no different from you and me. To understand people now, we must understand people then.

Hunter-Gatherers Until Recent Times

CHAPTER 8

Band Societies

T he sun had turned a faint red by the time South African anthropologist Louis Liebenberg brought our jeep to a halt near
the !Kung Bushmen camp at Gautcha Pan in the Kalahari of
Namibia (the ! represents a click sound). Louis spoke to a young man
named !Nani about a noxious beetle grub his people use for poison
arrows. Yes, !Nani said, he knew of some in walking distance; he would
take us.

The next morning we took !Nani in our car over a flat expanse broken by scrubby thorn trees and the occasional bloated baobab tree.
Clearly a Bushman's idea of walking distance differed from ours. After
guiding us by what seemed like quite a few kilometers, !Nani stopped
us at a patch of glossy-leafed shrubs. There he used a traditional Bushman digging stick to excavate the deadly pale larvae and showed us
how to squeeze their toxic juices onto an arrowhead.

!Nani and his fellow !Kung, along with other Bushmen and hunter-gatherers elsewhere, don't practice agriculture or raise livestock.
They depend entirely on the food available from nature, hunting
game with poison arrows and other simple implements and gathering
vegetable foods across a vast "walking distance."

For more than a hundred years, anyone seeking to understand early
humans has turned to the recorded evidence of how hunter-gatherers
have lived in recent centuries. Of special interest has been a pattern
wherein hunter-gatherers moved in little groups, known as bands.[1]
Each band traveled around to put up camps from which they sought
food and water. I call such nomadic hunter-gatherers band societies,

which, as this and the next chapter will address, typically consisted of a few bands.[2] I distinguish band societies from tribes, a word typically used, as I do in this book, to describe simple settled societies, most of which were dependent on horticulture, where plants are cultivated in gardens rather than plowed fields; as well as more mobile pastoralists who tend domesticated herds. (To confuse matters, "tribe" remains the word of choice for North American Indians, many of whom lived in band societies.) From the perspective of the study of human origins, horticulturalists and pastoralists are latecomers, less pivotal when it comes to grappling with the root characteristics of humanity. Agriculture is such a recent innovation that even the nations of the modern world need to be interpreted in light of how hunter-gatherer societies functioned.

Archaeologist Lewis Binford asked a native Alaskan man to encapsulate his itinerant existence in a band society. "He thought for a moment and said, 'Willow smoke and dogs' tails: when we camp it's all willow smoke, and when we move all you see is dogs' tails wagging in front of you. Eskimo life is half of each.'"[3]

As poetic as this old man's words were, and as important as hunter-gatherers have been to anthropologists, there hangs over the enterprise the question of whether peoples who have recently lived by hunting and gathering are accurate mirrors to our past. For centuries hunter-gatherers have had to either adapt to the presence of farmers and herders or be pushed by them onto harsh, unproductive land. We know they could have undergone profound changes before the first explorers documented their lifestyles.[4] Even shortly before: when the pilgrims arrived in America to found Plymouth Colony, the first Indian to greet them already spoke English (taught, apparently, by British fishermen). Nearly two centuries later, on their cross-country expedition, Lewis and Clark came upon tribes that already rode horses, animals that had been extinct in North America for thousands of years but were reintroduced by Europeans.[5] Clearly the Indians of early photographs had already reinvented themselves many times over.

Similar stories can be told about any hunter-gatherers. The Bushmen, a race of short, slim, reddish-tan, baby-faced people distinct from other Africans, are especially valued in studies of human evolution. They lived across a swath of southern Africa—in the same general area and a similar sort of desert and savanna habitat where humans evolved—with genetic evidence suggesting they split from other human populations in our deep past.[6] Nevertheless they were influenced by centuries of interactions with Bantu herdsmen who came from the north long before Europeans arrived.

Before Europeans made contact, Australia was one of the few major locations on Earth where local hunter-gatherers had few encounters with agriculturalists. Aborigines occupied the continent for 50,000 years following an early diaspora out of Africa. Arguably that makes Aborigines a reliable source of insight into the past, yet the northernmost Aborigines traded and intermarried with tribes from the Torres Strait Islands, who raised crops of taro and banana. Additionally, for a period starting in 1720, a fleet of Indonesian fishermen came to northern Australia to hunt sea cucumbers. The Indonesians took some Aborigines to visit their home city, Makassar. From them the Aborigines learned to make dugout canoes and shell fishhooks. They also garnered a taste for new songs and ceremonies, goatees, smoking pipes, sculpting wood, and painting skulls.[7] And of course like all peoples the Aborigines were not culturally stagnant; they invented the boomerang and conceived of spiritual cycles called the dreamtime, items and ideas found nowhere else yet shared widely across the continent.

When Europeans arrived, so did colonial diseases and military actions that decimated populations before anything but crude descriptions of their societies had been recorded; as one prominent anthropologist concluded, "The traditional local groupings of most Aboriginal tribes changed quickly and radically under the impact of European settlement."[8] For those studying how societies maintain their identities and separation from each other, that is a serious problem.

European contact with the hunter-gatherers represented a culture shock that is hard for us to conceive. It would have changed everything. Before Europeans appeared, hunter-gatherers likely had their own version of the geocentric view held by many civilizations until Copernicus proved Earth revolves around the sun. As the nineteenth-century explorer Edward Micklethwaite Curr told it, "Tribes living on the sea-coast probably think the world more extensive than those which dwell inland, and it is likely that those which occupy large tracts of desert country have more enlarged ideas on the subject than tribes in a productive, thickly tenanted neighbourhood." Of the inland hunter-gatherer societies, Curr wrote, "Their idea was that the world was a plane extending about 200 miles on every side, and that their country was its centre."[9] Ponder then what an Aborigine would have felt when European boats anchored off their beaches. It would have been as if Martians landed on the White House lawn. Their worldview, right down to many of their perceptions of their own societies and each other, would have been shattered, not gradually over years or even days but in one mind-blowing flash. Any once-divisive differences between Aboriginal groups would be exposed as slight and unremarkable. Although the societies of hunter-gatherers remained standing, their original sense of social identity and being distinctive from their neighbors became almost impossible to accurately reconstruct.

For all these reasons, I will speak of hunter-gatherers, and later of tribal societies, in the past tense, not to suggest the people themselves are gone, but that their original ways of life largely are. The few hunter-gatherers I have had the good fortune to spend a little time with like !Nani had already been required to stay in one place and stop hunting game. Nevertheless, we have some assurance that the hunter-gatherers studied in recent centuries can reveal something about how our ancestors conducted their business, at least in broad strokes. Band societies as diverse as the Arctic Inuit and African Hadza are alike in ways that suggest human nomadic existence has a blueprint—foraging widely for wild food was part of a package that, as a rule, included other social building blocks, to be discussed in the pages that follow. Along the way we shall investigate what societies represented for these hunter-gatherers, and why so many anthropologists have downplayed or overlooked the societies. What should become apparent is that hunter-gatherers lived in anonymous societies just as we do today.

FISSION-FUSION AND THE HUMAN CONDITION

The feature that distinguished band societies from other modes of life wasn't specifically hunting or gathering, which people still enjoy, if only when they bag deer or excavate truffles. Nor were these hunter-gatherers totally distinct for their nomadism. Pastoralist tribes like the Huns could disperse to camps part of the year to secure pasturage for their domestic animals.[10] What most set them apart was the pattern of movement of their members: nomadic hunter-gatherers spread out by fission-fusion, with people roaming with considerable freedom.

Fission-fusion nevertheless generally took a regimented form for the nomadic hunter-gatherers of recent times, as it presumably did for those living prior to the invention of agriculture as well. People mostly clumped here and there in bands. Each band consisted of on average 25 to 35 individuals comprising several, usually unrelated, nuclear families, often spanning three generations.[11] A person could visit other bands, yet tended to keep a long-term connection with one. Shifts between bands usually came about with little effort but not often, a far cry from the eternally fluid movements of chimpanzees and other fission-fusion species characterized by ever-changing parties.

Also uniquely, the members of a human band would separate each day into work groups—confusingly called parties—to find food. Each night everyone would return to the location that the band had chosen to put up camp, typically for a few days or weeks. This protracted commitment to a band, with each band having a home base that changed frequently, is unique to our species. The closest analog among other fission-fusion animals occurs in the gray wolf and spotted hyena during the weeks that the pack or clan brings back meat to pups in a den. In all three species, the base shifted occasionally both to keep enemies off their scent and to put the residents in reach of fresh hunting grounds.

The ecologist Edward O. Wilson has argued that having a well-protected home base is at the heart of our humanity.[12] In this view, early people were like ants foraging from a central place. I propose an adjustment to this view. Certainly, a band organized its camps more than a wolf pack does a den, and staying put for a while provided relief for the elderly, infirm, and young. Yet the camps were usually short-lived and not especially secure. Indeed, bands moved around in part to protect themselves. Some Ache and Andaman Islander hunter-gatherers, from Paraguay and in the Bay of Bengal, respectively, lived under constant threat, in this case from outsiders, and changed locations often—sometimes daily. A band's members could jointly fight off a leopard, although a baboon troop ganging up on the same cat would be pretty much as efficient; and one could argue that the multiple family fires set up around an encampment kept predators at bay more than would the single fire of a family on its own. But these were the extent of defenses at a band level: fortifications protecting everyone became prominent only for people settled in villages.

Our pedigree of living in nomadic bands that were part of a society-encompassing fission-fusion webwork of movements and relationships suggests spreading out was (and is) as consequential to human success as staying put. The time spent in camps allowed humans to hone their social and cultural lives based on extended face-to-face interactions in ways no other fission-fusion species can, but at the same time their mobility allowed an expansion of a society's realm beyond anything achieved previously by the societies of any other vertebrate land animal. This required an unsurpassed intelligence to monitor daily relationships with individuals in the same band, and to keep up, as well as people could, with the goings-on of others far afield.

Fission-fusion in other animals looks remarkably disorganized compared to that in band societies, but a chimpanzee (or bonobo) party and a human band aren't irreconcilably different. Certain chimps actually come close to settling into temporary home bases. Although most chimps are forest dwellers, some live in savannas, the habitat in which we evolved. Savanna chimps get little notice, though their social patterns more closely approximate those of hunter-gatherer bands. For one thing, they have to hoof it almost as much as nomadic hunter-gatherers. Primatologist Fiona Stewart estimates a community of 67 in Tanzania ranges across an area totaling somewhere between 270 and 480 square kilometers, the animals about as thinly spread as hunter-gatherers were in such environments.

Parties of savanna chimpanzees also have more stable memberships and locations compared to those of forest communities. Several factors are in play. The animals can't easily meander over to the next party, which could be quite far off. Their favored tree clumps are also few and far between, and the distance often encourages several chimps to stay around one spot for a few days before trying another site. Recent hunter-gatherers also liked clumps of savanna trees for their camps, not only because of the shade, but for the water and other supplies associated with them. In addition, savanna trees are seldom sturdy enough for the apes to bend leafy branches into a canopy bed the way they do in forests. Instead the chimps often bed on the ground, bundling grasses or folding saplings into structures that crudely resemble the temporary shelters that Bushmen construct of grass and sticks. They spend time in caves, as well, to dodge the midday sun.[13] Foraging from these rest sites, savanna chimps kill small primates with spears made from branches and shovel out edible tubers with sticks in something of the manner of hunter-gatherers.[14]

It is no great stretch to imagine that the common ancestor we share with chimps was capable of similar maneuvers, not to mention the long line of hominids that followed. The evolution of the social skills that individuals needed to stay with the same society members each night; the taming of fire by our predecessor *Homo erectus* 400,000 years ago; and cooperation in sharing food before bedding down would have given humans a haven worth returning to in regular fashion.[15] Proper camps would have gotten their start.

THE REALITY OF HUNTER-GATHERER SOCIETIES

There are shelves of articles and books by anthropologists who write of hunter-gatherers based solely on what happened in bands, while ignoring how hunter-gatherer bands identified with a broader society. Sometimes, when they acknowledge the connections between bands at all, their solution is to claim that the nomads didn't have societies to speak of.[16] This view often intimates that hunter-gatherer culture varied from one band to the next, with no wider human affiliations and nary a sharp boundary to be seen.

I find this doubtful. For one thing, all other human populations live in groups we call societies. Additionally, the hunter-gatherer bands documented over the last few centuries were affiliated with societies. Indeed, there is ample evidence that membership in societies was a crucial aspect of hunter-gatherer existence—one that is indispensable to understanding how the societies of today came to be. My term "band society" could suggest that one band constituted a society, and that was true at times, but societies generally extended across a specific set of bands. I am not surprised these societies have been missed or misinterpreted. Until primatologists like Goodall deduced that chimps had societies from the animals' spacing and confrontations with outsiders, the consensus was that fission-fusion animals were "open-group species," without social borders.[17] So it made sense that these human bands, whose movements were marked by fission-fusion as well, wouldn't have had them either.

Because the members of human band societies, as well as chimpanzee societies, rarely come together, finding out where a society stops and the next starts can be tough, but sharp separation in the memberships of societies are in place just the same.[18] Numerous accounts tell how hunter-gatherers of recent centuries felt secure in the presence of their "own kind."[19] Asked who they were, typically a hunter-gatherer would give you the name for a community encompassing several, often a dozen or more, bands spread over a wide geographic range. These band societies had populations ranging from a few dozen to perhaps a couple of thousand individuals.[20]

To support the notion that hunter-gatherers do not sort out into distinct societies, some anthropologists put forward the example of the peoples of Australia's Western Desert, a desolate region where resources are scarce and people sparse. The Aborigines across that region have been reported to lack any clear boundaries with their neighbors.[21] This idea is dubious. The anthropologist Mervyn Meggitt recalled a conversation with one of the Aboriginal groups of bands from the Central and Western Desert that lays out the Us-versus-Them view of the people of that region in no uncertain terms.

> "There are two kinds of blackfellows," they say, "we who are the Walbiri and those unfortunate people who are not. Our laws are the true laws; other blackfellows have inferior laws which they continuously break. Consequently, anything may be expected of these outsiders."[22]

Certainly the Western Desert peoples intermarried, but that's true for societies generally. It's also true that to survive, they might have had to be open to border crossings. Robert Tonkinson, an anthropologist, has written, "In areas as harsh as the Western Desert the need to assert a particular identity has to be balanced against the need to remain on good terms with neighbors."[23] Amen to that. Yet the peoples of the Western Desert fell into groups as distinct as those of Aborigines elsewhere.[24] Even if, by necessity, they exhibited an unusually generous cooperation with outsiders, societies don't dissolve when they are kind to each other. The border between France and Italy isn't as fraught as the one between North and South Korea, but it is present nonetheless.

We think of nations, which academics call states, as having governments and laws, and band societies have neither, formally speaking. Still, hunter-gatherers expressed the same comfort and trust around the people of their own society that we do at present, and in many ways their societies were analogous to our nations. The nineteenth-century historian Ernest Renan believed nations to be a modern phenomenon. Yet his definition of a nation as a strongly bonded people with a shared heritage of memories and a desire to identify collectively is an apt description of band societies, too.[25] Hence with some justification, "nation" has replaced "tribe" as the word of choice for many Native Americans. For them, the bonds are all important.

While the institutions of modern nations are efficient at inspiring passions toward our society on grandiose scales, the affections we discern as a national consciousness have a deep history.[26] As linguist Robert Dixon informs us, again about the Aborigines:

> A [band society] appears in fact to be a political unit, rather similar to a "nation" in Europe or elsewhere—whose members are very aware of their "national unity," consider themselves to have a

"national language," and take a patronizing and critical attitude to-wards customs, beliefs and languages that differ from their own.[27]

Like nations, and in fact all other human societies, band societies identified with an expanse of ground that they exclusively occupied. They were territorial—wary of, and often hostile to, outsiders entering this area. Nomadic they may have been, but the overall movements of band-dwellers could be just as confined as those of the people who came to depend on agriculture.[28] There were exceptions. Territorial-ity went away when nothing was gained from it. In the American West the tribes of the Great Basin wandered with relative impunity where pine nuts were so abundant that defending a supply was pointless.[29] Yet generally each society took over an area ranging from hundreds to thousands of square kilometers, with the partitions tacitly determined in a time before living memory.[30]

Unlike chimpanzees and bonobos, who can move anywhere within the space claimed by their community (with some biases to favorite spots by individual chimps), each cluster of people—each band—usually had primary use of only a portion of the society's territory, a space that its residents knew like the backs of their hands, and some-times inherited. A fondness for home didn't originate with people moving into a physical structure but rather with their connection to the land. The Australian anthropologist William Stanner brought this idea to life:

> No English words are good enough to give a sense of the links be-tween an Aboriginal group and its homeland. Our word "home," warm and suggestive though it be, does not match the Aborigi-nal word that may mean "camp," "hearth," "country," "everlasting home," "totem place," "life source," "spirit centre" and much else all in one.[31]

Some anthropologists call the land crisscrossed by each band a "territory," but use of the landscape was more flexible than that word suggests.[32] Like the neighbor casually knocking at the door to ask for a cup of sugar, any socially upright person could usually traipse onto an-other band's space. As long as permission was sought, the local band might share water or space to hunt if supplies were short elsewhere, or allow the visitor to stay around and chat with friends and kin. Just as individuals of the same band freely shared and borrowed from each other, reciprocity across the members of a society was routine.

There was more goodwill between bands than typically existed between the societies that contained them, whose claims were rigidly
enforced when necessary. Among the Bushmen, bands of the same
society occupied contiguous spaces, whereas a no-man's-land lay between different societies.[33] Similar unoccupied gaps separate the societies of other species, too—communities of chimpanzees, packs of
wolves, and colonies of fire ants.[34]

ANCIENT RACES

That an Aborigine could speak of "other blackfellows" reflected a
clear awareness among native Australians of who was who and at the
same time calls to mind the issue of race. Physical expressions of genetic traits that vary widely across populations have become valued in
humans in a manner without parallel in other animals. We associate
many nations, and certainly the groups within them, with a race. At
the same time we have been so distracted by the distinctive physical appearances of hunter-gatherers from different parts of the world, such
as the Aborigines or the Pygmies, that we have often ignored their
societies. We need to give some thought, then, to the racial identity of
hunter-gatherers, and how much it mattered to them.

Many sociologists and others claim races are socially constructed—
arbitrary products of the imagination.[35] This is sensible given that what
we often describe as races would have evolved as gradual shifts in physical traits across continent-wide expanses, with all sorts of intermediates
between. Yet in the context of societies composed of people of distinct
lineages that have come in contact, usually as a result of movements of
entire populations with different physical looks, race can have a robust
meaning. Societies of other species don't migrate enough for race-like
distinctions to crop up between neighbors in this way.

All Aborigines descended from the same migration of people, so
the members of neighbor societies on that continent have been similar in appearance, skin pigment included.[36] Aside from some contact
with Indonesians along the coast, categorizing people by well defined,
genetically inherited features wouldn't have come up for pre-contact
Aborigines, who would have been attuned mostly to social differences
from one society to the next. Native Australians became Aborigines
("blackfellows") and developed a sense of otherness and self-awareness as a race only after Europeans populated Australia.

In Africa, the crucible of evolution for the human mind and body,
the situation was different. Though the dissimilarities among the

peoples of Africa may not seem as black and white (literally) as many of the distinctions we take for granted today, some of Africa's neighboring hunter-gatherers were "as different genetically from one another as are the major ancestral groups of the world," as two experts put it.[37] This proximity of unlike groups could reflect the immensity of time during which human societies have diverged and moved around that continent to establish societies that ended up living near groups sharply distinct from their own. And so it was that for the past several millennia that the Pygmies have formed alliances with racially distinct farmers, while Bushmen have lived among the Khoikhoi pastoralists for perhaps 2,000 years and among the darker-skinned Bantu, who entered their lands over the past 1,500 years.

Yet despite the long contact with very dissimilar peoples, and certainly not blind to their resemblance to each other, the Bushmen never felt themselves to be a single unit, worthy of a name and acknowledgment of affinity, any more than America's original inhabitants thought of themselves as Indians before Europeans coined the term. Bushmen was a category that the Bushmen, who thought of themselves in terms of their societies such as the !Xõ and the !Kung, didn't recognize.[38] Even now they consider themselves Bushmen— or San, a pejorative name given to them by their Khoikhoi neighbors, meaning rascal—only when they leave the bush to take jobs elsewhere.[39] As with other human differences, a feature becomes a marker of identity only if the people themselves choose to see it that way—and that includes race.

Clear physical dissimilarities between broad groups living in proximity weren't exclusive to the hunter-gatherers of Africa. For instance, the Ache of eastern Paraguay had (and have) pale complexions compared to other native peoples of the region. Of course, slight differences existed within these races, too. People sometimes connected a certain appearance with a single group, such as the Ypety, a group of Ache that tended toward darker skin and thicker beards than other Ache. But for the most part the broad groups we think of today as races don't seem to be central to hunter-gatherer thinking. Only complete outsiders, usually European (to whom all hunter-gatherers across a region looked the same), lumped them together as one group—ignoring the fact that hunter-gatherers actually belonged to multiple societies.

Otherwise, sociologists are correct in identifying race as a recent invention. In this regard, race can be thought of as a rough appraisal of the overall appearance of vast numbers of individuals who have no clear common background. The modern fetish for races is an artifact of confounding people from a mishmash of sources into broad classes that hadn't existed before. The hunter-gatherers of a region often shared a far tighter genealogical heritage, but ironically those people seldom could have cared less about their racial commonalities. Some of these genealogical races were so scarce or limited in distribution that they were represented by a handful of societies: the Hadza are all of one group; those we call the Ache had four societies; Tierra del Fuego's Yamana, now culturally extinct, had five; and the Andaman Islanders were originally separated into thirteen societies. Meanwhile, "the" Aborigines, spread across the Australian continent, had societies numbering five or six hundred.

ANONYMOUS NOMADS

Commonalities with chimpanzees and other fission-fusion mammals notwithstanding, the societies of hunter-gatherer bands were anonymous—they were dependent on markers of identity rather than the members' personal knowledge of each other. Individuals were regularly spread out to such a degree that not all the strangers they came across belonged to other societies. As one anthropologist who lived among the Bushmen nearly a century ago put it, "The more widely separated bands of a tribe have no personal knowledge of or direct contact with each other."[40] Even today the Hadza, with a thousand members, think of themselves as a people despite their affiliations with small bands. Many individuals never come into contact with, or know, Hadza from the far side of their society's territory.[41]

This would suggest that hunter-gatherers conceived of their societies as being united around a common identity—language, culture, and other markers. The anthropologist George Silberbauer wrote in 1965 that calling the G/wi Bushmen a "tribe"—that is, a society—was merited "by the fact that there are also other features of their culture, besides language, which are held in common." Regrettably he never spelled out what these features were.[42] Finding myself struggling with the scarcity of information about hunter-gatherer identities, in the summer of 2014 I dropped by the home of retired Harvard professor Irven DeVore, a foremost authority on Bushmen. At the age of 80, sporting a thin white beard and flanked by tribal totems, Irv had come to look like a wizard. He pointed out that many relevant details about the interconnections between bands—the markers they shared—might have gone unasked or unrecorded, even if they hadn't already faded or changed since the people's exposure to Europeans.

One reason for this oversight may have been that many features of hunter-gatherer societies appear muted from the point of view of contemporary observers. Although we are still finely tuned to very small dissimilarities between people, and often exaggerate their importance, our world now revolves around supernormal stimuli, pushing our senses to intensities not experienced in the past. Many of our societal markers, from Big Ben to Times Square, are inflated for maximum effect. For hunter-gatherers the signage of daily existence was understated. Each nuance about the natural world and the people in it filled their senses. Distinctions between their neighbors that to us would be slight would have stood out as plainly as grass blades bent by a passing antelope.

The dialects among the Bushmen have been easiest to notice and are well studied. Bushmen today speak one of two dozen or so basic languages (or dialects thereof). Unfortunately, numerous other languages, and Bushmen groups that spoke them, have gone extinct.[43] However, there are many markers outside of language that distinguish Bushmen societies. Anthropologist Polly Wiessner has described certain goods—such as oracle discs used to divine the future, wood forks for puberty ceremonies, and aprons worn by women—as identifying them as part of a group of bands regardless of how little their people interacted.[44] Depending on their "band cluster"—their society—the !Xõ Bushmen spoke different dialects and made arrows that differed in shape. People from one !Xõ band cluster recognized the arrows from another such group of bands "as coming from !Xõ 'who are not our people,'" according to Wiessner. Meanwhile she found that another group of Bushmen, the !Kung, numbering 1,500 to 2,000, had an arrow style of their own.[45]

DeVore pointed out to me that some distinctions among Bushmen may have evaporated when they stopped painting and engraving on rocks, the reasons for this artwork long forgotten. The Bushmen also ceased making pots once they began trading goods with their Bantu neighbors centuries ago. Archaeologist Garth Sampson collected pottery shards from a thousand ancient Bushmen campsites in South Africa. He found unique pottery motifs in certain areas, and concluded that each area had been used by a different group of Bushmen, all of them now extinct. Concentrated in one tract of real estate, for example, were pots with comb stamps, a technique of pressing the edge of a mussel shell into the still-soft clay that must have been invented by the local society.[46]

Other hunter-gatherers had more distinctions between their societies. The classic scene in Western films where the cowboy sees Indians in war paint and tribal regalia and proclaims, "Here come the Apache," has a rough basis in reality. "A knowledgeable person can look at a pair of Plains moccasins from a hundred feet away and say, those are Ojibwa; those are Crow; and those are Cheyenne," Michael O'Brien, an archaeologist, told me. "The leatherwork is distinctive but the beadwork is what gives them away." Decorations on pots and tepee designs are also conclusive markers. This goes for war bonnets, too: feathers could be splayed flat or erected into a tube. Some tribes replaced the feathers with buffalo or antelope horns. The Comanche did away with the bonnets entirely, and instead wore the whole buffalo skin with horns attached.[47]

As for the Aborigines, "our laws are the true laws," the Walbiri people would say, a clear statement of how important customs and beliefs were in marking off one group from another. As for more subtle matters, historian Richard Broome wrote of the Aborigines that "even gestures can be misinterpreted, as winks and handshakes in one group are mere twitches or touches to the other."[48] Still, from what I can gather, sharp distinctions between Aboriginal societies (or ethnolinguistic groups, as anthropologists awkwardly call them) were few, aside from language. Hairstyles could vary; the Urabunna, for example, encased their coifs in a "net-like structure," while their neighbors inserted emu feathers.[49] But artistic expressions, such as in body scars (their version of tattooing), are thought to have changed only at a wider, regional level.

There is intriguing information on group identity for hunter-gatherers elsewhere. Among the Andaman Islanders, who for the most part fiercely avoided outsiders until early in the last century, were the Onges who painted their bodies, while the Jarawas tattooed themselves by puncturing their skin with slivers of quartz.[50] African Pygmies differed in the quality and cadence of their music, the dances accompanying their performances, and the instruments they played (and mostly still play).[51] The four Ache societies had different myths and traditions, as well as their own musical instruments and singing styles. What made one of those groups, the Ypety Ache or Ache Gatu, so extraordinary (and feared by the other three Ache societies) isn't so much that they ate human flesh, but that a people with only about 40 living individuals kept up an elaborate ceremony and lore around eating the deceased. As one observer described it more than 60 years ago, "The Ache [Gatu] consume their dead to keep their souls from invading the bodies of the living . . . carried away after the meal by the smoke that rises from the ashes of the skull. . . . The soul climbs into the sky to be lost in the upper world, the Invisible Forest, the Great Savanna, the land of the dead."[52]

Cannibalism aside, most of the differences I have pinpointed among hunter-gatherer societies might seem trifling, yet to them they were momentous indeed. The oddest beings a hunter-gatherer came across in his or her life would have been neighbors who were actually quite similar, and had very similar things. Still, their differences would have evoked a visceral reaction of anxiety or dread far in excess of what you or I feel during encounters with most foreigners in this day and age. After all, given the low density of most hunter-gatherers in antiquity, people would have encountered an infinitesimal fraction of the outsiders most of us are familiar with today.

The sparseness of nomadic hunter-gatherers affected relationships not just between societies, but also within them. While their societies were a step larger than those of other vertebrates, on a daily basis people carried on their lives among just the few in their band. That made each band equivalent to a close-knit neighborhood, even if a mobile one. Socially, then, nomadic life was something we would recognize. Economically and politically, not so much. All of these aspects, from enjoying each other's company to issues of workflow and group decision making, were fine-tuned and performed in combination to facilitate the success of these peoples in challenging landscapes.

CHAPTER 9

The Nomadic Life

To get some idea of how a band society functioned in the day-to-day, think of each band, with its two- or three-dozen residents, not merely as a neighborhood in the social sense. Having to operate independently, it also had to serve as a local manufacturing center. Picture not a steel works, but a minimal organized unit of production, something no city quarter can claim. The factory wasn't elaborate. The people had no need for complex or permanent infrastructure. As with animals in a small society, notably the simplest ant colonies, they put together simple dwellings, and anything else they wanted, from materials gathered on the spot. Forget pharmacies. In Australia's Northern Territory, I once watched an Aborigine treat a case of congestion like his hunter-gatherer grandparents did. The man pulled a tentlike nest of weaver ants from a tree, mashed the feisty swarm into a pulp, and held the pulp to his nose for a few raspy inhales. Trying it myself, I smelled the intense Vicks VapoRub scent of eucalyptus oil.

Few tasks in a band factory demanded multiple players. When something had to be made in steps, like a spear or lean-to, usually the same person (or maybe two) carried out the process from start to finish. Teamwork, when required at all, was seldom elaborate, although felling a giraffe or mammoth might have taken the coordinated efforts of several men, and perhaps butchering it required the ant-like participation of everyone available.

Division of labor by sex and age was the backbone of the factory. In a human band almost invariably the men hunted large game or

fished, while the women, often weighed down by breastfeeding children (which made hunting impractical), gathered the bulk of the
band's calorie intake in the form of fruits, vegetables, and small prey
like lizards and insects, and cooked dinner.

The rhythm of expeditions orchestrated by sex added a layer of
complexity beyond that of the movements of almost all other animals.
Outside of foraging gangs of certain ants, the most specialized outings
in the animal kingdom may be when a party of chimpanzees hunts for
primate meat, or when gray wolves or spotted hyenas gear up for the
chase.

As a factory, each band was assembled redundantly: there could be
multiple hunting and gathering parties, and the families performed
many chores on their own. Because they had to be mobile to keep
their stomachs full, the hunter-gatherers couldn't accumulate stuff,
and didn't have ownership in the way we think of it. People could possess just what they could lift—about 25 pounds of supplies is a figure
sometimes given, the weight airlines permit for a carry-on bag. (There
were a few exceptions. Inuits could carry more on dog-powered sleds,
the Plains Indians on travois made of nets between two crossed poles,
and other groups loaded canoes or dugout boats.) Left behind for
later return were fire pits, heavy stones to grind seed, and hunks of
rock used in making tools.[1]

What made the band an economic unit was the swapping of goods
and information between people. If someone could not make an item
when the need arose, then giving and lending were expected, with the
understanding that someday the giver would ask the receiver, or someone in the receiver's family, for something in return.[2] The exchange
most significant for survival of the factory was of food. The Batek, a
group I have visited in Peninsular Malaysia, saw food sharing not as a
benevolent act but as a reflection of the fact that all food belongs to
the forest, not to the person who found it.[3] A man who made a kill
would share the meat with the band. In the Ache and some Pygmies,
a hunter never consumed so much as a mouthful of what he caught.
This generosity contrasts sharply with the stinginess of chimpanzees
and the only slightly improved attitude of bonobos, who still never
split food among the many, yet was sensible for humans. Hunters can
kill far larger game than the whole band can eat at once, and animal flesh must be consumed immediately. With giving back the longstanding practice, a hunter who passed around meat one day could be
sure someone else would fill his family's bellies the next—the original
social security package.[4]

An upside of this factory life was that people weren't occupied with raising crops or struggling to take in excess food. They ended up with leisure time—a commodity that earned these nomads the moniker "the original affluent society."[5] The only thing stockpiled was time to devote to social relationships. The typical population of a band turns out to be ideal. Enough meat, produce, and goods were procured, processed, and exchanged on an average day to keep everyone fed and comfortable. Success became problematic when fewer than 15 or more than about 60 individuals resided in a band.

A band's size was self-regulating. The individuals and families, although interdependent, did as they wished. People could leave an overcrowded band to join another or go it alone for a time. Dispersal was an annual tradition each autumn for Nevada's Western Shoshone, when each family went its own way to collect the same nuts favored by pinyon jays. They split up not because food was scarce, but because it was scattered, which also gave them the excuse to visit friends elsewhere.[6]

JACK OF ALL TRADES

At present people rely heavily on experts, big (think Steve Jobs) and small (the watch repairman). By comparison, people in a band society followed the jack-of-all-trades strategy of ants in small colonies—and for the same reason. When a labor force is minimal, a dependence on specialists can spell disaster. This is especially true if one member dies with no trained replacement. That, and the bare-bones nature of ownership, meant everything about their nomadic way of life pretty much had to fit inside each member's head, even if peers, and especially the elderly, were useful at reinforcing what was correct.

This is hard to grasp now. No one person can make a pencil, let alone an iPhone or a car, from scratch. Expectations of what people should do by age or sex were the only job descriptions. Otherwise the sole specialized line of work in a band was as a medicine man. Even then, Aboriginal healers, whose training in that role could take years, were still expected to carry out all of life's mundane tasks themselves.[7]

The "one brain fits all" approach put a limit on the complexity of a band society and the labors its people performed. The baffling instruction manuals of modernity suggest passing on directions is no easy task even for details that are permanently recorded, and of course hunter-gatherers had no writing. The inventor of the boomerang, clever at carving wood as that Aborigine must have been, also had to find a way for the average Joe to construct one without much fuss, or the discovery would have vanished in a generation.

Although all members of a band carried the same survival kit, differences among individuals in creativity and skill would of course have been manifest. As philosopher Gunnar Landtman recognized in the 1930s, among the Bushmen, "some women are more clever and industrious in making beads than others, some men handier at twisting rope and boring pipes, but all know how these things are done, and none devote their life to them."[8] Still, evincing one's talents was a delicate matter. Because of the intimacy of life around camps, band societies rarely tolerated a showoff. There are endless stories of hunter-gatherers teasing the successful or talented person, statements like this one included: "When I kill anything, it is usual for our fellows, especially my Bushmen, to exclaim: 'how small!' The wounded one that escapes is pronounced wonderfully fat."[9]

Mockery aside, everyone would have known who the best hunter or gatherer was, no matter how humble others expected him or her to act. Recognizing expertise is in the blood of our species—even a three-year-old realizes that individuals differ in skill and knowledge, and children seek out the appropriate person to solve a problem by the time they are four or five.[10] This inclination likely originated deep in our evolutionary past. The societies of many species benefit from the varied talents of their members, and some animals gravitate to individuals proficient at the task, as chimpanzees do when learning to crack open nuts.[11]

Among !Kung Bushmen, unproductive people without a noteworthy skill were said to be *tci ma/oa* or *tci khoe/oa*, meaning "nothing thing," or useless. Men whose talents accrued benefits to others were called *//haiha* (*//aihadi* in the case of women), which crudely translates as "one who has things" (don't ask me how to pronounce the words).[12] So without specialists or jobs, band societies offered opportunities for dexterous toolmakers, dazzling storytellers, skilled mediators of social conflicts, and thoughtful decision makers. Gifted people who didn't gloat and took teasing in stride avoided the fate of the characters in Aldous Huxley's *Brave New World*, who had their creativity squelched. Instead, they were able to use their talents to enhance their quality of life in permissible ways, for instance by attracting a desirable mate.[13]

Some psychologists hold that individuality wasn't prized until the late Middle Ages.[14] Yet as long as their actions weren't anathema to the whole tribe, nomad hunter-gatherers had a wide latitude for making unusual choices—especially given their freedom to escape those who disagreed with them by moving over to another band.[15] Differences in how a task was performed could be condoned from one person to the next. A man might have a special way of fashioning arrows, if only to provide evidence a kill was his and thereby slyly show off his skill. This latitude for individuality must have been a wellspring for innovation, especially since the pool of talent included hundreds of people across several bands. Even if each family were obliged to fabricate an axe of their own, they might try to duplicate the methods of the one person most skilled at axe making.

Just as they had no specialists, hunter-gatherers seldom had special interest groups aside from their families and the hunting and gathering parties that went out each day. There were no political parties or fan clubs, no fashion followers, preppies, hippies, yuppies, or geeks— and certainly none that wore the equivalent of a team jersey. Cliques of the women who especially enjoyed knitting didn't call meetings to order; exclusive fraternities didn't make merry. The group that could be most closely compared to such an affiliation was the band itself, to which people gravitated based on social compatibility, and to some extent, kinship. But a band's members didn't ordinarily believe themselves to be distinct from or superior to the other bands of their society. Individuals could compete, yet while modern towns and cities have rivalries, I've seen no evidence to suggest bands did (bands never, for instance, went head to head at group sports).[16] Other than the personality mix of the residents, and perhaps proximity to a nice waterhole, there was no special advantage of being in one band versus another.

Overall, the lack of groupiness *within* a band society is another reason to suspect that the society would have been of inordinate consequence to its people. At present we devote tremendous effort to affiliations that shift depending on whether we happen to be attending church or practicing with our bowling team, in a protean aspect of identity referred to as the group self.[17] In a band society all this zeal would have been directed at the only membership most of the nomads had, nuclear family aside: their society. Among other indications that life in the society completely expressed their worldview, the bands didn't distinguish sacred beliefs (usually relating directly to nature) from other aspects of life. Likewise, rituals, entertainment, and education—all matters except family matters were part and parcel of their relationship to the society as a whole.[18]

As in present-day societies, the fact that individuals interacted with some members more often than with others was of little consequence: the focus of identity and affiliation with their entire society made the ties among all members as clear-cut, if not more so, than those en-joyed by citizens of a modern nation.

GOVERNMENT BY DISCUSSION

When the Canadian anthropologist Richard Lee asked a Bushman whether his people had headmen, the man slyly responded, "Of course we have headmen! In fact, we are all headmen. Each of us is headman over himself!"[19]

Among the jobs absent from bands were leaders. Anyone eager to make decisions for others was a problem for people who spent most of their lives in small groups day after day with little privacy. Ridicule and jokes were weapons deployed against those seeking to sway others, as often they were used to deflate other shows of skill or superiority. To be influential, then, flamboyance wasn't an option; keeping the economy of a band humming along required nuanced social skill, with a goal to, in the words one anthropologist, "persuade but not command."[20]

Humility, again, was key. The socially successful band member was a diplomat and master of the debating arts, delicately guiding the argument without being pushy, and ceding to those with more insight in a given situation. No wonder hunter-gatherers took delight in games less competitive than those of today. With a few exceptions, such as tugs-of-war or a pastime among Bushmen boys of throwing wands at hummocks to see whose bounced the farthest, winning wasn't the point.[21]

A person could of course lead in the limited sense of serving needs of the moment, as we all do from time to time. Such a leader might turn up when decisive action is required and there is no opportunity to discuss the matter. This occurs in animals, for example when a motivated lioness heads the charge on a zebra. Or an individual may have information of immediate value, as when a returning honeybee scout performs a "waggle dance" to inform other bees about exactly where to look for flowers.

Bands squashed any assertive display of pretensions or attempt to direct others by what's known as a reverse dominance hierarchy.[22] The majority would collude to put a stop to the egotist, the power-hungry, the showoff. Similar tactics are on display in inchoate form when chimpanzees or spotted hyenas gang up on the offending individual, but in a band these actions were effectual enough to subvert whatever rigid primate hierarchy our forebears inherited.[23] Reversals of dominance aren't infallible. All of us learn the hard way that successful tyrants collude, as when tough kids in league with each other wreak havoc on a schoolyard. But such power plays would have had limited success. Anthropologists like to say that hunter-gatherers could vote with their feet. When tribulations were rife, they left for another band. With no way to seize political control across all the bands, bullies could be safely avoided.

The failure of anyone to dominate, and the resistance of the group to being pushed around, created equality across bands. There are precedents for such egalitarianism among the animals—prairie dogs, bottlenose dolphins, and lions also have no leaders and little dominance. By comparison, life under the most caring alpha chimpanzee can be a rough go as his underlings vie for position.

What the lack of differentials in power or influence most brings to mind are the social insects. Their efficient hustle and bustle makes it easy to assume someone is in charge—but no. Think the queen heads a colony? She is at the head of no army, gives no orders to nest builders or nursery attendants. Rate her existence miserable. After one fling with sex, an ant queen is stuck belowground. Cranking out eggs is her only role. Meanwhile each worker operates on her own in the throng, or carries out tasks with those that happen to be near her, and eats, sleeps, and works as she chooses.

Are social insects egalitarian? Workers of the most colony-centric species don't tussle even when perilously short on food. Nonetheless they display a kind of reverse dominance in one instance of social discord. Patrolling workers of honeybees and some ants and wasps smash eggs laid by fellow workers who dare compete with the rightful egg-layer—the queen.[24]

Hunter-gatherer egalitarianism didn't mean parity to a tee. It didn't always apply to families—some dads have always ruled with an iron fist.[25] And while material wealth varied little, differing degrees of adroitness in diplomacy and other skills created disparities. In this the bands remind me most of lions, an egalitarian species with no dominance hierarchy, and yet the animals bicker at a kill. Equality means equality in opportunity, not outcome. Among humans it never arises spontaneously. In the words of social anthropologist Donald Tuzin, "For Americans at least, 'egalitarianism' has a gentle Jeffersonian ring to it, evoking images of ruggedly decent, buckskin-clad pioneers working together in harmony for the good of one and all. The truth is actually quite the opposite: egalitarianism is typically a rather savage doctrine, for it involves constant vigilance and intrigue among society's members as they struggle to stay equal to each other."[26] No wonder gossiping is considered such a primordial human skill.

Nor was equality expressed perfectly, with expectations consistently varying by sex and age in the areas of childrearing, cooking, and hunting, as we have seen. For the most part, however, each voice was heard, which gave Bushmen, among others, more sexual equality than exists in most societies today.[27] When an issue arose, every affected party spoke until a decision was reached by consensus in what must have been the primary source of entertainment before TV. Theirs indeed was the original "government by discussion," as the British prime minister Clement Attlee once described democracies.[28]

Overcoming discord was a primary concern. While band members had few formal means to regulate behavior, they shared beliefs about how people should be permitted to act. Today, we think of these as *rights*. In a way, identifying with those rules was the measure of citizenship: the duty to behave properly and participate in matters important to the group. "Our laws are the true laws," said the Walbiri man, and by this he meant their ethical code.

COLLECTIVE DECISIONS

Communal actions are determined by collective choices in other species too. A meerkat that can't find a meal gives a "moving call." That lone voice won't spur on the clan, but should a second animal echo it, the entourage heads to a new spot.[29] In this situation the dominant meerkat has no more power to sway the group than anyone else. Painted dogs sneeze to chime in about whether to go for a hunt. Honeybees and some ants use a similar technique when they select a new nest by a kind of voter turnout to each location under consideration. Who told you democracy was a human invention?[30]

The army ant is an exemplar of organization without oversight. In Nigeria I blundered into a carpet of army ants 30 meters wide, a horde setting out to seize quarry with their flesh-slicing jaws in a campaign of shock-and-awe. No ant chief directs the throng. Ant swarms rely on swarm intelligence, with each individual contributing at most a bit of information—perhaps signaling the location of prey or an enemy with a pheromone. Yet the overall pattern that arises at a mass level makes strategic sense in that millions of ants end up taking a productive course without anyone at the lead.

Social coordination in army ants scales up beyond anything shown by human bands. While pooled decision making can be effective for small human groups, our societies have grown ever more dependent on centers of power to carry out large-scale endeavors, and we are all the more vulnerable for it. A leader can be a society's Achilles heel. Because individuals often step up to leadership roles for personal gain, and indeed can be popular despite an arrogant lack of concern for others, a Hitler or Pol Pot may bring a society to ruin. Then again, a society might be thrown into equal disarray when the death of a well-meaning leader leaves a void. Thus, for a society that seeks to defeat another, regicide—the killing of kings—is a cheap alternative to war. With no commander to take out, you can step on ants till the cows come home and they will continue to invade your pantry.

Information, as well as the faculty to use it, is distributed across marching ants—and across hunter-gatherers spread out in bands. To some extent social media has brought about a return to this group decision making by enabling likeminded individuals to pursue sweeping collective actions without oversight and at almost no cost. In a historically important early example, a Philippine court was forced to impeach President Joseph Estrada in 2001 after Epifanio de los Santos Avenue in Manila was engulfed by protestors. What brought so many together was a rapid exchange of several million texts reading *Go 2 EDSA, wear Blk*. The very act of treating such a text as legitimate required trust that those forwarding it were coequals. That day saw the reverse dominance hierarchy firmly back in control in the form of what futurist Howard Rheingold calls a "smart mob."[31]

THE ADVANTAGES OF SOCIETIES FOR HUMANS: LIVING IN BANDS

Just about every boon of membership in an animal society applies to our species also, in how societies both provide for and protect their members. Consider the matter from the perspective of a band society. Membership granted the nomads a balance between the stability of a reliable home base (actually many, shifting at intervals for each band) and flexible mobility, with the band and the society serving different functions. A band was the unit of daily interaction, its intimacy on par with that of monkeys that stay in contact. Even in the band, the connections between members were dynamic, allowing for daily separations into hunting and gathering parties. Groups of people could jointly defeat enemies or predators, obtain food (and widely share it, unlike most primates), and tend the children. In the latter task, which bonobos and chimpanzees do poorly at best, women took chief responsibility but men participated as fathers, as did grandparents: a role for the elderly uncommon among animals, though female bottlenose dolphins too old to reproduce aid in child care. Most remarkably, social learning was pivotal, and intricate, for humans: children garnered lessons not just from mom and dad but from other adults, absorbing the rhythm of life in their society.

In contrast to the rapport afforded by a band, the societies that hunter-gatherers belonged to were about identities—inclusion and exclusion. Even though each band carried out its quotidian chores alone, the society offered a security blanket, supporting reliable market and marriage relations beyond the local fireplace. Hence while people across all the bands were less intricately and constantly reliant on each other than we have become in nations, their inclusion in a society was of immense survival value, providing the recourse to share goods and information and act together as problems and opportunities came to light.

Hunter-gatherers in their band societies minimized petty altercations and stayed egalitarian by dispersing in factory-efficient bands. How then did humans come to take advantage of the social glue afforded by their shared identities and forge civilizations?

The process began with the hunter-gatherers themselves. As it turns out, my description of their societies, while in line with much of what has been written about them, has been incomplete: I have not conveyed a sense of the possibilities open to them. Human fission-fusion can be reconfigured to permit people to distribute themselves in a diversity of alternative ways, from simple village to urban throng, in the service of feeding and defending themselves. No less than with the farmers that followed them, how hunter-gatherers lived varied by circumstance. Resources permitting, a society could settle down even as its members continued to forage from nature. Their aversion to accepting the authority of leaders could be put aside, and displays of talent at a particular task could become a justifiable point of pride or even a necessity. Both the benefits and strains of participation in a society would have intensified, as the coming chapter will establish.

CHAPTER 10

Settling Down

Between Mount Eccles and the sea in Victoria, Australia, on a lava plain laid down by a volcanic eruption about 30,000 years ago, are the archaeological remains of hundreds of dwellings. The structures cluster in groups of a dozen or so, some so large they are partitioned into apartments. People by the thousands settled across that expanse in these small villages, members of settled tribes that jostled, fought, and forged lasting alliances.

The region around the villages was transformed into a vast, managed waterscape, with streams and rivers variously dammed and diverted to create a labyrinthine yet integrated drainage system. The waterways, which extend for kilometers, are ancient, many dating back 8,000 years, with the system reaching its full glory 600 to 800 years ago. The canals were used to harvest wild game—a species of eel—with traps reaching a hundred meters long and constructed in some places of stone walls up to a meter high. The people also carved out artificial wetlands in which the young eels could thrive until they were large enough to eat, and caught the fish in such abundance that the excess could be preserved and stored for the off-season.[1]

Like all the Aborigines elsewhere in Australia, the people at Mount Eccles lacked domesticated food. This entire elaborate infrastructure was the brainchild of hunter-gatherers. And yet the homes look to have been permanent, and some may have been occupied year-round—the descendants of the original residents claim this was so. Indeed, the lesson of the Mount Eccles Aborigines is that, even before societies took up farming, people had the option to reside in what I call a settled hunter-gatherer society.

The social lives of the hunter-gatherers who clustered in bands present us with a riddle. Many aspects of their existence seem antithetical to modern life. Most of us gladly follow a leader we respect and some of us strive to be leaders ourselves. Nomadic hunter-gatherers regarded both choices with disdain. We subscribe to a social hierarchy, not only fostering status distinctions, but also looking up to the powerful and prestigious, admiring those of breeding and affluence, and spending inordinate amounts of time tracking the private lives of stars and presidents. Karl Marx saw the history of societies in terms of class struggle, yet hunter-gatherer bands had no classes to struggle over. From whence, then, came the value that civilizations have accorded to accumulating wealth, if hunter-gatherers owned little and were willing to give it all away? Where did our personal ambitions come from and how far back in time did they exist? Finally, as it is clear that we aren't hardwired to be generalists in the manner of ants in small colonies, from whence our desire to be different from each other, for example by excelling at a narrow career path?

Settling down forced people to draw upon a set of implements from their mental toolkit distinct from the ones hunter-gatherers employed in bands. Although conditions we take for granted—such as social inequality, job specialization, and acquiescence to leadership—were far from ubiquitous among settled hunter-gatherers, each passing generation spent at one place made those conditions more likely to come to the fore. Because of this elaboration, anthropologists generally describe settled hunter-gatherers as complex, in contrast to so-called simple hunter-gatherer band societies. But fission-fusion offered its own complications. These included the laboriousness of tracking down food that could be spread thinly, as well as in finding good campsites and the struggle to maintain social parity. For this reason I will avoid labeling either as simple or complex. Rather, I focus on the constancy and compactness of their living places as the ultimate cause of the complexity (or simplicity) of hunter-gatherer lives.

Many descriptions of human prehistory assume that food domestication made way for the power and role differentials of civilizations to come into play. True, agriculture tipped the balance. However, just as chameleons change color as conditions demand, humans reconfigured their social lives—transitioning from equality and sharing to deference to authority and hording, and from roaming to setting down roots—as suited the situation. This can be hard to believe because we have come to think of hunter-gatherers much as they would have thought of us: as an alien life form. Yet human cognition nonetheless remains adjustable across all the social options that had once been available to hunter-gatherers.

GATHERING TOGETHER

People can be content when right on top of each other, thanks in part to anonymous societies that allow us to tune out others. Most animals have very a narrow tolerance for their spacing from society-mates. Despite their need to stay with the troop, baboons keep a cautious distance from all but their closest allies, whereas army ants constantly stand practically on top of each other. Pinyon jays are more flexible, thriving in dense flocks part of the year and paired up on nests when rearing chicks. Paradoxically, even most fission-fusion animals don't take advantage of the full range of dispersal possibilities, since, in species other than humans, wolves, and elephants, the individuals only put up with a few others nearby at a time. When Goodall stocked her research station with enough bananas to keep her community fed, the chimpanzees didn't all settle at that spot but rather underwent a meltdown that drove them completely apart (a topic for later). For humans, that would be no problem. Even people who have until recently lived in hunter-gatherer bands can take to city life (though as with all people who grew up in a radically different society, the cultural shifts can often be arduous). Meanwhile, the tourist immersed in Grand Central's multitudes can board a train to Blue Mountain for a pleasant walk with barely a human in sight that same day.

Some of the versatility of the human expression of fission-fusion was on display when hunter-gatherer bands converged in aggregations of as many as a few hundred. These occasional social gatherings were staged during a season when resources were bountiful enough to sustain everyone for a few weeks. During this time, people massed together much like monkeys in a troop, except lodged in one place. The chosen site could be a watering hole or other productive location. Each year some otherwise roving Andaman Islander societies rebuilt residences they kept near the ocean, and bands would rendezvous to fish for a couple of months. The pole-and-palm-frond structures reached well over ten meters in diameter and several meters in height, making them larger than my Brooklyn apartment. Inside, each family had a sleeping platform and fire. The communal domiciles were maintained for so many generations that each was associated with piles of refuse in some places thousands of years old and more than 150 meters in circumference and 10 meters high.[2] For another example, every autumn in North America, bands of certain tribes gathered to frighten herds of bison over cliffs.[3] No chimpanzee community enjoys each other's company this way, as one grand assembly; the apes parcel out their social engagements in smaller doses.

In some fundamentals, life during a band get-together was little changed from normal, with each band often keeping its status as a "neighborhood" by camping a bit apart. Yet the throngs were the reunions of the day, enlivened by gossiping, gifts, songs and dances, and, much as occurs when elephants aggregate, with males on the prowl for conquests.[4] While the most common instances of socializing between bands were visits by lone travelers or families to those they knew personally, these assemblages were greased by the people's shared identity. Surely, this must have given gatherings a significance in buttressing group practices and behavior, if just by circumstance. I've seen no evidence decisions were made that affected the collective. Still, the activities must have conveyed a sense of common purpose.[5]

The gatherings were a step in the direction of permanent settlements. But while they may have reaffirmed people's affiliations, this type of assembly, in which the itinerant hunter-gatherers simply stopped moving, was doomed for several reasons. Nearby sources of food would be cleaned out, offal and wastes piled up (a problem that the nomads managed poorly), and biting insects had a field day. Worst of all, with so many personalities right on top of each other, jealousies and resentments came to the fore. After all, what reunion doesn't reunite a few archenemies?

Indeed, the foremost reason people have never scaled up the egalitarian lifestyle of band societies is that, like most mammals, we squabble a lot. Just as rock concerts can devolve into social entropy, large gatherings could end in brawls. Murders peaked at this time.[6] Bands retired to their homelands, with some musical chairs as people joined a mate or friends living elsewhere. "Free societies are societies in motion, and with motion comes friction," the novelist Salman Rushdie has said. And so it was literally for roving hunter-gatherers.[7]

THE RAINY-DAY MENTALITY

In some cases bands developed practices that made their wild foods productive enough to approach the yields of agriculture. Even without

sowing seed or domesticating the food, certain modes of extracting
from the environment made staying put for a time easier.[8] In 1835
the Australian surveyor-general Thomas Mitchell recorded a vista "extending for miles" on fire-cleared plains by the Darling River that had
"the agreeable semblance of a hayfield." The Wiradjuri people reaped
the millet with stone knives and heaped it to dry in what was clearly
a time-honored process. Despite claims about what constituted the
Paleo diet, the grain was ground and baked into bread.[9]

In some instances hunter-gatherers fiddled with nature to improve
its productivity. The Ache of South America established plantations of
guchu beetle larvae, which mature to a hefty ten centimeters in dying
Pindo palms, by felling and sectioning trees for the beetles to find.
The meanderings of the bands had to be choreographed so everyone
got back in time to harvest the plump insects. The delicacy was so relished that if the Ache had found a way to safely raise enough of them
at one spot, my guess is they would have taken up residence right then
and there as beetle farmers.[10]

To drop the itinerant mode of life required bounty on an ongoing
basis. That included any excess that could be stored to tide people
over times of dearth. Among the insect societies operating this way are
harvester ants that jointly build well-protected underground pantries,
where they keep seeds fresh for months. However, most vertebrate
societies fail to show delayed gratification by means of a group effort.
For instance, each pinyon jay caches its own seeds and attacks any
member of its flock who steals from it.

Describing the gathering practices of a Bushman society, the anthropologist Richard Lee explained, "The !Kung do not amass a
surplus, because they conceive of the environment itself as their store-
house."[11] Nevertheless, cultural practices that yielded a lasting food
supply could support long-term home bases. Both Bushmen and
Hadza took a small step in that direction by drying meat (though not
enough to keep them at a place long).[12] Inuits did one better by putting seal carcasses on ice—for them the outdoors represented one
giant refrigerator. When the Western Shoshone split into families every autumn to enjoy pinyon nuts, they gathered the excess in baskets.
During the winter doldrums they came together to consume them,
turning a seasonal bust into a social and culinary boom.[13]

Few environments offer the opportunity to accumulate food in
bulk from nature year after year. Even when the resources are present,
the risk of committing to a site is great since the option to move elsewhere should things go awry can vanish when neighbors also decide to

settle down. Examples include the Mount Eccles Aborigine tribes and the Jōmon of Japan, who lived most or all of the year in settlements. Some of these villages were small and simple, like those of the few New Guinea hunter-gatherers, who caught fish and ate the pith of wild sago palms on land too swampy to raise the pigs and yams favored by their horticultural neighbors.[14] More elaborate were the settled hunter-gatherers of North America: the Calusa of southwestern Florida, the Chumash of the Southern California coast and Channel Islands, and, by far the best studied, the tribes of the Pacific Northwest, which ranged from the tall forests of Oregon north to the stunted landscapes of coastal Alaska.[15] All these tribes mass-collected and stashed aquatic foods to keep themselves fed through thick or thin.[16]

The Pacific Northwest was densely populated by hunter-gatherers when Europeans came to North America. Much as the Bushmen and Aborigines had never seen themselves as a group, so too, the peoples of this region had no term in their languages that encompassed everyone living a settled lifestyle and that distinguished them collectively from the roving bands living further inland. The reason for this semantic omission may be that among the settlements there were populations from varied Inuit and Indian backgrounds, each made up of multiple land-holding tribes that had taken to this way of life. Most lived off the ocean, but a few dependent on salmon found places to permanently congregate along rivers.

Some sites in the Pacific Northwest were occupied for centuries by a couple hundred to nearly 2,000 individuals. The most developed of these were impressive indeed. Their longhouses and other residences could be enormous, the biggest on record extending 200 meters long and 15 meters wide. That's as spacious as the homes of many modern celebrities, though several families shared the building; a small settlement might occupy one longhouse, while a large village spanned several structures.

The societies were neatly set apart by markers of identity, more exuberantly so than for people living in small bands. The differences were striking and well documented in the Pacific Northwest. Most stunning were the labrets, ornamental plugs worn in perforations in the cheek or lower lip, ranging from ivory disks to multicolored beadworks. Labrets made their appearance 3,000 to 4,000 years ago and revealed much about the social and economic standing of the wearer, though their primary and original value was to connect people with a tribe.[17] In the far north, for example, the Aleut tribes were distinguished as well by their tattoos, nose pins, necklaces, and the choice

of animal fur for their parkas. Wherever foreigners might be encountered, whether at a friendly meeting or in battle, there was no marker
more crucial for the Aleut to wear than a tribal hat that resembled the
beak of a bird, painted with brightly colored decorative motifs.[18]

LEADERSHIP

Much of the intricacy of settled life bears on handling the personal
discord and logistical issues that routinely broke up groups of nomadic hunter-gatherers. That this was a problem for the nomads is
supported by the observation that individual hunter-gatherer bands
around the world, across habitats as divergent in the resources on offer as tundra and rainforest, kept their size to a few dozen individuals.
Social failure, not some inescapable issue around hunting and gathering, would explain this regularity.[19] The bands of some Bushmen
societies, for example, grew unwieldy and dysfunctional every two or
three generations.[20]

To circumvent this dysfunction, people in settlements came to
tolerate those with a knack for making decisions they could abide
by. There are animals that show few differences among members in
power and influence and do fine—ants operate through a kind of
collective intelligence. People aren't ants, for which markers are sufficient to keep a society together—transcontinentally in the case of
Argentine ants. Social conflicts have existed for all humans, band societies included. But where did the human proclivity to either lead
or follow leaders come from? Certainly it reflects an upbringing in
which we defer to parents who know more than us, and who expect
us—indeed, often coerce us—to behave in a manner appropriate to
our society and station, and yet it has come to extend throughout our
lives. Today we are surrounded by authority figures from teacher and
boss to sheriff, the president, and Congress. For societies to function
today, people must know their places under different circumstances
and behave accordingly, whether to lead or to follow.[21]

In many vertebrates, a dominant animal can remind us of a leader.
Dominance affects who interacts with whom, and how. Still, while expressions of power and control of resources can be important components of high status, they don't necessarily put the one at the top
of the heap in charge. Among most species the "top dog" is likely to
push others around without mobilizing anything significant for the
group. Some alpha individuals have a certain influence: naked mole
rat queens nudge and nip at the workers, supposedly to prod them to

work. Following the most powerful, highest-ranked animal can give others safe cover if something goes wrong—the alpha ringtailed lemur typically sets the path taken by her troop for the day. That's not the case in bands of horses, though, where the mares, and not the lead stallion, set the route.[22]

What people more typically think of as a leader, however, is someone who takes a substantial role in directing a society's affairs. The alpha male and female wolf do just that: in addition to setting the pack's direction, they get everyone moving and initiate hunts on prey and attacks on foreign wolves.[23] In an elephant core the oldest animal earns her matriarch position not by pushing anyone around. The others defer to her in apparent recognition of her accumulated wisdom on such matters as which outsiders are friends.[24] She also intervenes when tensions surface between her underlings and comforts the hurt parties after, something dominant chimpanzees with political savvy do as well.[25] Even so, adroit negotiators in these species have limited impact. They cannot lay out long-term courses of action for the whole society the way monarchs and presidents do (albeit admittedly such leaders today rarely carry out such agendas without the approval of others).

Leadership in this strict sense is rare in nature. Yet we have seen that even people don't require leaders. In hunter-gatherer bands everything from daily activities to long-term plans was a matter of discussion. When more than a few dozen people resided alongside each other for the long term, however, this egalitarian approach became untenable. In the first villages the easy nimbleness of moving away from pushy individuals or acting together to knock them down a peg was lost: at best one could walk to the far end of town.

Under these circumstances many factors played a role in determining who became a leader. Magnetic personalities garnered support but would come along rarely in a band society or small settlement. We don't see many John Kennedys step forward in nations of millions. Still, having even a modestly talented person in charge can offer advantages when acrimony is high. At such times the effort to be an authority figure pays off, as does willingness to bow to such a person. The two must go hand in hand for leadership to work out, with people putting aside disagreements to support an individual rising to the occasion.[26] Then as now, humans would have gravitated to those who made themselves the center of attention and responded to issues quickly. Part of that faculty, first honed in band societies, was public oratory, a skill vital to the incipient stages of leadership. In what's called the babble effect, chatty people have always held a certain sway, though

Bushmen youths, in their egalitarian bands, are said to have stifled the urge to follow them.[27]

So while hunter-gatherer settlements didn't put together anything like a government, they did have influential people, if not clear leaders. As an example, the chiefs of the fisher-people around Mount Eccles were treated as nobility, able to declare war and claim the best spoils.[28] The New World ruler closest to a king in stature was the Calusa chief, who kept the peace seated on a stool, an unassuming throne by current standards, in a building that, according to one historian, was "said to hold 2,000 people without being crowded."[29]

The Chumash and Pacific Northwest chiefs, although ostentatious, were less overtly assertive of their power.[30] They had to act cautiously compared to the army-backed heads of large agricultural societies, relying less on coercion than persuasion and recompense, such as feasts to encourage people to meet obligations. Leaders have always been virtuosos at political maneuvering and at guarding their own interests.[31] Yet the chiefs often presented themselves as exemplary members while displaying some of the humility, integrity, and steadfastness that were expected of people in a band. These are still admired qualities in leaders today, perhaps a holdover from egalitarian times. Indeed, by convincing people to work together, chiefs ensured that elements of the egalitarian mind-set stayed intact. Even then their sway was limited. In a never-ending push and pull between leader and followers, small settlements supported chiefs upon whom the people exerted control to some extent.[32] Pacific Northwest chiefs sought the support of informal councils that had a say in the pedestrian aspects of village life. This was leadership by committee, with the committee taking up the role handled by the entire band in a nomadic society.

BACK AND FORTH BETWEEN WAYS OF LIFE

The troubles that make leaders necessary need not be internal to the society: some threats come not from discord between the members but from outsiders. History's most revered leaders—George Washington, Abraham Lincoln, and Franklin Roosevelt are some American examples—came along when citizens wanted someone they could trust in times rife with conflicts.[33] This could be another path to the emergence of leaders among hunter-gatherers, as well. Early accounts by European settlers tell of hundreds of Aborigines in pitched battles that presumably took organization, and according to some descriptions these had strong headmen. Bushmen embraced leaders when making a united front against enemies, too.[34] That was so for the =Au//ei living in what is now Botswana (the = and // represent click sounds). The records concerning these Bushmen from the early decades of the nineteenth century recounted the =Au//ei as seasonally occupying villages surrounded by defensive palisades. To obtain sufficient food while so entrenched, the people innovated methods for culling game by driving herds into pits that were a maze of yawning chasms.[35] The =Au//ei of that time were warriors prone to blood vengeance who burned wagons, stole cattle, and collected tribute from other Bushmen.

That a society adapted to leaders and village life didn't mean the change was permanent. During the final decades of the nineteenth century, =Au//ei reverted to roving, leaderless bands. By 1921, headmen had reemerged with the reduced role of orchestrating armed clashes, this time without the people settling down. The leaders gained influence during periods of encroachment by armed tribes such as the cattle-rearing Oorlam.[36] When the incentives to fight diminished, the leaders fell out of favor and the people reverted back to an egalitarian band life. One might assume it would be their *original* way of life, except there is no telling how far back in time the requirement for self-defense might have arisen. The dynamic =Au//ei culture suggests that changes in social organization come and go, leaving hardly a trace.

Famously, nomadic American Indians had leaders, too: senior individuals whose roles continue to the present day. It is impossible to say how often those hunter-gatherer tribes, with their portable tepees and wigwams, were led by chiefs prior to the arrival of Europeans, who brought with them horses and guns. What's clear is that when dangers were ongoing, nomadic societies embraced greater social complexity. Among the Plains Indians, warrior associations—the West Points of the day—put select men through rigorous training to prepare for combat.

At the other end of the gamut from the =Au//ei, who spent most of their time as nomads, hunter-gatherers widely thought of as fully settled were just as malleable. Even the Pacific Northwest Indians, the gold standard for settled peoples who foraged from nature, didn't always stay put. Villages either moved or disbanded as suited them. Some longhouses were seasonal residences, with families traveling by freight canoe to homes elsewhere. There's evidence of temporary camps, too, suggesting people would go out to hunt game, perhaps like sportsmen pitching tents today.[37] Probably few hunter-gatherers lived at the absolute extremes, as perennial homebodies or incessant wanderers.[38]

LIVING WITH DIFFERENCES

Once people had settled down for long enough, with property no longer restricted to the items each family could haul around, possibilities opened up for expanding technologies. The revolution in social order that ensued affected the very structure of day-to-day existence.

Many of the technologies generated by settlements improved food production. The Pacific Northwest Indians, for example, built massive seaworthy canoes, equipment to bulk-process fish, and watertight containers for prolonged storage of dried seafood. The range of devices diversified until fish could be caught in a number of styles of nets, or clubbed, speared, harpooned, brought in on a line, or trapped in weirs. The wide assortment of intricate gear permitted a few to collect much of the food for all. That was important given that the concentration of bodies in one place often made it impractical for everyone to fan out and procure food individually. Bringing home the bacon—or salmon—became one of many social duties to benefit from specialized know-how.

A person's interests and talents, say, in knapping arrows or weaving cloth, therefore found a more positive and open reception than they had in bands. In the Pacific Northwest some occupations became hereditary, with parents passing on their specific skills. The growing need for expertise meant the mechanics of every aspect of the society's overall functioning was no longer within anyone's grasp—which also meant that the sum total of the knowledge in the society was expanding. Just as staying in one place assured that social complexity stopped being limited by the physical baggage members could carry on their backs, settling meant social complexity was also set free from the cultural baggage all must carry in their heads.

Since the pioneering work of Emile Durkheim, sociologists have regarded specialization as a force for enhancing the cohesion, or solidarity, of societies.[39] This it certainly is, though the cohesiveness he had in mind wasn't entirely restricted to people in settlements. The jacks-of-all-trades in a band society likewise relied on an exchange of goods that contributed to the people's solidarity. While trading among bands could occur between people blessed with different talents, the clear-cut jobs in many settled societies accentuated the effect such that people grew ever more interdependent, in a trend that continues right up to the present day.[40]

Job differences would also come to simplify interactions with little known others and total strangers as they grew more common in larger settlements. We can figure out how to behave toward someone merely by identifying the person's role—based on the police uniform the person is wearing, for example—without knowing anything else about him or her. Similarly, a worker ant interacts appropriately with a soldier of her colony, whether or not the two have been in contact before.

Mammals other than humans seldom take on special jobs other than caring for the young and certain temporary tasks like sentry duty. One could argue that the quality of life for chimpanzees and bonobos might be enhanced if one ape expert at finding nuts turned her nut supply over to another dexterous at cracking them open. Nothing about this is complex: many harvester ants have one type of worker that collects seeds and larger workers that chop them apart for all to eat. Yet the closest any vertebrate comes to this separation of tasks is the naked mole rat, whose relatively large societies support the ant-like division of labor into queen, kings, workers, and soldiers.

In our species, specialization informs not only how we interact with strangers and others based on their social positions, but also how we identify with other people in general, including their connections with all kinds of special groups. For nomadic hunter-gatherers this social differentiation within the society tended to be less important, beyond sex and age differences. The people of many Australian societies, however, belonged to groups known as skins and moieties. Children were assigned to a skin next in sequence from that of one of their parents, and to a moiety based on their connection to an ancestral being and ties to a plant or animal species, connections that determined how they socialized and whom they could marry.

Among settled hunter-gatherers, specialist groups would often proliferate over time: secret societies that offered ancient rituals and hidden truths, shaman societies, and others. A likely scenario is that this multifarious identification spun out of the primal affiliation to the society itself to create scores of collectives of lesser urgency, stature, and duration. It's curious how seldom other animals do the same within their societies, beyond the social networks of females that serve especially to raise and protect offspring in mammals from dolphins and hyenas to some primates. Nor does any one of these animals think of itself as being associated with similar-minded companions who, say, really relish fruit. Whether chimpanzees feel solidarity when on a hunt or battle together, as people do, is hard to say.

Groupiness in human settlements would have reduced internal competition and divided social stimulation into chunks that were manageable and gratifying. The notion of optimal distinctiveness from psychology helps explain this. People have the most self-esteem when they achieve a balance in their sense of inclusion and uniqueness. That is, they wish to be similar enough to others to feel a part of their groups, but they also want to be different enough to be special.[41] While membership in a large group may be held dear, it fails to fulfill a desire for specialness; this can motivate people to set themselves apart from the crowd by connecting to more exclusive groups. Nomadic hunter-gatherers had small enough societies that this was seldom an issue. Aside from membership in a few groups such as moieties and skins, everyone's quirks and personal social ties were sufficient for them to feel unique in a society of hundreds. There was no need to express differences by identifying with a job or club. In fact, it might have been discouraged. As settled societies grew larger, though, humans felt a greater urge to distinguish themselves. For the first time, the fact one would want to know about anyone was, *What is it you do?*[42]

FEELING SUPERIOR

A second, perhaps more dramatic shift was the emergence of status differences between society members, above and beyond submission to a leader. People in bands and in some of the simplest villages considered meeting short-term needs of finding food and shelter a satisfying goal and thought of life as hard only when this minimum wasn't met. Compared to the mostly utilitarian items in a nomad's meager travel pack, largely interchangeable and readily loaned or exchanged, settled hunter-gatherers had a greater number and variety of things. While ownership was a fuzzy concept in a band society, settlements could in time concentrate resources in ways that individuals were able to control. Oftentimes, materialism took off extraordinarily. People not only possessed but inherited property, some of it unavailable to others.[43] The Pacific Northwest Indians could inherit anything, even the right to sing a song or recount a tale.

Ownership became a sign of status in an expanding hierarchy of wealth and influence. In what was the death knell of the ethic of everyday sharing enforced by band societies, chiefs in particular were self-aggrandizers. Most of the time, chiefs would solidify their standing by taking advantage of community productivity to horde part of the surplus for themselves. The cycle of disparity was continued after the chief passed his status and riches to his children, or someone else of his choosing (most chiefs were male). Wandering bands would be nonplussed by both the behavior and the objects, which had little to do with a chief's labor, other than that of managing debts owed him. Pacific Northwest chiefs put their economic clout on conspicuous display at ceremonial feasts called potlatches, calculated investments for political gain. There, the chiefs would dazzle by giving away or even destroying fine food and material goods that had taken years to accumulate, making such persons "rich for what they dispensed and not for what they hoarded."[44] So wrote the American anthropologist Morton Fried.

People in settlements came to refocus their goals from maximizing their free time to earning power and esteem. In the Pacific Northwest, admired artisans, sought out for their masks, home decorations, and totem poles, were among the few who plied their trade full time and were compensated well enough to earn a rank just below the aristocrats.

The Pacific Northwest nobility, lacking the power to force their will on average citizens, had a fallback option to get jobs done: slavery. The slaves were captives or their descendants, part of the plunder taken in ambushes of other tribes. As tribes grew in intricacy, they adapted their markers accordingly, modifying them to display people's position in the hierarchy of prosperity and influence in addition to their tribal membership. Only slaves, who had no rights and were not considered members, wore no labrets.

I am amazed how readily people who lived as equals in bands adjusted to inequality. People accommodated to slaves, aristocrats competed for positions of authority, and overbearing chiefs were deposed, and still there are no records of revolts among the general population of these societies. Perhaps this was because, even as the elite took disproportionate control of resources, everyone was protected and fed. In any case, the uprisings that might have materialized in hunter-gatherer bands to quell the power-hungry grew harder to arrange. It wasn't just that more individuals, with more opinions, were involved: support was also harder to rally. Smart chiefs could draw on the blessings of hangers-on and toadies as well as other elites—a military officer and a priest backed the Calusa chief. A further factor no doubt was that once lives depend on status differences, a known property of human psychology comes into play: the disadvantaged see those on top as deserving their position.[45] This tendency may have evolved from power hierarchies like those in chimpanzees, with our acclimation to differences in stature one of the ancient implements in our species' mental toolkit.

SETTLEMENTS AND POWER DIFFERENCES
IN PREHISTORY

Before some point in our past, fission-fusion must have been the sole life choice for our ancestors. I say that because chimpanzees, bonobos, and humans have in common fission-fusion into small groups. The simplest explanation for this shared trait is that the ancestor from whom all three species sprung had this mode of life, too. With spreading out the only game in town, the nightly encampments of early proto-humans must have served as incubators for the social dexterity we would need to take up residence at one place. But that split between us and the other two apes occurred 6 million years ago or so—long before our ancestors were anything one could call human. How far back in the human evolutionary lineage did settlements appear?

Putting down roots, along with the concomitant development of the social features often connected with settlements such as leadership and inequality, has been portrayed as a watershed step for humanity. It's true the full potential of the settled life has been achieved since the development of agriculture. I see no reason, however, why these customs couldn't have sprung up, as the situation merited, with all the versatility shown by the =Au//ei, from the dawn of humanity.[46] The constant efforts expended by people in bands to keep on a level playing field suggests that egalitarianism wasn't the original condition of humankind but rather an option recently perfected. After all, even the affable bonobo is only somewhat egalitarian: while bonobos may not abide bullies, let alone leaders, they can be competitive and often at best tolerate many others in the community. Clashes over power and resources are a part of the human heritage we have always expressed. If people have been open to adopting the customs of specialization and status for a very long time, why then did so many hunter-gatherers in recent centuries live in egalitarian bands rather than in settlements? I expect that hunter-gatherers hiked less and settled more before agriculturalists laid claim to the world's choicest, and most fertile, real estate.

That Aborigines could make bread from wild grasses that looked, to the eyes of outsiders, like fields of cultivated grain shows that differentiating between settled hunter-gatherers and farmers is, essentially, nit-picking. Indeed, while anthropologists traditionally lump tribes like those of the Pacific Northwest together with band societies under the rubric of hunter-gatherer, it was the reliability of the local harvest, rather than whether food was domesticated, that mattered most.[47] Shifts from nomadism to settlement and from hunting and gathering to agriculture were gradual. The hunter-gatherers of the Fertile Crescent stayed put in villages for centuries before gradually domesticating the sheep and wheat that they ate. Once people had agriculture, however, they could increase food production far beyond the yields from nature. Many settled hunter-gatherers were, from the perspective of population growth and the cultural extravagance that can accompany it, a civilizational dead end. Domesticating the aquatic life that nourished most of those societies was impractical, with no way to control its reproduction to feed expanding societies, or to allow those societies to spread far from their sources of wild food. By contrast many domesticated crops and livestock can be taken from their ancestral habitat by locating or creating environments to suit them. Shepherds found pasturage for their sheep and farmers terraced and irrigated the land until their communities lived everywhere. Not that agriculturists were always driven to scale up their production. North America was occupied in part by corn-planting farmers, but even after centuries those tribes were hardly more elaborate than those of Pacific Northwest hunter-gatherers.[48]

Reliant as we are on domesticated food and biased toward a way of life in which anonymous societies have been amplified to massive scale, we are prone to think of our ancestors as progressing from simple to like us. Yet transitions from hunting and gathering to farming were not a given. Whether in villages or bands, few hunter-gatherers saw raising food as a move forward. "You people go to all that trouble, working and planting seeds, but we don't have to do that. All these things are there for us," an Aboriginal woman was reported to have remarked to a white settler. "We just have to go collect the food when it is ripe."[49] A traveler to the Andaman Islands in the mid-twentieth century wrote that the natives there reacted similarly to the prospect of planting coconuts, asking why anyone should "care for a tree for ten years to get its nuts when the island and seas around are teeming with food for the taking?"[50] So it was that the Chumash hunter-gatherers traded fish for the corn grown by their farming neighbors, not once taking up shovels and hoes themselves. Even the African Pygmies who lived for centuries as seasonal laborers for farmers and therefore understood agricultural techniques intimately never took to sowing seeds full time. As it would turn out, giving up hunting and gathering was no advance in quality of life. After the advent of farming, people grew smaller, weaker, and more sickly as they struggled to nurture and harvest crops—conditions that wouldn't be reversed until the invention of the plow and harnessed oxen.[51]

Taking to cultivation at all but the smallest scale of simple gardening had another drawback that no early farmer could have predicted: it could ensnare a society in a plant trap.[52] A trap, because the option of going back to hunting and gathering full time faded away once an expanding society committed to agriculture. Sure, hunter-gatherers such as the Lakota and Crow and some small South American tribes once farmed, but gave up on the practice.[53] Yet once a society grew to a huge population, or was packed in tight with other agricultural societies, the numbers of people would be too great to be supported by native foods and starvation would be guaranteed.

Pre-agricultural settlements at concentrated and dependable food supplies were the original testing grounds for civilizations, their people having taken small steps toward nationhood in the modern political and organizational sense. There are excellent grounds, however, not to lose sight of the full variety of lifestyles that existed among hunter-gatherers, roving bands included. All the hunter-gatherers of recent history were like us—fully evolved humans who exhibited a wide range of social possibilities we have largely abandoned but remain entirely capable of today. This is what makes both nomadic and settled hunter-gatherers relevant in understanding modern people. It is also why I will return to them often in presenting topics extending from psychology to international relations to the ways people structure their societies.[54]

The continuing biological evolution of the human species notwithstanding, I see little reason to expect there were radical changes in our social potentials over long tracts of the distant past, either. Early peoples must have already evolved the versatility to spread out in small bands, or settle for stretches of time into enclaves, even if these were modest by today's standards. Both modes of life assembled around markers of identity, making for anonymous societies that were simpler than, but in the fundamentals little different from, what we have now. For answers to how our anonymous societies came to be, whether set in place or mobile, we must trace what likely happened much farther back in prehistory, before our days as hunter-gatherers in the mold of Bushmen or Aborigines, back when humans were neophytes on Earth.

The Deep History of Human Anonymous Societies

CHAPTER 11

Pant-Hoots and Passwords

T he gated community of Pinnacle Point lies at the edge of the resort town of Mossel Bay along South Africa's coastal garden route, its manicured golf course extending to a precipice above the Indian Ocean. Descending the flank of this bluff are wooden stairs custom-built by a local ostrich farmer, commissioned by archaeologist Curtis Marean of Arizona State University. Part of the way down, hidden by a tarp, is a surreal world. To the right behind the tarp is a shallow cavern where researchers sit at rickety tables, busy at laptops. Blocking a sunny ocean view to the left is a hillock of sediments, illuminated by floodlights and pin-cushioned with little orange flags. Other scientists slowly carve steps into this mound. Three surveyors manning high-tech instruments stand between the computer data experts and the excavators, every few seconds calling out, "Shooting . . . taken!" to hypnotic effect as they map the coordinates of each newfound item. Together, the archaeologists are working to uncover every trace left behind by the people who visited here off and on, spanning an era from 164,000 to 50,000 years ago. Most of the remains are simple artifacts made of minerals, stone, or shell, representing some of the best information we have about how our early ancestors lived.[1] I visited Pinnacle Point curious about the findings. Was there evidence the humans of that remote time were cut from the same cloth as those of the modern era?

Because both we and our closest extant relatives, the chimpanzee and bonobo, live in societies, we can be reasonably confident the people of Pinnacle Point did, too. The evidence from this site and elsewhere, scattered and incomplete as it is, points further to the conclusion that those humans could have made the transition to anonymous societies deep in our species' past and provides strong clues as to how this may have happened. Ultimately however, the answer may lie not in the archaeological record but rather in what we said to one another, before we even learned to speak.

Five to seven million years ago, our predecessors split off from an ape whose other descendants would evolve into the bonobo and chimpanzee. The successive descendants of these forebears were a varied lot. Human minds automatically simplify any such a profusion by arranging it in a line from primitive to complex. This linear thinking distorts the facts. At most points in the past several million years, multiple humanlike species prospered concurrently to form a bushy family tree. All but one of the branches came to a dead end, leaving us as the sole survivor.

The earliest of the species, the australopithecines and their progenitors, resembled other apes to the novice eye. Our genus, *Homo*, originated about 2.8 million years ago. Some of these early humans, such as *Homo erectus*, left Africa, and later species like the Neanderthal of Europe and Southwest Asia and the "hobbit" of Indonesia, *Homo floresiensis*, evolved elsewhere. But like our earliest ancestors, *Homo sapiens* is African in origin. Everywhere else in the world we are an invasive species, just as Argentine ants are invasive to California and Europe.

PUZZLING OUT THE PAST

How early *Homo sapiens* lived and—even more indecipherable from the evidence—how they identified themselves and others presents an enigma almost beyond solution. Further obscuring our view is that for a good part of the period of residency of our species on this planet, including much of the timespan when humans camped at Pinnacle Point, *Homo sapiens* was in trouble. The harsh arid climate over many centuries kept our numbers down. DNA data indicates that at one point just a few hundred people remained, fewer than many of today's endangered species.[2] It is humbling to think how close we came to failure.

The archaeological evidence is all the more scarce because hunter-gatherers had no reason to overbuild objects that would withstand the ravages of time as we do today, by manufacturing bottles that will last an eternity for sodas gulped down in minutes. The survival of their art in deep caves makes plain that those places held a significance, probably spiritual, to the people. However, we only have those works, some up to 40,000 years old, because the conditions in the caves proved ideal for preservation. Luck can smile on the archaeologist: even chimpanzees create an archaeological record. The hammers they use to crack open seeds, identifiable as tools by their accumulation under trees where their surfaces were worn down from repeated bashing, have been traced back 4,300 years.[3]

Archaeologists must heavily depend on stone tools to understand early humans, too, but judging from recent hunter-gatherers, these would have been only a small part of their traveling kits. Most of the artifacts they left behind have vanished. Among the Australian Aborigines, for example, paintings drawn on the desert floor with colored sands were gone with the next wind. Similarly, the twigs, teeth, bone, thorns, and foliage employed in ceremonies would either have rotted away or have become indistinguishable from other detritus. So while we can surmise the people's favorite spots to bed down at Pinnacle Point, we don't expect relics of their bedding, and we can only guess as to whether they made baskets or wove cloth. Many of the crucial attributes of the societies, notably how people managed their social relations, leave little, if any, trace.

Even settlements in prehistory would be hard to detect if early homesteads were rudimentary, or made of ephemeral materials. The stone walls of the canals and homes near Mount Eccles have been mainly reduced to rubble, in spite of the fact that some were occupied a mere two centuries ago. However, vestiges of huts a few tens of thousands of years old, which look to be suitable for long-term stays, have been unearthed in Europe. Structures possibly dating back hundreds of thousands of years, well before humans evolved to their present form, are reported at sites such as Terra Amata, France. Some claim these are remnants of edifices made of tree branches braced with stones. If this is true, those dwellings would have been big enough to house many people.[4]

The most recent revolution for hunter-gatherers wouldn't have been a settled life, but the creation of artifacts that persisted long enough for archaeologists to find evidence for such settlements, or at least for places people would often gather. The earliest known enduring piece of architecture, and a monumental one at that, is Göbekli Tepe, on a ridge in the southeastern Anatolia region of Turkey. Construction began there at least 11,000 years ago, before any plants or livestock had been domesticated. Proclaimed by one archaeologist as "a cathedral on a hill," Göbekli Tepe is the oldest known religious site.[5] Arranged in circles on a slope are T-shaped limestone monoliths, three meters tall and up to seven tons in mass, engraved with stylized animals: spiders, lions, birds, snakes, and other mostly dangerous species, all made improbably with simple flint tools. The antelope that frequent the area must have drawn hunter-gatherers to this corner of the countryside from midsummer to autumn. Archaeological explorations have uncovered evidence that Göbekli Tepe was a center for feasts featuring the earliest known bread and beer made from grain reaped from wild grasses.[6] To create such a formidable structure, the site's builders must have lived nearby for all or part of the year. Though similarly ancient dwellings have yet to be found nearby, other researchers working a few hundred kilometers to the south have unearthed substantial homes and elaborate headdresses for hunter-gatherers going back 14,500 years, well before Göbekli Tepe. The Natufian settlements prove that village life, and the disparities in status associated with it, arose long before people succeeded at domesticating a thing. Other substantiation of differences in wealth indicative of village life includes elaborate burials near Moscow going back 30 millennia; the clothing on the bodies is adorned with thousands of ivory beads, which would have taken years to produce.[7]

The scarcity of still earlier archaeological finds has been used to assert that Stone Age people never settled down, and, moreover, they had little or no art, music, or rituals, no complex weapons, nets, traps, or boats; perhaps they couldn't even speak. Some claim that humans evolved the brainpower for abstract thought and complex reasoning only recently. That would seem to make the first *Homo sapiens* little more than bumbling cartoon characters.

There was more to them than that. The Pinnacle Point team has excavated a considerable quantity of cultural material over the past 15 years. During my two-day visit, they dug up shells that were remnants of a meal, a hearth on which those mollusks were cooked, bits of a red pigment called ocher, and a quartzite blade the size of a pocketknife. Throughout the strata are bladelets made by heat-treating silcrete rock, so toothpick thin that their only function would have been as spearheads or arrow darts.

Simple stuff, perhaps, but we can at least say the coastal visitors had a sense of esthetics. As early as 110,000 years ago they started bringing helmet shells and dog cockles to the cave, pretty objects scrounged from the shoreline by individuals reminiscent of beachcombers today.[8]

The archaeological record indicates that life improved for *Homo sapiens* over those early ages, and remarkably so in the past 40,000 or

50,000 years. The number and sophistication of artifacts increased, among them the outstanding cave paintings of Lascaux—of which Picasso reportedly said, "We have learned nothing." Additionally, tools became better designed and more diverse, and also left a greater imprint in the archaeological record.

Indeed, many of the implements carried by modern hunter-gatherers can likely be traced back to these times. In the past decade, an array of artifacts that recent Bushmen would have considered necessities of life have been unearthed by archaeologists in a South African cave where they had been buried for 44,000 years.[9] Among the preserved items are sticks used to dig up beetle larvae and tubers, bone awls, wood segments notched for counting, beads made of shell or ostrich egg, arrowheads (at least one of which was decorated with ocher), resin for attaching arrowheads to the arrow shaft, and an applicator to apply poison to the tips of the arrows. Any Bushman from the past few centuries would know these objects very well.

At the same time that recent Bushmen would have recognized those items, they doubtlessly would have seen them as somewhat foreign, just as they did the tools made by a different Bushmen society living in their own time. In fact, the ancient material goods so resemble those of the Bushmen of a century ago that it seems logical to conclude that their owners would have likewise made them a touch differently from place to place, and just as quickly come to associate those variations with the particular Bushmen societies of their day.

The Pinnacle Point artifacts offer plenty of hints that early people might have taken an interest in the stylistic flourishes commonly associated with more recent human societies. Starting 160,000 years ago, ocher (iron oxide) was carried into the caverns to char in fires. Heating is an almost certain sign the intent was adornment: it turns ocher the vibrant red of blood. Hunter-gatherers the world over, including North American Indians like the Chumash, painted their bodies with ocher designs that bespoke their identity. Many Africans, Bushmen included, still do. One hundred kilometers down the coast from Pinnacle Point, researchers have excavated a 100,000-year-old ocher workshop complete with grindstones, hammerstones, and pigment stored in abalone shells. Also in Blombos cave were 71,000-year-old hunks of ocher scratched with geometric patterns and snail shells with holes drilled in them so they could be strung like beads.[10]

Nonetheless, the artifacts and living conditions at Pinnacle Point and Blombos might have seemed abysmally primitive even to a Bushman of 44,000 years ago. So stark are the differences between those

periods of prehistory that many anthropologists have argued that the
clear cultural changes that unfolded for humans starting between
40,000 and 50,000 years ago must have been predicated on a substantial, and rather sudden, evolutionary transmutation. This claim is
implausible. It makes little sense to believe that cognitively "modern
humans" sprang into being during that 10,000-year interval, long after
the emergence of our species. This idea is equivalent to thinking that
because the people of the early Industrial Revolution led squalid and
unsophisticated lives compared to those of us in the modern era, the
mental endowments of humans from the eighteenth century were fundamentally inferior to ours.[11] Two scientists put the matter succinctly:
"The archaeological record tells us only what people did in the past,
not what they were capable of doing."[12]

Pointing to scratched bones and shell necklaces as Stone Age versions of the national flag, archaeologist Lyn Wadley has proclaimed
that humans became behaviorally modern the moment they committed to storing abstract information outside their brains.[13] The problem is of course that we can never prove what our ancestors thought
about ocher designs or styles of beads and arrowheads. Were these
informative to them or the equivalent of aimless doodles? Still, if a
certain nonutilitarian item turns up repeatedly in the Pinnacle Point
caverns, such as a specific kind of shell, it may have been valued as a
marker: significant things tend to reappear, the way cats do in Ancient
Egyptian art. Even so, for an artifact to designate a society, it wouldn't
just reoccur; it would be specific to a group. Too few archaeological
sites exist to resolve the pattern for early people.[14] The best we can do
is infer that a certain object could have served as a marker because it
did for recent hunter-gatherers.

That the first Bushmen likely had societies differentiated by markers opens up the possibility that anonymous societies could have an
older genesis, perhaps extending to the birth of *Homo sapiens* or even
earlier humans. A research team in Kenya's Eastern Rift Valley has
uncovered evidence going back 320,000 years of complex technologies and what could be interpreted as symbolic behavior. The archaeologists propose that the obsidian tools and ocher showing signs of
grinding to make dye found there might have been used by the people
(possibly *Homo sapiens*) to decorate each other and mark out group
identities. That these materials were precious to their owners is clear.
The hunks of the ocher and obsidian had been hauled to the site, the
obsidian coming from locations up to 91 kilometers away.[15]

Caves would have been far from the first and only places our fore-bears expressed their identities. I have no doubt the people at Pinnacle Point embellished trees with carvings and boulders with drawings long weathered away to proclaim *this is ours.* Since the birth of *Homo sapiens*, Africa may well have been awash with signs of societies hardly less plentiful than the flags that flutter across continents today, whether placed to warn, celebrate, or express a reverence for the land. The reason for my confidence is the ease with which humans could have evolved to depend on markers, as studies of other primates suggest.

EVOLVING MARKERS

Cultures and the symbols attached to them now saturate human so-cieties to such a degree that it's hard to imagine the earliest people could have had societies without them. But symbolic cultures and large populations aren't compulsory for anonymous societies to exist. The identities of ant colonies, no matter what their size, are laid out as chemistry, plain and simple, without symbolic information (that we know of). Their communal odor need only distinguish *us* from *them* (with things getting a little more complex for ants that distinguish be-tween enemy colonies). The same goes for pinyon jays with their kaws, sperm whales with their clicks, and naked mole rats with their scent.

The simple markers of the first human societies needn't have con-veyed anything abstract like patriotism or the people's connection to the past. Such qualities could have been added later. Once we stop assuming that markers must carry profound meanings, the dawn of anonymous human societies becomes easy to envision.

What the first markers may have afforded was a reduced like-lihood of error about who goes where. Every society dweller faces risk from misidentification. Errors can go both ways. They may involve confusing a potentially dangerous foreigner for a member and be-ing attacked for it, or concluding, perhaps in a pinch, that a member doesn't belong and mistakenly going on the attack. Both blunders can be avoided if the individual in question gives a clear indication that he or she is not a threat.

Underlying the development of such a marker might have been the drive to match the behavior of others in one's group. Our early ancestors, like many animal species, would have excelled at social learning. This talent can generate culture, that is, socially transmitted information taken in aggregate, including traditions such as those of meerkat clans that sleep in later than their neighbors, or dolphins and whales that pass on tactics for fishing to their offspring.[16] People learn a Pledge of Allegiance in some parts of the world and become skilled with chopsticks in other places.

Copying isn't restricted to species that are social or brainy—wood crickets menaced by spiders learn to hide by watching experienced crickets.[17] Still, social learning can be of supreme importance in species with societies. Monkeys raised in a troop trained to prefer corn dyed pink to corn dyed blue will shift to blue if they change membership to a troop that prefers it, one study showed, even when both colors are readily accessible.[18] The fact that youngsters duplicate their elders results in variations among chimpanzee communities in how the apes employ stones to crack nuts, fish for termites with modified twigs, use chewed leaves to sponge up water that's just out of reach, or hold each other while grooming.[19]

Compared with human conventions, however, variations in chimpanzee cultures are simple and few, and as far as anyone can tell, they don't matter for social acceptance. Chimps don't register leaf-sponging techniques to keep track of who belongs where; nor do dolphins monitor differences in fishing strategies. There's nothing to suggest that a chimp deviating from the local tradition of, say, holding hands while grooming is noticed by its companions, let alone avoided, corrected, scolded, or killed.[20] Beyond sometimes ostracizing a disabled individual, chimps don't look at unfamiliar behaviors with alarm, which is to say they haven't taken the step of perceiving their differences as markers. The female who transfers to a new community apparently adjusts to local habits without suffering consequences for any bad manners she unwittingly exhibits.[21] The community comes to adapt to her presence as an individual; this acceptance isn't based on whether she assimilates to their ways.

That's true except for one possible, and very intriguing, exception: the loud call chimpanzees make to keep in touch with each other, the pant-hoot.

In certain species even loose and temporary groups adopt the same vocal signal to serve the needs of the moment: birds, for example, can match each other's calls during the time they spend together.[22] Sometimes this vocal matching may approximate the sort of mirroring people do when they synchronize with each other's speech patterns and mannerisms as a sign of sharing common ground—even a monkey prefers humans who adopt its behavior.[23] For chimpanzees, not only do the animals distinguish the voices of individuals—just as monkeys can, and indeed, just as people know whether Tom, Dick, or Jane is speaking—but they also fine-tune how they give their pant-hoot by listening to others, until the exact same pant-hoot sound is shared across a community.[24] Indeed, the pant-hoot "accent" turns into an indelible part of each community's repertoire.[25]

Even though the pant-hoot differs from one community to the next, the apes don't seem to use this accent to identify whether individual chimpanzees are part of their community the way we detect a person who isn't from here by his or her dialect. Instead the pant-hoots are thought to serve primarily as what I call a group-coordination signal, to help assemble and mobilize the society members (as vocalizations are used in most bird flocks) and to monitor the locations of chimps from other communities. This is a common function of vocal cries in species where society members claim ownership of a site or coordinate behavior at a distance. As an example, greater spear-nosed bats employ a screech specific to each roosting group, which guides companions away from foreign bats to the fruit-bearing trees of their territory. Once there they continue screeching, one assumes, to assert their property rights.[26]

Chimpanzees know the pant-hoots of neighbors well enough to anticipate the consequences of an encounter. Pant-hoots from their own community elicit a pant-hoot in response. Those from a foreign community well known to them provoke an advance—even an attack should the listeners judge from the sound that they have a numerical advantage. Unfamiliar pant-hoots more often initiate cautious retreat: nothing distresses the apes more than a community of chimps previously unknown to them.[27] PET scans show a response by the chimp brain that intrigues, too. In a pattern yet unexplained, pant-hoots differ from other chimp calls in that they fail to activate the posterior temporal lobe.[28] Since this region of the brain is associated with emotions, foreign groups, perhaps, get a chilly reception.

A group-coordination signal is a marker of a sort, one used to coordinate activities and lay claim to space, rather than necessarily to discriminate individual members from outsiders. We have them, too. Nowadays national flags and monuments act as effectively as the vocalizations of apes or scents deposited by wolves or ants to indicate the domain of each society. Converting a group-coordination signal such as a pant-hoot into a marker for recognizing fellow members is a cinch, however, and could have been driven by the simplest of reasons. By a slight customization in how such a call was given and received, the members of a society would be able to confirm whether an individual they had trouble identifying was in fact one of them. The pant-hoot—or anything like it that could be presented as needed to make one's group identity known—would become a proof of membership, a simple marker serving as a password.[29]

THE PASSWORD

That early human markers were passwords is a hypothesis, albeit a simple one. Our other close relation, the bonobo, has a vocalization called the high hoot that each community seems to employ much the way chimps use the pant-hoot.[30] That makes it reasonable to think our common ancestor with those apes had a similar call that could be adapted as I described. People seldom depend anymore on passwords to recognize each other, though the soldier approaching his platoon in wartime is wise to give a sign of reassurance he's friendly. Along a similar vein, a Yanomami tribesman returning to their village in South America's Orinoco basin rainforest will cry out the word "friend."

In fact, the experts may be underestimating the chimpanzee in this regard. Though chimps don't pant-hoot as a regular greeting upon meeting each other, the call could still act as a makeshift indicator of community identity. Primatologist Andrew Marshall told me about a chimp who failed to pronounce the pant-hoot of his zoo community. This social misfit was barred from grooming sessions and wasn't permitted around food until everyone else had had their fill. Eventually the other males drove him into a moat, where he drowned. It's impossible to prove that his odd call was the reason for his ill treatment, but the idea rings true.

The initial human password could have been marks on the body made by ocher, equivalent to wearing a national flag. Since marks can easily be missed, I think a sound broadcasted like a pant-hoot was more likely, however. In that case dialects would have existed before words. Language, prior to the emergence of writing, is another trait that leaves no trace in the archaeological record. No one can be certain when our species began to speak, yet it may be that a password had to evolve first to support the profound levels of trust needed before productive conversations could be had.[31] Mastering the sound would have become imperative in childhood, more so than copying such behaviors as seed-cracking techniques had been before. It would have been essential for newcomers to the society, as well, although both the young and immigrants would have had to be cut some slack until they got the password right.

My conjecture is that the first passwords materialized as our ancestors began moving from the forests out onto the African savannas. Judging by their digestive tracts, teeth, and tools, our progenitors ate more meat than chimpanzees and bonobos, and hunting or scavenging savanna game animals would have required roaming vast tracks of land.[32] While chimps might see some reclusive community members only occasionally, for people in the most far-flung band societies, years might have gone by before they happened upon members of the most distant band. Even if those individuals had met before, lapses in recognition as time passed, owing to both changes in physical appearances and fading memories, could have led to precarious moments of uncertainty. A dependable password would not just avoid this dilemma but, in time, would allow for members who had never met and, perhaps, were completely unknown to each other.

For animals relying on individual recall, we would expect that movements across extremely wide expanses might cause a society to come apart, because long-separated individuals would have forgotten each other, if they weren't complete strangers. Human societies avoided this fate when the members could depend on a password to distinguish both acquaintances and unknown fellow members from foreigners.

Because markers permit individuals to forget others while retaining their societal connection to them, and even to be comfortable around outright strangers, they allowed societies to expand not only in space but in population. Chimpanzee communities top out at a couple hundred, those of bonobos at somewhat less. There's suggestive evidence that the societies of our ancestors pushed beyond these sizes very early on. Based on estimates of brain volumes, two anthropologists concluded that the social networks of *Homo erectus* and early *Homo sapiens* individuals were already bigger than those of chimps.[33] This would indicate their societies were already well in excess of 200. Another study based on archaeological evidence projected that societies had already reached populations of several hundred individuals by the time the genus *Homo* emerged 2.8 million years ago[34]—when the human diet had first begun to include more flesh. Both results suggest markers are a venerable part of our heritage, having come into play long before the first complex artifacts, like cave paintings, found their way to the fossil record.

There's no way to know if our ancestors' societies could already include members who were unknown to each other early on, when markers were limited to a simple password, or if being around unfamiliar others became possible only after markers grew in sophistication and diversity. Whichever was the case, from that point on early humans were freed from the requirement that they be acquainted with other members, regardless of how often they came across each other or how far apart they called home. With people at ease for the first time with complete strangers, I would describe our species as "released from familiarity."[35]

LIVING BILLBOARDS

A password, given as needed, would have been a crucial development in human evolution, one that has been previously overlooked. Societies of a few hundred, though huge by the standards of most mammals, are absurdly tiny compared to those of today. At the time, however, they would have been a breakthrough as our forebears faced ever more imposing rival societies of their own kind—as similarly large band societies have done up to the present era. Nevertheless, to truly secure a society's membership, what had been one signal had to evolve into an entire system.

A problem with having a single marker is that it leaves societies vulnerable to cheating or mistakes that might enable outsiders to tip-toe in—or perhaps even confound the memberships of entire societies. Nature is rife with examples of things going awry when species depend on signals that are too easy or too few; think of the spider that freely enters a nest of hostile ants after dousing its body with the ant's password, the colony odor. Still, I don't imagine a human society has ever taken over a foreign competitor by the stealthy copying of its identity. What would have made such a move impossible is that our initial passwords were eclipsed by more and more varied and inimitable markers—intricate rites, for example. Cracking the elixir of molecules comprising an ant's scent is far simpler than faking the mixture of markers people register about each other.[36] So while no marker needs to be costly, over the passage of evolutionary time, certain markers, and definitely a society's entire package of markers, grew exceedingly hard to duplicate. It came to be, then, that even without shouting the word "friend," people could make their identity quickly obvious and unimpeachable.[37] Even if one individual misses the telltale signs that a person doesn't belong, someone else will catch the error. This reliance on the collective to pick out intruders is found also in insects: a foreign ant that makes it by the first nest entry guard will be detected by one of the sentries it passes next.[38]

Universal among people, and probably among the first markers to follow passwords other than perhaps simple gestures, are those on the body. Humans evolved to be living billboards for their identities. Our naked skin and head hair, anatomical features that distinguish us from other primates, represented an early canvas for expressing ourselves and learning about others at a glance, both in terms of who they are as persons and as society members.[39]

In Gabon, I once showed a book of pictures of various African tribes to a crowd that had gathered, voices abuzz. The images drew curious stares and some heated exchanges; fingers landed on jewelry, plumed hats, and any other detail that struck the locals as bizarre. The bits of the photographs that got most smudged, however, were the unfamiliar (and thus patently absurd) hairstyles. Why did humans evolve mops of hair amenable not merely to the grooming we see among other primates, with their short fur, but also to being elaborately trimmed and sculpted, if not because that styling can be accomplished in culturally distinct ways?

Hair has had to be styled through every age of humanity if just to keep it out of the eyes. In ancient China, unmanaged hair "invariably denotes figures outside the human community: barbarians, madmen, ghosts, and immortals," according to one historian.[40] The terra-cotta warriors in the tomb of Qin Shi Huang, the founding emperor of the Qin dynasty who died in 210 BC, are precisely rendered with coifs connoting their ethnic origins. Among the Kayapo of Brazil, one anthropologist has observed, "each people has its own distinctive hairstyle, which stands as the emblem of its own culture and social community (and as such, in its own eyes, for the highest level of sociality to have been attained by humanity)."[41] Certain North American Indians cut their hair short while others let it grow to the ground; some had a row of bangs across their brow; others shaved the crown (or, notably the Mohawks, the sides) of their heads. Hair could be parted or braided various ways or wrapped in beaver pelts or formed into a hornlike lock.[42] Even people like the Bushmen, whose kinky hair never grows long, take the time to pretty it up.

The bare skin of our species is no less suitable to flaunting our identities. Darwin thought that bare bodies evolved to make females sexier, an idea a male chimpanzee might refute. What hairlessness did give people was a surface on which they can etch, sketch, paint, perforate, tattoo, or swath to define who they are. Personal customization may have started with ocher crayons prepared on a fire to generate differences in style that anthropologist Sergei Kan aptly describes as transforming the natural skin into a social one.[43] Outsiders so dislike the facial tattoos of women in northern Myanmar that marriage to their own tribe is all but assured.[44]

The Iceman, a 5,300-year-old mummy discovered in the mountains between Austria and Italy in 1991, bore 14 sets of tattoos made by poking soot into cuts on his back and ankles.[45] Our bodies would have been available as billboards long before the Iceman died. How far into the past our unruly beards and mops of hair go no one can say, but human skin has been all but hairless for 1.2 million years, taking us back to the early days of *Homo erectus*.[46]

The cartoon caveman (usually a man) of everyday imagination cared not a whit for his appearance. Rating ourselves above nature, we imagine this brute to be crude, unkempt, and unwashed.[47] This persistent stereotype, which can extend to contemporary people of other cultures, expresses belief more than reality. Social primates groom each other for hours—scruffiness is a sign of bad health. Tailoring hair is the human version of their primping and cleaning, with the motive not so much the bonding of groomer to groomed but for the groomed person to consummately broadcast his or her identity. Before vanity mirrors, we would have been completely dependent on others to help us with our appearance, notably to manage "the bits around the back," as primatologist Alison Jolly once put it. A caveman wouldn't have been merely tidy, but elegant. As Jolly notes, "Figurines of 25,000 years ago show women naked with beautiful braids."[48]

In recent millennia tinkering around with our skin and hair have become just part of our fetish to turn ourselves into walking advertisements. Around the world the signage on our bodies would extend from head to toe with deformations of the skull; lengthening of the neck, earlobes, or lips; chiseling of the teeth; decoration of the nails; and shrinking of the feet. The earliest rationale for clothing wouldn't have been modesty but the further embellishment of the human figure—one more way to own our identity.

What may have begun as occasional announcements of identity would have multiplied into an ensemble of markers integral to peoples' moment-by-moment experience of others. Their reliable presence eliminated the constant need to keep track of who was who. This ongoing broadcasting of identity would have become mandatory. As societies grew sufficiently populous and dispersed, humans needed continual reassurance about everyone's attachment to the group. Society-specific routines would no longer have been copied at a whim but enforced: a person who behaved or was clothed unacceptably now would shock even his or her friends. Comparisons with outsiders would have been part of the equation. Though not everything foreign had to be maligned—one society could covet another's goods, for example—a sign of allegiance copied from a foreigner might be enough to get a person banished.

The ability to conceive of fellow society members in the abstract allowed people to relax around anyone in a crowd who fit their expectations without taking the slightest notice of them as individuals. It could be tough going to keep a marriage partner or significant friend happy if you had to mull over each unfamiliar face that passed by.[49]

CULTURAL RATCHETING

The nitty-gritty details about our social lives became a core part of our suite of markers.[50] Humans came to have cultures: rich and complex systems that we teach each other and that differ from one society to another. Cultures went far beyond serving to identify where people belonged; they kept their members safe and fed and in general held deep significance for those members in a way no chimpanzee would recognize.[51]

All kinds of social traits eventually emerged as an outcome of our ancestors' creation, modification, and diversification of cultural behaviors. By this means people incrementally innovate and improve on everything they do, a unique human trait called cultural ratcheting.[52] Ratcheting ramped up in the last 50,000 years and especially in the past 10,000 until it went on not just from century to century but from year to year. Some refinements came to be shared across society boundaries, like the trendy cell phone models of today. Along the way, people turned anything they like into a marker for their identities to create a nexus of emblematic complexity next to impossible for outsiders to replicate.

Though we can't imagine a world without this perennial novelty, much of it is a product of modern consumer capitalism. Such constant tinkering is not obligatory to human life. The sparse archaeological footprint of early societies suggests that innovations were exceedingly rare for most of the span of human existence, with minor variants yet few steps forward. The experiences of 99 percent of the human generations that have ever lived were pretty much the same as those of their parents and the parents before them, their societies varying by shifts too slight to register in the dust of ages. Case in point, the ratcheting evident at Pinnacle Point was only a hairsbreadth above the zero rate of change for the nut-cracking chimpanzee, which has used similar rocks as hammers for thousands of years.

Early humans may have innovated little because they were few. Cultural elaboration takes an ample population. Despite my earlier assertions about the self-reliance of hunter-gatherers, recollection of how to do things wasn't *entirely* stored in each person's head. We constantly remind each other about everything imaginable. The responsibility for remembering is distributed across everyone—call it a collective memory. Lacking books or Internet, our ancestors relied on each other. The more they communicated, the less they forgot, reducing the onus on each person to know how to accomplish each task down to the last detail. Human learning is imperfect and skills can deteriorate over time.[53] With enough people in contact, however, collective memories can extend widely and efficiently, not just from band to band but across societies that learn from their neighbors.

Collective memory wouldn't have gotten us far earlier than 50,000 years ago, when humans were scarce. In the way that job specialization can endanger a small society, overdependence on the knowledge of others has its risks when people are few. Basic survival skills could disappear by sheer bad luck. This is named the Tasmanian effect. Many anthropologists think the Aborigines there forgot such skills as fire-making and fishing after being isolated 8,000 years ago by rising oceanic waters that made Tasmania an island.[54]

Neanderthals suffered from low densities, too. Their brains were heftier than ours, yet their societies simpler. This simplicity has often been taken as stupidity, but because the game they hunted in the harsh north supported a very sparse population, Neanderthals might simply have been stuck in the same sort of rut as early *Homo sapiens*.[55]

Another impediment to progress when people were few may have been the echo principle, named to convey how the past echoes through the ages.[56] Durable goods can be discarded but never entirely forgotten. Evidence of prior generations is everywhere, extending people's collective memory into the past. A figurine or axe wedged in the soil was almost as likely to have been made millennia ago as yesterday. Never losing sight of their predecessors' designs, early humans could have kept returning to them.

With the goods of populous modern civilizations both swamping and outshining the castoffs of distant eras, this practice is almost lost on us. Arrowheads were common on the ground in Colorado when I was a kid, and yet even though we snapped them up as souvenirs, we had little motivation to figure out how to make them. But for early

Africans, and even later for the first humans to colonize Europe, who got by at low densities too, the echo principle may have been a life-saver. By studying the scraps of past ages, talents such as crafting tools and figurines didn't indefinitely vanish from their lives. This may explain how many artifacts were replicated with so little variation over immense spans of time.

Rather than a fundamental brain upgrade, a more plausible explanation for the proliferation of material goods 40,000 to 50,000 years ago was a surge in populations at that time. The population explosion was attributable to a congenial climate across Africa and the spread of humans throughout the Old World, then or slightly earlier.[57] The resultant boost to the collective memory caused both practical technologies and other aspects of identity to burgeon.[58] More than ever before, humans would also have come into contact with foreigners. Struggles to keep themselves straight from those outsiders would explain why variation in objects likely to express memberships, such as in necklaces and art, took off at this time. Such markers can spread fast. Anthropologist Martin Wobst reasons that tailoring objects to convey the message *we made this* would start a chain reaction. Once one society added, say, a motif to its pots, the unadorned wares elsewhere instantly signaled that their makers *didn't* belong. As a result, people responded by innovating alternative styles that put the spotlight on their own identities. Presto: new versions of goods cropped up in territory after territory, those markers fueling the cultural enrichment of societies just as they had originally facilitated the growth in their populations.[59]

Because our identities emerge in part as a response to contact with alien groups and the things they make or do, I propose a rule: the more societies interact with different competitors, the greater the number, complexity, and conspicuousness of the markers they display.[60] When societies pack together, then, their peoples are likely to grow more unlike each other to avoid confusion and to protect themselves. This explains the varied labrets distinguishing the numerous tribes of the Pacific Northwest. Reflect on the high recognition factor of adornment and ritual across New Guinea, an island densely occupied by more than 1,000 tribes known for their colorful and elaborate costumes, compared with the relative uniformity of their sparsely populated Aboriginal relations next door in Australia.

The brilliant cacophony of art and decoration, language and activities that distinguish societies the world over would grow increasingly elaborate. The origin of all this diversity traces back to a fundamental

shift to anonymous societies that occurred at the inception of our species, or earlier. To recapitulate, the markers our societies use would
have evolved by gradual steps from behaviors that perhaps, as I've
imagined here, were originally similar to those still seen in chimpanzees and bonobos today. First, there would have been a password. Subsequent markers would have involved the use of the whole body as a
canvas for expressing membership in a society, but these would have
left little trace in the archaeological record. Several tens of thousands
of years ago more complex societies prospered when human populations swelled and interacted sufficiently for people to collectively
remember, produce, and improvise on far more complicated social
traits, partly to set themselves off from their neighbors.

The path leading from the individual recognition societies of other
primates to the fully human anonymous societies, with all their cultural extravagancies—unknown in the world of ants, with their simple
markers and preset social lives—was a long one. It vastly predated civilizations. The evolution of anonymous societies has been part of an
enormous rewiring project extending from the cerebral cortex to the
lower brainstem. Much of the necessary neural circuitry grew out of
what would initially have been a rudimentary interplay of stimulus and
response to markers and the groups sharing them. Since then our re-
vamped brains have come to associate our representations of individuals and societies with an unruly republic of emotions and meanings
that animates our conduct from one moment to the next, and across
the years. The interplay of such behaviors, largely still unaddressed by
evolutionists, is being laid bare by the discipline of psychology.

Functioning (or Not) in Societies

CHAPTER 12

Sensing Others

S pend time among hunter-gatherers and you realize sleeping through the night is a modern idea. On my trip to Namibia, I listened as Bushmen conversed under a chiseled Milky Way in voices elaborate as birds, full of clicks, twang, and twitter. Their huts were barely visible in the flicker of family fires as they shared both traditional tales and events of the day with gusto and moments of dramatic acting. When the sun was up their conversations centered on daily business. Nighttime was story time, with tales conveying the big picture of a proper social life and affirming people's connections with the greater society.[1]

Years later their animated storytelling came vividly to mind as I stood behind Uri Hasson, a professor of psychology and neuroscience at Princeton University, while he hunched over a computer displaying a series of brain scans. Hasson had been observing the cerebral activity of people watching a movie. The snippets he played were open to interpretation: some viewers might suspect the husband in the film was unfaithful while others might conclude the wife was a liar. When Hasson looked at scans of the viewers' brains, he found they differed accordingly. If, however, audience members talked to each other as they watched a video, their cerebral cortexes synchronized: the same portions of their brains lit up when they began to follow the story line with a unified point of view. Hasson calls this mind-to-mind coupling "a mechanism for creating and sharing a social world."[2] The pleasure the Bushmen expressed that starry night must have come from such a melding of minds.

I've asserted that every society is a community that has to be socially constructed in the imaginations of its people. This applies to the details of how the society operates, too. Humans turn everything they do into a kind of story, and interpret their lives in light of what it says. Further, that story expands through people's constant interactions into a society-wide narrative in which everyone plays a part. From the moment of birth we enter the web of expectations this larger story presents, with rules and expectations about work, money, marriage, and on and on. The full narrative brings life to a society's most beloved social markers and imposes structure and meaning to the world by creating an integrated framework in which people function. Passed down through history and slightly reshaped by every generation, the narrative affects how we conceive of our society and the line we draw to divide ourselves from outsiders.[3] It isn't the specifics of how we are governed so much as the psychology that underlies this narrative, and how it affects our identification with others, that sets the line. The formation of human identities, and our reactions to those identities, guide our lives in ways scientists are trying to understand.

Thus far in this book, we have approached our investigation of the origins and evolution of societies in phases. We began in the animal kingdom, considering how various species create societies and what they gain from their memberships. Next, we investigated the versatility of the original societies of our species, finding that hunter-gatherers had well-defined societies set off from each other by markers of identity, which have persisted as the organizing principle of societies since the origins of humanity. To ferret out where those identity signals may have come from, we then explored the remote past, where we saw that markers likely began as simple passwords. But the story has grown much more complicated than simply recognizing the identities of other people. As I will describe throughout this section, the human relationship with markers, and the groups they establish, evolved along with a rich underlying psychology.[4]

DYING FOR A FLAG

"Humans follow a flag like an experimentally imprinted duckling, a ball," says Irenäus Eibl-Eibesfeldt, a pioneer in studying human action through the lens of how animals behave.[5] Evidence suggests that learning markers and using them to categorize people, places, and things is instinctive—organized in advance of experience.

Even though the inspiring stories we share like the raising of the American flag at Iwo Jima add to the meaningfulness and importance of markers, knowing those meanings isn't imperative for us to be aroused by anything that acts as a potent marker. Nor does such a signal need to be connected with a person to set off an impassioned reaction: either its presence (a stirring anthem) or its absence (imagine a town where Americans shoot bald eagles) engages the limbic system, the brain's emotional center. Those circuits can go off like a firestorm when a powerful marker is evoked—an act of violence becomes all the more horrendous when it also involves the destruction of a national monument.[6]

Given the right onlooker and context, even the simplest object or word is enough for a strong emotional response. As an example, picture an equilateral cross with arms hooked at ninety-degree angles. Seeing this apparently banal design can put a Holocaust survivor into a faint. Yet pausing to think about the symbolic implications of the swastika isn't required for the horror it may cause. Once an anguished response is set by a kind of Pavlovian conditioning, casting it aside is no more possible than stopping our gag reflex at outlandish ethnic food: take a moment to visualize walking cheese, once popular in Corsica, wriggling under maggot-power. Even so, at the time they were in widespread use, swastikas made Nazi hearts swell with the same rapture we expect to experience when we stand before our national flag at the ball game.

People do love a flag. Today, even "mild-mannered Danes go nuts over their national colors," says historian Arnaldo Testi. Indeed, he goes on, "in democratic, secular republics, the flag acquires an even greater, near-sacred, civic importance, if anything, in the absence of other public binding icons, like king or God."[7] This mania ties to our most essential groups: to fight or die for the flag is a personal pleasure and honor.

How a mere color pattern, shape, or sound can precipitate fervor or fear in the human brain is poorly understood.[8] The response emerges in childhood, and no wonder. In the United States, flags are displayed in classrooms, and the Pledge of Allegiance is often recited in preschools (though children can now opt out). By six years old, a child perceives flag burning as bad; pride in country comes soon after.[9]

The omnipresence of signs of national identity accords everyone a similarity of experience that primes our feelings even when our minds are elsewhere.[10] Faced with adversity, symbols turn into beacons that

galvanize us to action. A recording of the national anthem became a
hit single in 1990 after the United States committed to the Gulf War;
American flag sales exploded in 2001 the day after the 9/11 attacks.[11]

HOW BABIES CATEGORIZE PEOPLE

Much of the psychology of interactions between human groups concern how we respond to the markers directly associated with individual people. The efforts of Hasson and other psychologists show that
our brains are constructed to socialize, and recognizing the identity
of others is a part of that process. An essential first step is to notice
other beings at all. Suppose you are playing checkers on a computer,
thinking your opponent a machine, and midway through the game
you find out that a person, not a computer program, is behind the
opponent's moves. On cue your mental activity would shift to areas
of the brain reserved for interactions with other people, including
your medial prefrontal cortex (at the front of the cerebrum) and your
superior temporal sulcus (a furrow in the temporal lobe). This would
occur even if the information was wrong and no person was there.
Same game, same IBM, yet you switch to another mental state—the
state that is normal when we attend to people, and infer what is on
their mind.[12]

As social beings, we depend on our awareness of others. A mental
acuity we share with the chimpanzee and bonobo, honed by the upbringing of all three species in fission-fusion societies where individuals come and go, is our adeptness at registering slight differences—and
commonalities—between persons.[13] Psychologists have come to fascinating conclusions about how people attend to one another, both as
individuals and as members of groups. Almost all the research concerns not independent societies, but rather races in urban settings.
Hence the examples I give reflect that bias.[14] Nevertheless these attributes would have originally evolved back when hunter-gatherers experienced societies more uniform than those of today, as responses
to the markers of their own and other societies. It makes sense then
to assume that societal markers such as language or adornment will
yield similar results. (Why there should be a near equivalence of human psychological responses to races and societies will be a subject
for later.)

Recognizing human groups begins very early in life and cannot be
suppressed.[15] From molecules transmitted through the amniotic fluid
and later in the breast milk, infants get a taste for what their mother

cats, including any strong flavors, such as garlic or anise, preferred by her ethnic group.[16] One-year-olds observing individuals who speak their language anticipate that they will favor similar foods, and that people from other backgrounds will have a different diet. By the age of two, this expectation becomes cemented as a preference: children that age and up are partial to anything eaten by the members of their own group, whether it's a fried scorpion or a tuna melt.[17] This is an expression of a partiality for the familiar, and therefore what's safe, as anyone knows who's watched a baby burst into tears over the slightest strange thing.

Even a three-month-old hones in on faces of its own race.[18] By five months this preference extends to those speaking the parents' language, and with their dialect; a child grows up finding strange accents exhausting to comprehend and is sensitive to differences in speech.[19] Between six and nine months of age, babies become good at categorizing people of other races based on facial cues, but after that they become worse at distinguishing individuals of races other than their own.[20] This precedes the loss of adeptness at learning foreign languages after the age of five. I anticipate a baby's response to equally striking markers of identity such as hairstyle or dress will show a similar decline.

Eibl-Eibesfeldt is right in that the response to membership badges could be as much a reflex for a human baby as it is for a chick connecting to its mother (or to a ball, if its misfortune is to think a ball is mom). That imprinting occurs for each species in its own time window. You may wish to be reassured that such instincts are more hardwired in chicks than in humans with all their smarts, but the distinction isn't clear-cut. Flexibility is required for the survival of all creatures, not just humans. A chick must adapt to its mother's appearance after she molts, which gives her a new look.[21] Should mother be eaten for dinner, the chick can imprint on another hen. You need not be a brainy vertebrate to handle such eventualities. Even ants sometimes adjust to the identity of workers of other ant species, and learn to treat them as nest-mates, if the foreigners are introduced into the colony before they mature.[22]

We may be more flexible than most animals in recognizing the members of our societies as well as the ethnicities and races we are born into within them, yet the difference is one of degree. Some vertebrates can even hone their skills on other species. A human baby exposed to monkeys by the age of six months becomes adroit at distinguishing monkeys as individuals, and monkeys reared by people demonstrate similar lifelong talent at telling humans apart.[23] I wonder if a youthful exposure to the wrong species explains the vervet monkey observed living and mating with baboons in the 1960s, and some other oddball companionships like bears playing with tigers.[24]

In short, before a baby can read, before it speaks or comprehends language, memberships such as race and ethnicity are distinctions it attends to effortlessly, without any input from elders. But even though we are hardwired to pick out basic group memberships like society and race, how we think about them is far more multilayered than merely identifying features the way ants detect a colony scent. People discern the categories of living things, humans included, as being manifested in their very core, as we shall now consider.

THE HUMAN ESSENCE AND "ALIENS"

Starting at about three months old, infants begin to attribute an *essence* to every life form, an elemental *thing* at its core that makes it what it is and not something else.[25] This mental construct is registered in the cosmologies of hunter-gatherers, who, as animists, were convinced that spirits organize the world. They saw animals and plants, and each other, as suffused with essences. Paraguay's Ache, for example, thought a child was given its essence by the man who provided meat to its mother while it was still in the womb.[26]

People ascribe essences to species based on how they look and behave. Yet the essence that underlies those features, and not the features themselves, is what matters. That's how the brain handles discrepancies. A child has no problem with the notion of a chair turning into a table when its back is chopped off, but knows a living thing is different: a butterfly remains a butterfly even with its wings torn off; a swan raised by ducks is still a swan.

We act as if the attributes of swans are embedded in their very atoms. Geneticists speak of the swan's DNA to support this view; however, genes are mutable and allow for transitions, say, when one swan species evolves into another. With their belief that essences are stable, people don't allow for such intermediates. In experiments, children perceive an image morphing from lion to tiger to be one species or the other and not both, even if each child differs in where he or she sees this switch occurring as the cat changes appearance.[27] Essences thoroughly lodge living things in the categories where we think they belong.

Here's where I believe societies, certainly as they existed for early humans, differ from most groups in modern life. Joining a book club or bowling team is optional, and no one would presume anyone would be constrained to choose a spouse from either of these groups. Nor do we expect architects and lawyers to riot or declare war on each other, as anthropologist Francisco Gil-White points out.[28] This is not to deny that some people have stronger allegiances to their faith or the New York Yankees than to their country.[29] Yet a Yankees jersey, goth apparel, or job uniform can be discarded without undermining who we think a person is: the sports-obsessed or counterculture youth is *going through a phase.* By contrast, starting at the age of three, people see their society, and their race or ethnicity, as essential and inalterable features of identity, fixed permanently by an essence, the way species are.[30] More so than many betrothals, a national or ethnicity identity stays with us till death do us part. For us, they are, in a word, natural.

On this basis the single human species is splintered into many. Outsiders—the people of other societies, and nowadays other ethnicities and races—are treated as if they were distinct creatures. Their markers are passed on to their offspring as dependably as the traits distinguishing a swan from a duck, so that their memberships persist through time. Even so, because societies and ethnicities are perceived to run deep—to be *in the blood*—we can be open to the possibility that an individual who looks like a member of one group is part of another, the same way we can accept, appearances to the contrary, that whales are mammals, not fish. To think an odd person belongs, we must be convinced he or she was born of members of our group and that his or her kids will be members of that group, as well. We can attempt to conceal our pedigree by immersing ourselves in the lifestyle of another group, but however elaborate the cover-up, anyone who knows better will feel our essence remains. Purging ourselves of our essence is well nigh impossible, an idea as absurd as a baby swan training to be a duckling.

What of individuals who marry or are adopted, like a swan by ducks, into another group? While such family members can be treated lovingly, their passage to full acceptance can be long and hard. As reveals of their inner essences, our eagle eyes still detect differences, for instance in the descendants of immigrants. Even people who are products of intermarriages over several generations may not blend in fully.[31] This also explains why, though Americans use the term *biracial* to label someone born of parents of two races, the child of a white mother and an African American father, no matter how light-skinned, will be understood in the context of white-dominated America as African American—a point of view once known formally as the one-drop rule, where the slightest trace of black ancestry renders the person black.

ORDER FROM CHAOS

Regardless of modern obsessions with racial groups, actual physical differences aren't the only metric of identity. The man with a drop of "black blood" can be as white as any Caucasian. And to the bafflement of outsiders trying to understand the interminable conflict of Israelis and Palestinians, the people involved look much alike: Haredi rabbis and Islamist imams even sport the same beards. Genetics proves they are of a single, narrow stock, a similarity that the people themselves would have a hard time conceding.[32]

Yet what physical differences exist are hard to ignore; in fact we tend to magnify them in our heads. Over the past 12,000 years our rules of social engagement have adapted first to narrowly defined groups, such as the homogeneous people of the neighboring hunter-gatherer society, and later to races as we think of them today—races that now apply broadly to people fitting loose physical descriptions that we boldly assert are so clear-cut we code them by color: white, black, brown, yellow, red. That the categories are artificial can be shown by how readily they shift. Early in the twentieth century few Americans considered Italians white, let alone Jews, Greeks, or Poles, the distinctions made in the United States at that time suggesting a probably exaggerated virtuosity at detecting fine-grained differences among humans. Meanwhile the British of the time referred not just to Africans, but to Indians and Pakistanis, as black.[33]

Compare those crudely defined color categories of race to what I call "genealogical races" such as the Bushmen or Ache, each of which, while divided into multiple societies, could be traced back in time as one reasonably tight overall population. Though our aptitude at picking out different human morphologies may be ancient, it's unlikely that a sensitivity to human variation originally evolved to respond specifically to racial traits varying across populations of our species such as skin color, most of which arose as fine gradations across vast distances—continents even.

Certainly categories of race or ethnicity are rarely fashioned from one set of human features alone. Just as the child makes sense of an image of a lion that changes into a tiger by seeing it as one or the other but not as both, the mind sees faces intermediate between groups (races, in most studies) as a match to one or the other. Yet we are also sensitive to other nuances of identity that render categories less ambiguous and more sharply defined than objectively they really are. When researchers concocted a face to be intermediate in appearance between two races but gave it the hairstyle typical of one of the races, the person was categorized as of the race associated with that hair. Put a guidepost like an Afro or cornrows on the head, and subjects confidently proclaim the person black.[34]

That's how Rachel Dolezal, a woman of European descent who had pale skin and straight blond hair as a kid, passed as African American, becoming the president of Spokane's NAACP chapter. She gave special consideration to the style and texture of her coif while keeping her skin tan to look as if she had the requisite drop of "black blood." Few questioned her self-stated identity before her duplicity was exposed in 2015.[35] Once it was disclosed that no drop was there, however, her black essence vanished, rendering her apparent racial markers superficial and irrelevant.

Hairstyles label us so strongly that if a hairdo associated with a dark-skinned race crowns the head of someone with racially indeterminate features, we read his or her skin as darker than it is.[36] The skin tone shift happens in much the way we perceive the same line to be longer or shorter if we draw arrowheads pointing inward or outward at its tips.[37]

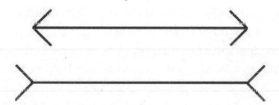

The optical illusion works because the brain dislikes confusion—including in social categories. If someone's identity confounds us, our improvisational gray matter pieces together where she fits in from other information we consider useful. Along with hairstyle, the Jewish skull cap, the Sikh turban, and the black garb of a Greek widow are showy examples. More muted signs could be used by hunter-gatherers, for whom physical appearance would have seldom differed by much, and may not have mattered when it did. By this mental procedure people turn artificial (constructed) categories into social realities.

Earlier I brought up how markers speed up the recognition of groups. What's remarkable is how swiftly and unthinkingly the information is absorbed. We register a person's ensemble of markers with no deliberation. Alex Todorov, an exuberant Bulgarian with a dramatic mop of hair and an office near Hasson's in Princeton's psychology department, has shown that a face glimpsed for a tenth of a second, a span of time too short to enter our awareness, is subconsciously assessed in terms of the person's emotional state, sex, race, and (Todorov assures me) ethnicity and society, too.[38] Our nimbleness with markers must reduce our cognitive load—the conscious effort we expend. And as Solomon Asch, a pioneer of social psychology, recognized in the 1940s, we can no more prevent these impressions than refrain from hearing a melody.[39]

Should markers, even in aggregate, come up short, signs of identity have occasionally been forced on people. Jews were compelled to wear a yellow badge in France during the Middle Ages and the Star of David in Nazi-controlled Europe. The rationale given by authorities was that the badge served to indicate shame, but surely a reason to coerce visible marks was that people of an ethnicity were now impossible to misidentify. Proof that words harm as much as any stick or stone, even word of mouth can mark people, with rumors about family backgrounds the downfall of many under anti-Semitic regimes.

When push comes to shove, people are liable to be so anxious to avoid mistaking unfamiliar outsiders for members of their society that they slip up, concluding that someone who comes across as odd doesn't belong, even when he or she does. This tendency is so predictable psychologists name it (rather clumsily) the ingroup-overexclusion effect.[40] On occasion the consequences of exclusion from the ingroup are cataclysmic, as a World War II–era news report indicates:

> Mistaken by Nazi administrators for deported Jews, a trainload of German refugees from Hamburg were exterminated by Gestapo officials in the "death chambers" of a Jewish concentration camp near Lwow, the Manchester Evening Chronicle reports today. . . . When the train arrived, the hungry and exhausted German refugees looked not much different from the hungry and exhausted Jews who have always been brought to this camp to their death in gas chambers. The Gestapo guards lost no time in stripping the newcomers naked and in sending them into "gas chambers."[41]

SUBCONSCIOUS ALARMS

How do our minds process the identities of the people we meet? Once we distinguish a person from an animal, a computer, or other object, the brain switches to taking in information about that individual, assessing, among other factors, whether he or she may be a threat. Such judgments are indispensable in an uncertain world, a matter of life and death for hunter-gatherers coming upon someone unexpectedly or for troops in a war. But the evaluations occur even in less fraught situations as part of the background activity of our nervous systems.

The mental reaction that ensues depends on who the people are and how familiar we may be with them or their group. If the person is unknown to us but is identified as being one of our own or from a trusted other group, our familiarity with his or her markers of identity make the stranger seem less strange.[42] Whether or not we make the effort to individuate the person—treat him or her as an individual—we have some reassurance about the person's beliefs and behavior, though we might still be on guard if someone seems shady or unpleasant, or is simply having a bad day.

Foreigners—and definitely those from a disliked or unfamiliar society—can raise subconscious alarms regardless of their behavior. At a minimum we may feel discomfort in their presence. Among Bushmen, even a trace of a foreigner, like an oddly made arrow found embedded in the ground, induced anxiety since the conduct of its maker couldn't be predicted.[43] One researcher studying the G/wi Bushmen noted the "reassurance and lessening of tension that is seen when a stranger is recognized as a fellow G/wi."[44]

The difference between a stranger (an individual we don't know, whether he or she is part of our society) and a foreigner or "outsider" (a member of a different society, whether we know him or her or not) is a big one. The two are clearly distinct in our species: we can be friends with the foreign exchange student in our class, while we have never laid eyes on a mysterious next-door neighbor. Yet psychologists often confound strangers and foreigners. Insofar as the English language muddies this distinction, it is unfortunate. The word "xenophobia," for example, is often applied willy-nilly to our negative reactions toward strangers and foreigners alike. The fact is human minds have likely evolved to respond to strangers and foreigners differently, and foreign strangers most strongly of all.[45]

Our instantaneous and unthinking categorization of people has obvious adaptive advantages. As we have seen, it would have alerted us to others whose actions couldn't be predicted and reduced tensions toward those who resembled us. Our evaluations not only occur beneath our conscious radars but also affect us to the bone. For instance, most participants in a study who watched a video of people being stuck with a hypodermic needle perspired less and showed less activity in the bilateral anterior insula section of the brain when the person being injected was of a race different from theirs, a neurobiological response indicating diminished empathy.[46] Even a chimp will display this selective connection to others, yawning only in response to the yawn of another individual that belongs to its own community.[47] This reaction betrays a basic ingredient in human interactions: we act as if the essence of our own people is superior, as if we are more human than other groups. When facing someone who doesn't fit, we disengage. In extreme situations, the brain activity of a person viewing an individual he or she considers foreign looks the same as that of a person viewing an animal. Once people are typecast as outsiders, the mind abandons nuance and may expel them from the human category entirely. Taken together, these reactions scaffold the rickety edifice of human stereotypes.

CHAPTER 13

Stereotypes and Stories

S tereotypes are mental shorthand, an unavoidable outcome of partitioning our experiences into categories that make the world comprehensible. Without such expectations about regularities we would be chronically surprised, not anticipating the lily's scent or the hot sting of a bee.[1] Walter Lippmann, the journalist who gave "stereotype" its modern meaning, wrote about them this way: "We notice a trait which marks a well-known type, and fill in the rest of the picture by means of the stereotypes we carry about in our heads."[2]

So it is that by categorizing an object as a chair by absorbing its characteristic legs and seat, and thereby anticipating that object will serve our desire to sit, we drop into it without expending a thought (perhaps one day being surprised when, not noticing it's made of paper, we, and the chair, crash to the ground, our chair-stereotyping having failed).

Of course, we don't usually associate stereotypes with furniture. When applied, as the word usually is, to people, stereotypes are socially shared, simplified predictions created by the overloaded brain to evaluate others. Some of our predictions are inoffensive. We give money to a café barista without thought. Indeed, we treat him or her, most blatantly when distracted or in a rush, more like a machine to get us caffeinated than a person. The barista can be fine with that, treating us in turn as just another in a long line of customers. Yet we also create stereotypes about the people of societies and the ethnic groups within them, for example anticipating their actions, including how those others will act toward us, or prejudging what they will think about the people of our own group.[3] The essences we imagine in others correspond not only with verifiable human markers, but with the baggage of our beliefs and biases about the people who have them. These generalizations can harm by yoking the people of a certain society or background to a package of undesirable and largely outright incorrect assumptions.[4]

Our personal biases are more widespread than we might like to believe. One of the powerful tools for exposing our stereotypes is the Implicit Association Test. The test consists of a series of randomly paired images flashed on a screen, one a word and the other of a person, most commonly with either a dark- or a light-skinned face (though two ethnic groups would work as well). The onlooker is asked to connect a type of face with a certain kind of word. For example, one person might be told to press a button only if the word shown alongside a dark face has a positive connotation, such as *peace* or *joy*, and not press it if the word is negative, such as *violence* or *disease*, and do the opposite for light-skinned faces.

It turns out that just about everyone is a racial profiler. Almost all Americans perform the task faster and with less effort (with fewer mistakes) when asked to match dark-skinned faces with unfavorable words and light-skinned faces with positive words. This is true even for those who find stereotypes objectionable and take pride in being unprejudiced. The result comes as a complete shock to most subjects of the experiments, who are seldom aware of their own prejudices. More shocking still, according to the social psychologists who invented the test, is that "awareness of the hidden biases did not seem to help us to eradicate them."[5] Practice doesn't improve the results.

In the musical *South Pacific*, the composer-librettist team of Richard Rodgers and Oscar Hammerstein put to song the conventional view of how these prejudices first enter our heads:

> *You've got to be taught to be afraid*
> *Of people whose eyes are oddly made,*
> *And people whose skin is a diff'rent shade,*
> *You've got to be carefully taught.*
> *You've got to be taught before it's too late,*
> *Before you are six or seven or eight,*
> *To hate all the people your relatives hate,*
> *You've got to be carefully taught!*

This is erroneous: no boy or girl has to be methodically taught such things. Rather, learning stereotypes is an extension of a child's developing mastery of detecting patterns—the primary job of any immature creature striving for independence.

For the child, distinguishing categories of people is just the start. A child associates those categories not only with patterns of what he or she sees people do, but also with what others say they do, and further, whether those actions are evaluated as good or bad. This discrimination doesn't require a motive. People don't have to feel that the others are adversaries or competing with them somehow. Such biases emerge easily even when the groups are totally arbitrary. For example, kids randomly assigned to one of two groups predict that those in the other group do more bad things.[6] In effect, groups—and especially enduring and significant ones like nations and ethnicities—are represented by mental files that fill up beginning in childhood with rules of thumb ranging from the reasonably inoffensive, such as *Italians eat pasta*, to such problematic conclusions as *Mexicans do menial work*.

"Racism isn't something that happens to the child, it is something he does," argues the psychologist Lawrence Hirschfeld.[7] By the age of three years, children don't just recognize races; they perceive races as more than skin deep and freight them with stereotypes.[8] This transpires inexorably, as the child assimilates the attitudes of others, not exclusively the child's parents, to assemble a point of view. Actually, mom and dad often have surprisingly little influence.[9] Even the progeny of progressive parents absorb the biases of their society. Young people are little prejudice machines, displaying negative attitudes of the same magnitude as adults but often doing a worse job at covering them up.[10]

Figuring out where they themselves fit in is important to children, too. To do this, children prioritize their time by observing others. Rather than studying their reflection in a mirror, they pay most attention to those of the same race and ethnicity as their parents or caregivers.[11] When adopted by another ethnicity or raised by a foreign nanny, however, children seem to figure out their own background and how they themselves are expected to act from how other people treat them.[12] That a child accepts the identity thrust on him or her by others is a symptom of the intense desire humans express to find their place. No doubt if your infant brain had been transplanted into the head of a child in a Senegalese jungle or a housing complex in Macau you would have grown up exactly as at home in that social setting as you are with your own identity today. Children can even absorb two cultures, albeit as they mature they are likely to see themselves primarily as a member of one or the other.[13]

Although no specific prejudice is set by genetics, that our proneness to be biased is innate has a troubling effect: once formed, stereotypes resist revision. This unfortunate fact, a leading group of psychologists has concluded, "may explain, in part, why conflicts among different language and social groups are pervasive and difficult to eradicate."[14]

SNAP JUDGMENTS

In all these matters, snap judgments are the norm. Alex Todorov, the Princeton psychologist, finds that in the same tenth of a second or so it takes us to stick a person into a category, we have already come up with opinions on his or her trustworthiness, among other stereotypes. An outsider to our group incurs the quickest and most superficial evaluations. He or she excites the anxiety-triggering amygdalae, the same portion of the brain activated when we slap away a bee, heart racing. Our amped-up, quick-and-dirty response to whatever we deem a potential threat is "resistant to change and prone to generalization," one neuroscientist writes.[15] What amazes me is how we make such verdicts before a person registers in our conscious mind. If we continue to examine a face for more than the blink of an eye, the only result is to justify conclusions already formed unconsciously. In effect, in that critical first moment of perception, we don't see a person. Instead, we register our stereotypes about him or her, erecting a simulacrum in our mind's eye that overrides most qualities of the real individual.

Any departure people show from our stereotypes about them is a measure of their individuality. Valuable information, wouldn't you think? Nevertheless, we give only people from our own society or ethnic group serious consideration. For those we grant such special privileges, we absorb personality details beginning in the fusiform gyrus, the section of the temporal and occipital lobes devoted to face recognition. If we interpret an ambiguous face as black after it's been "tagged" with an Afro, for instance, we are more likely to form a more complete impression of the individual if we are black ourselves. This reaction intensifies should we be especially likeminded with the person, say in our political leanings: in response, neural firing shifts to the lower part of the medial prefrontal cortex, the same region that activates when we think about ourselves.[16]

The price of this automated filtering system is the bottleneck it places on understanding outsiders. Not that we can't decide to put in extra effort, say, by learning such social niceties about an outgroup person as her name. However, given how clumsily we individuate those not of our "kind," Martin Luther King's dream remains an uphill battle: well intentioned people often fail to judge individuals from another group solely by the content of their character.

In a sense, in the instant of encounter, we automatically judge the book by its cover, taking more than a cursory glance inside only if the cover passes muster. What's more, we recollect people more accurately over the minutes, days, and weeks after meeting them if they are of our group. If they are not, any specifics we may have discerned about them fade from memory. One result is the sad fact that many false imprisonments are based on the eyewitness testimony of accusers who are of a different race than the accused.[17]

What would seem to be a positive sign for group relations within and between societies is that sufficiently powerful signals can supplant malignant prejudices. For example, if a man of a different race or ethnicity wears the jersey of our sports team, the jersey can become paramount in how he is evaluated: given our shared allegiance, we may forget his outsider status.[18] Unfortunately, the change will be temporary, and only affects the fellow wearing that jersey. See him or someone else of his ethnicity sporting another shirt and our ingrained biases snap back into place. Attempts to suppress our own negative views can backfire through a "don't think of an elephant" effect, where we end up thinking of something—such as an elephant—even more after we try to force our attention onto something else. We end up activating our prejudices even more strongly.[19] Ironically this occurs because the energy we burn in attempting to overcome stereotypes can wear us out, with the result that we can slip up and behave worse.[20]

One might think the problem could be mitigated if different ethnic groups or societies got to know each other better. However, as can be seen from the racial prejudice against African Americans throughout American history, even where black people live in large numbers, frequent contact isn't sufficient to override the effects of stereotyping. Psychologists Kimberly and Otto MacLin compare the problem to traffic in Manhattan, where pedestrians see thousands of cars but, never having a reason to own a car, fail to distinguish one from another. At best, they may know the markings that label a car as a taxi.[21] To distinguish foreigners as *persons* rather than in terms of a stereotype, we can't merely be exposed to them as a New Yorker is to cars. To stretch the "apples and oranges" idiom in a new direction: fairness to outsiders means going beyond the easy-out of seeing them as the equivalent of apples and ourselves as all sorts of interesting varieties of citrus. Attending to outsiders in such detail doesn't occur effortlessly, or often.

Only bigots and naïve small children flaunt their biases. As adults, most of us either rationalize any prejudgments we are aware of and the anxieties they elicit, or we submerge them so deeply we fail to notice their existence. Yet the Implicit Association Test tells us that we don't need to be radically xenophobic to show prejudice toward outsiders. Even among well-meaning people perceptions unavailable to conscious inspection still influence responses ranging, in the words of two social psychologists, "from derogation to degradation to disconnection."[22] Just about every store owner, employer, traffic cop, and passerby distinguishes between ingroup and outgroup members, reacts to certain outgroups with heightened vigilance, and is less likely to come to their aid. Because the same people are more apt to feel concern for those they flesh out as individuals—a penchant known as the identifiable victim effect—they may also have trouble treating outsiders fairly because of the strain required to distinguish one from the next. Yet most people haven't the foggiest clue that they misbehaved. Such dynamics play out in every interaction between societies and between the ethnicities within them, and the impact on interpersonal relations is beyond calculation.[23] All these biases stand up to contradictory information. We exclude from our prejudices the friend who is a citizen of a nation we dislike without that person improving our assessment of his or her people—our buddy is an exception to the rule.[24]

Despite the negatives, the human aptitude for dealing with utter strangers serves to simplify positive interactions both within and between societies in a way unachievable for species whose members must know each other to function socially. People can register someone as foreign but still enter into a productive relationship with him or her. A modest adverse reaction can be benign, keeping individuals engaged yet at a formal distance.[25] This is only partially good news. Beyond our everyday biases toward treating our own people subtly better than others is the danger, given a provocation or opportunity, that any disgust or jealousy can be purposed to justify actions against entire classes of outsiders.[26]

Social pressure to conform can turn stereotypes into a self-fulfilling prophecy. Individuals tend to try hardest to shine at skills that the stereotypes they have internalized about themselves suggest they will be good at, with the prevalent clichés in the United States being that black people excel at sports and Asians at mathematics, undermining whatever other potentials they had and thereby further supporting the stereotypes.[27] Whether talent differences generate stereotypes or stereotypes foster differences in talent is an open question, but individuals can face backlash from their own group and peers should they break from expectations about group behavior.[28] This suggests that our biases aren't limited to outside groups: we expect people like us to act a certain way too.

What this boils down to is that we act in a biased manner even when we believe we don't. This impulse must have evolved to nudge people to serve their group's best interests even when their relations with a foreign people were good. When ties failed, then as now, undisguised discrimination would rise to plain view.

REMEMBERING, FORGETTING, MEANING, AND STORIES

The research makes clear that when it comes to other people, we respond automatically, our positive or negative feelings and biases triggered within milliseconds of an encounter. I expect that our responses to the markers themselves are every bit as automatic, and that only when called on to explain ourselves afterward do we rationalize our reactions by spelling out what a marker symbolizes to us. That's not to claim that its meanings are insignificant. The concepts and stories we grow up with touch upon many aspects of our identities and guide how we interpret our place in a society and the world. In the last chapter I explained that humans turn everything into a story. The vital cultural particulars that people chose to convey—for example, the account of Betsy Ross sewing the stars and stripes on the first American flag—give those details an emotional weight that improves their later recollection and continuity. Such stories resemble stereotypes in that they save us mental effort by cutting through a morass of information to remind us about what really is important in our connections to others.

The stories of a society may tell of the aspirations of the people and accounts of their past. A story can be recent, perhaps recounting the success of that year's Olympic athlete, but the most important ones endure over the generations: Italians still commemorate the Roman Empire, and the citizens of India can surely recount the contributions of such early dynasties as Maurya. The rendition of a society's birth especially can be a source of inspiration and pleasure whether a person's forebear signed the Declaration of Independence or he or she has just become a naturalized citizen. No origin story is a straightforward recounting of events, however. Creating what amounts to a collective historical consciousness is delicate business. What matters isn't truth but *tale*, one that conveys a proud past and bravery in a crisis for the sake of the group and its values, where the abiding question that all men and women confront in their lives, "who am I?," turns to "who are we?"[29]

After Vietnam broke away from the Han dynasty of China in the tenth century, the scholars of the time set down accounts of the country's history, which served that function nicely. For centuries afterward, a founding dynasty known as the Hung Kings was considered part of the Vietnamese heritage. The archaeological record has revealed that those kings were medieval fabrications.[30] In the first half of the nineteenth century, revolutionaries did the same for the Han, an ethnic group that has come to compose the majority of China's population. They invented a streamlined history and ancestor for them all, the mythical Yellow Emperor.[31]

Happy is the nation without a history.[32] So said the seventeenth-century thinker Cesare Beccaria. I interpret his claim to mean that a complete and accurate rendering of the past undermines the bonding function of collective memory. Think of history as the contents of a box of mementos found in an attic, from which we draw what we desire and bury what's best forgotten. Whether the narrative that results from storing the keepsakes together is error-free, a complete fable, or something in between, any challenge to the story is frowned upon or forbidden. A well-crafted history serves the interest of putting the people in the best light and gives shape to their future—although an artful leader can manipulate the contents of the attic box to build a passionate following.[33] At the same time people sear into memory every detail of the tales that they share, as the historian Ernest Renan reflected, "Forgetfulness, and I would even say historical error, are essential in the creation of a nation."[34] So we have Turkey's obstinate refusal to acknowledge the Armenian genocide and the American account of its war of independence, which in the eyes of the British has many faults, such as greatly underplaying the role of French support of the rebel cause.[35]

Hunter-gatherers, lacking written accounts and mechanisms for uncovering upsetting facts, were masters of selective recall. They relished stories, but they were more devoted to explaining nature than they were to recognizing their forefathers' achievements. Not all societies treat history prominently, and least of all people living in bands. While the necessities of daily living, such as how to make a fire, were conveyed as assiduously as possible, forgetfulness was absolute when it came to bygone days. For preliterate societies the past was seldom an unfolding saga worth memorizing. Rather, they saw time as at once endless and twisted into cycles similar to the phases of the moon, virtually ensuring that history, as the philosopher George Santayana predicted, would repeat itself.[36]

This makes for an intriguing contrast with our current fascination with history. By all accounts band societies lived and breathed in the present tense. One of the few instances I've read of hunter-gatherers telling stories of long ago were Aborigines recounting the arrival of Indonesian fishermen three centuries earlier.[37] When I asked about this lack of interest in former times, anthropologist Polly Wiessner proposed that the past gained significance—and I imagine, too, a set chronology associated with themselves and their land—only after people created political systems that had to be justified and conveyed through the generations, such as that set out in the US Constitution.

"This land was made for you and me," sang Woody Guthrie. This sense of joint ownership of a space, as with the things in it, would appear to be vital in a person's connection to the society. I've told you how hunter-gatherers were almost always territorial, as people are generally. More than the founding-father tales central to the ethos of nations, what most resonated with hunter-gatherers was this attachment to the land, and its sacred sites, such as the reverence of Bear Butte mountain in South Dakota to the Cheyenne. This intimacy conveys how people can be willing to die for their homeland as a question of cultural survival.[38] It is *here* that their knowledge of the world applies, where their sacrosanct ways permeate the soil. In fact, stories and space can be linked. Those with expert recall assign their memories to locations in an imaginary place or landscape. Both memory landscapes and real landscapes are encoded in the hippocampus.[39] Australian dreamtime falls into this memory tradition, its potent stories connecting to places in such detail that Aborigines could reconstruct the lay of their land without maps.[40]

Citizens today retain this conviction of sharing a national territory even as many of them have come to possess, as individuals, just a tiny piece of earth within it. Even if "a nation is the same people living in the same place," as James Joyce's fictional character Leopold Bloom observed, no one has to walk every inch of their society's territory to feel a spiritual connection.[41] Its borders are as fixed in the mind's eye as the boundary in identity that separates Us from the Them who live on the other side. Indeed, the expanse of ground too wide for anyone to ever see in its entirety is no less concrete in the people's imagination than the vast community of compatriots, most of whom they will never meet. And so emotive pronouncements about the homeland grace our national anthems: there are Fuji's "shores of golden sand and sunshine"; Bulgaria's beauty and loveliness—"ah, they are boundless"; the Chilean countryside as "a wonderful copy of Eden." Meanwhile Guyana is "set gemlike and fair, between mountains and seas," and "among the lands" Lesotho "is the most beautiful." People can adore a foreign land, but few merge this feeling with the warm glow of belonging and deep connection to stories that ties them to their country of birth.[42]

Not that people require a territory to tenaciously identify with their group. Perennial wanderers who originated in northern India, the Romani, often called gypsies, are a durable ethnicity that retain a common culture despite being landless since spreading over Europe a thousand years ago. Still, without a home, or at least no claim to a homeland, a group, ethnic or otherwise, can seem impotent.[43] Hence the emotional resonance of the epic stories of dispossessed Jews or Palestinians seeking their own plot of ground.

Stories, and the territory they are often anchored to, are the great binders of societies. The minds of the Bushmen, at their camps, must have linked up firmly as they told each other stories of mutual significance, building what one expert proclaimed to be the group mind.[44] Whether taught at school or around a fire, stories handed down through the generations, and the land and people described, frame our common upbringing and destiny: they remind us either that we are in this together or that the others—*they*—are in this together in ways that may not serve our interests. Counterintuitive aspects of a traditional tale, whether Homer's Cyclops or Moses parting the Red Sea, are memorable and impossible to confuse with the beliefs of others. A story's themes can be so pointedly improbable that outsiders inevitably find them absurd, yet people accept their idiosyncratic facets without question.[45] The role of such myths isn't to convey logic but rather to elicit emotion and bond us to the place we live, and to each other. How a story is presented is important, too. Spoken aloud in the hushed tones or uproarious chants of the devoted, not only are stories likely to be remembered, but mangling the account can be a travesty or sin. Properly recounting a story becomes a common practice of our upbringing and part of our very souls.

A people's connection to their land adds to the emotional pull of the membership they share, just as their narratives and stereotypes about themselves and others create an identity that gives life meaning. As we shall learn, what makes each of these elements formidable is that they all can be leveraged to maintain our superior self-image. They can also be weaponized against groups that compete with us, or offend us. Alas, this is what makes them so frightening.

CHAPTER 14

The Great Chain

A mong hunter-gatherers, the Jahai of Malaysia call themselves the *menra*, or "real people." This is also the meaning of *Dana-zaa*, the name the Canadian Beaver Indians give their kind, and *mihhaq*, the word used by the Kusunda of Nepal. "Even the gentle San [Bushmen] of the Kalahari," E. O. Wilson tells us, "call themselves the !Kung—*the* human beings"—true, yet even there the designation "human" applies not to every Bushman but merely those Bushmen who belong to the !Kung society.[1]

It should come as no surprise that the people of a society can harbor a sense of their superiority, though rarely do we stop to contemplate how extreme, and at times even outrageous, this feeling can be, as the critic Ernst Gombrich has nicely spelled out.

> I know a wise Buddhist monk who, in a speech to his fellow countrymen, once said he'd love to know why someone who boasts that he is the cleverest, the strongest, the bravest or the most gifted man on earth is thought ridiculous and embarrassing, whereas if, instead of "I," he says, "we are the most intelligent, the strongest, the bravest and the most gifted people on earth," his fellow countrymen applaud enthusiastically and call him a patriot.[2]

That we cultivate and maintain this confidence in our supremacy, and the way we register the relative status of those in other societies, are human universals. Surely Americans didn't invent the idea that their nation is the greatest. "Somewhere there may be a rationalist country that asks its children to withhold their emotional evaluation of their homeland until they've carefully compared its inventions, heroes, and breathtaking natural wonders with those of its competitors," social psychologist Roger Giner-Sorolla says. "But that country is probably not where you or I grew up."[3] This outlook is embodied again in national anthems, which voice almost identical themes. Among them, the adulation of national history and heroes, expressions of pride in the people's work ethic, commitment, and courage, and in the peace, security, liberty, and—the most overused word of all—the freedom a nation offers. For many people the very idea of plausible points of comparison between their society and any other one is absurd, even when the similarities are clear. Social minutia can be enough to arouse a supercilious attitude toward others who in other respects may be all but indistinguishable, a phenomenon that Freud named the narcissism of small differences.[4] A society's members seldom have to be sold on why their ways are best: they absolutely know how things should be, and for them a life arranged that way is worth living.

A sense of superiority is just part of the picture. Compounding the situation is the unnerving ease with which people are capable of devaluing strangers. If extraterrestrials were to land on Earth and witness human behavior, they might rate people socially inept: without considering the matter, we bounce between treating an animal as a person, and a person as if he or she were an individual human being, a coffee dispensing device, or an inferior kind of animal. Given the practical payoff of a skill at *humanizing* animals (hunters, for example, often imagine what a deer is thinking to predict its next move), it's a shock to discover just how readily people *dehumanize* other humans.[5] The names many hunter-gatherers give themselves reflect how humans treat outsiders differently from, and usually inferior to, their own people or other groups they know and trust. Even the names many modern nations apply to their own citizens, such as *Deutsch* for the Germans and *Dutch* in the Netherlands, derive from the word in their languages for "human."

The architecture of minds built to perceive human groups as having their own essences doesn't grant equality to them all. We see our society and those of other people as falling into a hierarchy along with other living things, a notion codified in the Middle Ages as the Great Chain of Being. Typically the royal We surmounts the chain (surpassed only by God and the angels). Other humans follow in a descending order, some of them, Aristotle announced, "as much inferior to their fellows . . . as beasts are to men."[6] This hierarchy continues its plummet through the natural world, with "some animals," as Orwell wryly wrote in *Animal Farm*, "more equal than others."[7]

The scale wasn't fashioned in an ancient Greek ivory tower; people intuited the universe this way before words were scratched on parchment.[8] More than likely it's a basic feature of our psychology. Research shows that children think of people as superior to animals, and of outsiders as closer than their own group to animals.[9] Furthermore the proneness of hunter-gatherers and tribal peoples to describe themselves as *human* suggests this type of thinking is typical even among societies that are small and have much in common with their neighbors (far more in common than you, with your wider knowledge of the world, likely share with a Tibetan yak herder today).[10] Their use of epithets meaning *nonhuman* or *animal* also suggests those people felt entitled to treat at least some outsiders categorically as other species, an outlook that would naturally affect their relationships.

RANKING OTHERS

Psychologists have a lot to say about how people rank outsiders. The status we assign different groups is associated with beliefs we hold about how, and how well, their people express emotions.[11] First, a bit of background. Six emotions are often considered basic—happiness, fear, anger, sadness, disgust, and surprise. These are discrete, hard-wired, physiologically distinct mental states expressed first in infancy and recognized worldwide with slight cultural adjustments.[12] We combine these feelings into secondary emotions, culturally influenced expressions of ourselves in relation to others. We use secondary emotions to interpret each other's intentions and respond to what others think about us as persons or as representatives of groups. Many of these complex sentiments are positive, such as hope and honor, but there are negative ones like embarrassment and pity. These shared secondary emotions are important to bonding the members of a society, who are motivated by the satisfaction of pride or patriotism and the avoidance of shame or guilt to act together in costly situations.[13] Secondary emotions require learning and more brainpower than basic emotions and emerge later, while a child is absorbing its identity.[14]

People widely intuit that all humans express the basic emotions, and that we share these as well with some animals (picture a happy dog wagging its tail).[15] We think of secondary emotions differently, treating them, along with such qualities as refinement, self-control, and civility, not just as unique to humans but as less developed or questionable in peoples other than our own. We especially doubt the emotional complexity of those near the bottom of the Chain of Being, whom we see as driven almost entirely by the basic emotions of animals.[16] We assume such groups lack self-control as well as the capacity for nuanced feelings, and that they can't be dealt with rationally. We discount evidence that they express regret—a secondary emotion—for any transgressions we attribute to them and doubt their sincerity.[17] Because we connect our august standing with moral caliber, we may associate such pariahs with immorality: they do not (and, we often think, cannot) follow proper ethical codes. Their presumed moral failings thus place them outside the boundary of fair treatment—our scope of justice.[18] To top that off, the dehumanized people elicit few secondary feelings from us in return, a coldness that worsens matters by seeming to confirm the other side's assessment of our own emotional disabilities.

Our unequal treatment of outgroup members is aggravated by miscommunication. Much is lost in translation not only because of possible language differences but because of a lack of adroitness at reading even the basic emotions on a foreign face.[19] While we observe all manner of details about people like ourselves, in interactions with outsiders who make us uncomfortable we detect only the most extreme expressions (notably anger and disgust). Hence research has shown that American Caucasians are faster to register rage on black faces than on white ones and, when presented with a racially indeterminate person, are more likely to conclude the face is of the black race when it expresses anger.[20] Individuals most susceptible to associating an ethnicity with danger—namely, racists—make the quickest, shallowest assessments. The darker the skin, the more dangerous the man.[21] On June 27, 1994, *Time* magazine published a mug shot of O. J. Simpson, accused of murder, that had been tweaked to blacken his face, as became obvious when *Newsweek* published the same image, undoctored. The outcry forced *Time* to pull the issue from newsstands.

Even figuring out when a foreigner is truthful can be hard. In a recent study of Turks and Americans, participants failed to catch each other's lies.[22] Likewise the cliché of the inscrutable Chinese, whose emotions, unreadable to Europeans, are assumed to be undeveloped, while Chinese citizens, acting by other rules, find Europeans overreactive. The social implications are real. I have seen tourists rant away while missing the irritation on the face of an "inscrutable" native of China or Indonesia, where under ordinary circumstances emotions are understated.[23] In part to counter the passivity Europeans and

Americans see in them, some Asians have had the fold in their eyelids removed surgically.[24]

Two aspects of our perception underlie our emotional reactions to outsiders. One, dubbed warmth, is a measure of their trustworthiness, judged with the blazing speed of a first impression. The other, appraised slowly and determined by our recognition of the group's position in the hierarchy, is competence. Competence is a measure of their strength as a people, and hence their capacity to act on their views about us—that is, whether they will be able to help us or cause us harm.[25] We respond differently depending on these instinctual assessments, pitying members of groups we see as warm but incompetent (Cubans, for instance, in a study of Italian subjects); envying competent others who are perceived as cold toward us (Japanese and Germans, in the same study of Italians); and craving interactions with competent groups that treat us warmly. Finally, disgust tinged with anger—contempt—is an at-times undisguised reaction to those regarded as hostile and inept (a response many Europeans have to the Romani).[26]

The disgust brought on by the Romani and other denigrated people place them firmly among the vermin at the nadir of the hierarchy. Examples can be dredged up by the thousand. The name given to Ache hunter-gatherers by their farming neighbors is Guayaki, or "ferocious rats."[27] Proclaiming that "not every being with a human face is human," as one political theorist of the Third Reich put it, the Nazis likened Jews to leeches and snakes.[28] During the Rwanda genocide of 1994, the Hutus compared the rival Tutsis, a mildly different racial group of slightly taller people, to roaches. The tactic of equating black people with apes began with the first European contacts with West Africa. We like to think this egregious view is ancient history, yet the Implicit Association Test demonstrates that the characterization remains embedded in American minds.[29] These aren't throwaway comparisons equivalent to disparaging someone for acting like a pig or complimenting the person for being smart as an owl.[30] Registering a people as disgusting and subhuman allows us to deal with them the same way. Such was the fate of the war prisoners at Abu Ghraib, photographed in humiliating ways by American prison personnel in 2003 and 2004. At best their misfortunes are disregarded or treated with schadenfreude—pleasure at their pain.

Disgust is a complicated emotion. It is directed not only toward certain humans and animals, but toward anything contaminated or unhygienic. This includes such outlandish ethnic dishes like walking cheese, as well as anything—or anyone—reminding us of our unclean animal natures.[31] Disgust toward people and disgust toward unsanitary things appear to be essentially interchangeable, a product of activity in two areas of the brain: the insula, a deeply folded portion of the cerebral cortex, and the amygdalae, a pair of almond-sized clusters of neural tissue buried in the temporal lobes, part of the brain's rapid-response system. Meanwhile the medial prefrontal cortex, a lump of brain tissue involved in human interactions, fails to activate, as if we were merely confronting an object.[32] I suspect the Gestapo's mis-identification of the trainload of Germans accidentally sent to the gas chamber for "delousing" resulted in part from the travelers' repellant appearance after a long journey in crowded, less-than-hygienic con-ditions. That the American downtrodden have been obliged to use separate drinking fountains and toilets makes sense if you think clean-liness is next to godliness and believe that the badly sullied lie at the filthy and godless end of the scale.

Immigrants, often belittled as unclean or as vermin, are especially vulnerable to extreme xenophobia. Indeed, disgust may have evolved in humans as a way to keep diseases from entering a society by enforc-ing a bottleneck for contact with outsiders. Such infections as avian flu and Ebola can trigger the same fear today.[33] A territory closed off to outsiders functions as an island to which some parasites have a hard time making a journey.[34] Diseases became unintended weapons in con-quest. Colonialist societies often transmitted illnesses to which the na-tive population had no immunity. The Arawak contacted by Columbus were among the many American tribes who succumbed to smallpox.[35]

SURVIVING AT THE BOTTOM

Whether or not marginalized peoples succeed in building relation-ships with more powerful neighbors, their low status puts them at a psychological disadvantage as well as at an economic one. Not only do they feel inferior, they can be downright depressed about themselves, as a !Xõ Bushman's explanation of the modern-day beliefs of his peo-ple discloses:

[The God] *Gu/e* first made all the people alike, later he divided them into different peoples, some to work for others. Nevertheless, when God first made man he started by creating the white man and then the black one. With the pot scrapings he had left he made the Bushmen, that is why the Bushmen are so small and why they have so little sense compared to other people. *Gu/e* did the same thing when he created the animals. First he made the large ones, then the smaller ones and with the pot scrapings, the smallest creatures.[36]

Yet even people battered by outsiders and with poor self-esteem strive to appreciate their own distinctiveness by finding things specific to their group to value and give their lives meaning.[37] From what I know about the !Xõ, for example, I wouldn't be surprised if they take what pride they can today from their superior knowledge of nature.

That the values and identity of a society can be protected in the face of demonstrable foreign superiority underscores that the very idea of what it means to be human can be open to interpretation. Each society or ethnic group puts a favorable spin on its standards and defining features. Thus, without completely dehumanizing the people of other societies and ethnicities, we can ascribe to them, and to ourselves, differing qualities.[38] The Chinese perceive their nation as admirably communal, while Americans take pride in individualism. We can recognize foreigners as clever, more ambitious, or more expressive than ourselves, in which case we may decide that our straightforwardness, contentment, or reserved nature is appropriate, and excusable, deportment. By calling our shortcomings "only human," we keep ourselves as close to the top of the hierarchy as we can. We rationalize that while those others have a lot going for them, they fall short in categories where we shine. Perhaps they end up working like machines, are hopeless at the things in life our group finds important, or are wanting in moral principles.[39]

PREJUDICES AMONG ANIMALS AND IN EVOLUTION

A susceptibility toward seeing disliked outsiders as disgusting vermin suggests that prejudices against humans and creature phobias may share an underlying psychology. Once I spent a day watching Barr Taylor, a psychiatry professor at Stanford, treat a patient with arachnophobia. Dr. Taylor introduced her to spiders incrementally, one body part at a time, starting off by drawing a head, adding the abdomen, then each leg. What had been frightful grew routine as the woman absorbed its details.[40] In this reading of the issue familiarity breeds understanding, or at least tolerance. A failure to properly register outsiders as individuals could betray a similar discomfort with people we find unintelligible—a squeamishness about looking them in the eye.[41]

Equating outgroups with vermin may have antecedents among the other primates. Male macaques readily associate troop members with fruit and foreign monkeys with spiders in a simian version of the Implicit Association Test.[42] Conceivably, then, dehumanization predates language. A reasonable proposition is that our ranking of societies and ethnic groups in the Great Chain emerged out of the dominance hierarchy that organizes the relationships of individuals *within* the societies of many animals. For a macaque or baboon, every moment is dictated by status differences in which dominants pester subordinates. As such, perceiving foreign societies and their members below the hierarchy of its own troop would be cognitively simple, and a plausible precursor to the dehumanization of groups.[43]

Still, among species other than ours, it's hard to conceive of the elaborate prejudices that humans show. One reason for this is biologists don't know whether any other animal has the secondary emotions that seem to be a necessary underpinning for human appraisals of outsiders. But putting that question aside, taking stock of a foreigner's warmth, competence, emotional depth, or status, shallow as those evaluations might be, only makes sense given the chance to build a relationship. That's impossible, for example, in chimpanzees, whose sole recourse on encountering a foreign animal is to flee or bludgeon it to death (some encounters with lone females excepted).

The emergence of anonymous societies no doubt shaped our perceptions of foreign societies as species. Markers in effect are traits we broadcast to establish ourselves as bona fide human beings.[44] The showcasing of our identities using our bodies as a canvas possibly got its start from our ancestors' reactions to Neanderthals and other now-extinct offshoots of our family tree, whose weak chins, projecting faces, sloping foreheads, and behavioral anomalies marked their own bodies as distinct. If so, our response to foreigners that once were truly distinct species eventually transferred to different societies of our own species.[45]

That we assess outsiders this way, and see in them what we fear, may have been a smart response in our distant past, especially before 40,000 or 50,000 years ago when contact with outsiders was rare, knowledge about them sketchy, and life spans too short to depend on intimate learning. Generating stereotypes (even if little more than hunches to stave off an overload of uncertain information) gave our ancestors a quick-and-dirty guide to predicting whether foreigners might benefit or hurt them. Reliance on prejudices would be less risky than figuring out the individual personalities of those whose morals, motivations, and modes of expression were strange and whose ties to them were shaky or nonexistent.

The philosopher Immanuel Kant asserted that our moral concern should take in the human race.[46] It should be clear by now that the mind doesn't readily accommodate such a point of view. Situating ourselves at the summit of the Great Chain of Being may explain how we can hear about deaths from wars overseas every day and still enjoy dinner.[47] One upshot of judging other people's mental and emotional capacities as impoverished, and their ethics as lacking, is that we block ourselves from interacting with them by ordinary means. Committing atrocities, facilitated by psychological assessments of outsiders as inferior, may represent a considerable payoff in pure evolutionary terms. Once the deed is done, seeing those we wronged as less than human turns into a self-preservation mechanism for assuaging guilt, allowing us to move on.[48] Generations can pass before we recognize our actions as heinous, should our selective memories allow it at all. In fact, just as we doubt expressions of regret in others, saying we (as a society) are sorry is a surprisingly recent idea, and doesn't come easy.[49] Buried in the 2009 Defense Appropriations Act was an apology "for the many instances of violence, maltreatment and neglect inflicted on Native Peoples by citizens of the United States." It didn't make the front page.

Thus far, I have described social stereotypes as characterizing the individual members of a society or ethnicity. We have yet to tackle how our biases also channel our perceptions of a group considered in its entirety. Moving forward, I will explore how a society can seem to be a single thing, how it can respond with one voice, and how it can carry out actions as a united front.

CHAPTER 15

Grand Unions

The greatest displays of power in prewar Nazi Germany were the rallies at Nuremberg. Each year hundreds of thousands of enraptured people—eventually the number climbed to nearly a million—filled an arena bordered by 130 searchlights bright enough to be seen a hundred kilometers away. During fervent speeches wave after wave of troops marched in goose step under stupendous flags and banners to the thunder of Wagner.

The spectacle made plain that the Nazis placed the Germans at the summit of the Great Chain of Being. But for people watching the newsreels from elsewhere in the world it also played into a human predilection to conclude that members of outside groups—Germans in this case—are of a kind. The extravaganza was a show of such solidarity that its countless participants, all looking and acting alike and identifying with the same symbols, seemed to blur in the minds of those terrified of the Nazis into a swarm of a dystopian uniformity that people normally associate with insect colonies. Indeed, George Orwell called nationalism "the habit of assuming that human beings can be classified like insects and that whole blocks of millions or tens of millions of people can be confidently labeled 'good' or 'bad.'"[1]

Even when groups don't overawe the way the Nuremberg throng did, the members can merge together like so many ants because their shared unfamiliar appearance and ways of doing things make it a struggle to tell them apart. This is in sharp contrast to the people in our own group. Today the problem extends both within and across societies. *They all look alike*, Caucasians say this about Asians, who in turn say the same thing about Caucasians. It's a view that misses the trees for the forest since both races show the same variability in features.[2] *They can't be trusted*: this prejudice was once as sweepingly common among hunter-gatherer societies as it is today.

Once we shoehorn others into categories, the chanting horde of a hostile country or the cheering supporters of a friendly one meld in the mind's eye as if the individual people were so many cells in a Frankenstein-esque body, a flesh-and-blood entity with a personality and ambitions of its own, able to sustain itself for the long haul. Psychologists say the group has a high entitativity, the quality of being perceived as an entity.[3] Overbearing displays of symbols such as the flags and music at Nuremberg add to the effect. They make foreigners look not just more stitched together but more competent (and thus a greater possible threat) and coldly intimidating, and therefore untrustworthy.[4] Our sizing up of a group's strength and prospects for destructive competition with us in turn influences whether we think of it as an enemy, an ally, a needy dependent, and so on.[5]

While the creators of the fanfare at Nuremberg undoubtedly took pleasure in impressing the outer world, the core of their game plan was to strengthen the German people's identification with the nation. For the participants, of course, the whole business was intoxicating. You can predict that the Germans underwent much the same reshaping as they did in the minds of foreigners, except with a different outcome. While most of the time people find the recognition they crave by expressing their connections to a society through their contributions as individuals, and seeing one another as persons, the joy of being peas in a pod wins out when they feel part of a unified crowd. The participants' impulse to express their individuality is put aside and their differences forgotten, leaving the sensation coursing through their veins of being both like each other and one with a united ensemble.

Conceiving of our society as a concrete entity is a normal experience nurtured by any national celebration, with its parades, fireworks, and flag waving.[6] The perception begins in infancy with the urge to belong, a comfort with like others that transforms into strong bonds with our society as we age.[7] That sense of being part of something greater leads us to inflate how much we have in common with other members, whether we individuate them or not. It also hones the distinctions we perceive between our society and those far more homogeneous outsiders, who appear all the more identical when they frighten or anger us—while our inattention to detail further enhances the impression that each group is united.

Those in power contribute to this entitativity. In extreme cases, as with Hitler during the Third Reich, the leader becomes a symbol and an embodiment of the identity of everyone else, as formidable as any totem pole. People may similarly look upon a foreign leader as representative of his or her society. This view enhances the stereotype bias that prompts us to treat outsiders as carbon copies. The leader becomes the original from which the copies were stamped.

Our belief in this carbon copying requires no proof that the foreigners really are alike. If we assume all persons of a nationality or ethnic group are the same and we come across a hostile one, we are apt to expect them all to be hostile. As a first approximation, people are taken to represent their society, much as their leader does in the international arena. This is especially true for any personality traits we dislike.[8] That's no different from how we generalize about other threats. It is adaptive to reason that all bees sting if one does. Slapping bees is expedient. Our response to societies, treated as species, can be equally simplistic, and sometimes just as deleterious to outsiders. Better safe than stung.

SOCIETY BECOMES SELF

In considering the sense in which societies are entities, it may be helpful to step back for a moment and overview the extraordinary range of human connections, from the most personal to the most abstract. Different styles of interaction work well in groups of different sizes, and these styles and group sizes have been important from early times.[9] Most intimate are relationships between pairs such as a wedded couple or a subordinate and boss. Then there are the relationships within the handfuls of people who perform a task together. For hunter-gatherers this meant parties that gathered plants or hunted game, and now it applies to workplace collaborations where decisions have to be made smoothly to achieve a goal. *Where can we dig up these tubers?* became *how do we market this car?*

Next up were the original hunter-gatherer bands of two- or three-dozen souls that today correspond in size to many classrooms, business departments, and hobbyist clubs. Conflicts can be harder to manage in these larger groups, but people are still tied together by a sturdy knowledge of individual others. A few hundred to a few thousand had an allegiance to a band society, and today we see corresponding numbers in social groupings like church congregations, conferences, and schools, institutions that are pivotal for exchanging information and resources across a broader community. At this tier, collections of people extend well beyond individuals who can directly interact, or even know each other. Most of the anonymous crowd identifies instead in some symbolic manner.

It would go too far to say that a classroom is equivalent to a hunter-gatherer band, or certainly that a group as ephemeral as a conference is equivalent to a band society. Yet our contentment with the intimacy of the classroom or with the more abstract joy of being at a conference alongside scores of fellow enthusiasts may echo how our psychologies interfaced with others in early bands and societies, respectively. Regardless, for most individuals the societies themselves, no matter how populous, retain their primary hold, their interwoven markers setting up the conditions for people to see each other, and foreign individuals, as part of permanent, unified wholes.

So it is that people find significance in both their fellow members and their society, which is perceived as an entity amounting to more than the mere sum of its members.[10] We see ourselves as playing a role in our society's continuance, carriers of its celebrated history, traditions, laws, and social mores—all those markers that bind us into the future.[11] It's fair to say that people feel they live on by way of their progeny and also through this union: love of country becomes a vicarious path to immortality.[12] The early ethnographer Elsdon Best reported on one expression of the phenomenon among native New Zealanders:

> In studying the customs of the Maori, it is well to ever bear in mind that a native so thoroughly identifies himself with his tribe that he is ever employing the first personal pronoun. In mentioning a fight that occurred possibly ten generations ago he will say: "I defeated the enemy there," mentioning the name of the tribe.[13]

Thus if our nation's soldiers die in war, our response is distress or anger or fear. If our national team triumphs at the Olympics, we are thrilled not purely out of joy for the team, but because of a sense we won the game ourselves.[14] Uniting around other citizens comes with a surge in pride and awareness of strength and glory. This sort of ethnocentric love some claim is a function of oxytocin, a hormone that dampens the amygdala anxiety response while increasing our empathy for similar others—positive feelings also associated with the flag.[15] Moments of such solidarity, and oneness, can represent the highlight of a person's life. Americans felt this affinity at that moment *we* landed on the moon, or, for the Brits, the *we*-ness that came with crowning a king or queen. We forget our differences when buoyed by the identity we share, and, in an effect called a group emotion, our intense, and overlapping, feelings draw us closer.[16] Ask a person how she feels, and she might say quite happy; however, if she then hears about an act of terrorism against her country, she will likely report extreme sadness or anger. Intrinsic to human self-esteem, group emotions focus pride, attention, and energy at a national level, so that one's society inherently becomes part of the self.

When reactions do spread, at times like a contagion, the response is heightened the closer together individuals are, whether excited monkeys converging on a fruiting tree or thrilled fans cheering across a sports arena.[17] Yet modern-day crowds aren't necessary to cultivate group emotions and bonding. Togetherness on a more modest scale will do. For roving hunter-gatherers, their connection with their society must have peaked when the bands gathered in one place. Members would have confirmed their solidarity not merely by swapping goods and building camaraderie, but more fundamentally by identifying with the assembled community. The get-togethers would have offered a sense of the sacred in the form of collective pride and patriotism. People across bands reaffirmed their unity through feasting, stories, songs, and dances.[18]

The pleasure of carrying out these activities, and of ceding one's individuality to the group, may spring from an impulse to act as we think we must to be accepted. This is seen in how we mirror not just the vocal tics and gestures but the emotions of those we respect, which we tend to do most with those of our race or ethnicity.[19] These actions are unintentional, induced in part by mirror neurons in the premotor cortex that fire whether we engage in an action or watch another carry that action out.[20] The behavior has genetic underpinnings: newborns imitate the sadness, fear, or surprise of others.[21] Mirroring others, and emotional responses at a group level, spring up in other animals, too, like the excited monkey troop. Chimpanzees match the playfulness or anger of apes in a video.[22]

ACTING AS ONE

Not long ago I was in Tokyo enjoying the crisp January air at the Teppozu Inari Shinto shrine, recognizable by its century-old torii gates and arched rooftops, when more than a hundred individuals, stripped down to white loincloths and head cloths, gathered in the inner courtyard around a pool filled with ice, and began to chant. About 30 of them climbed in, sunk to their knees until the freezing water reached their waists, and replicated the motions of rowing a boat as their voices alternated from a melodious prayer to a two-note chorus and then to words spoken in a rumble. After a few frigid minutes the devotees climbed out to be replaced by the next set of people.

The ritual is known as the Kanchu Misogi winter purification festival. I stood watching with Panos Mitkidis, affiliated with the Center for Advanced Hindsight at Duke University, where psychologists investigate human thought processes, including, in Mitkidis's case, how people act in unison under trying conditions. Mitkidis had worked up a sweat despite the chill while his team gathered data as the day's events unfolded.

Rituals consist of replicated sequences of actions that have no obvious practical utility in and of themselves. The rallies of some mammal species are the behaviors closest to rituals that we see in nature. Gray wolves leap on each other and yip, while spotted hyenas go through bouts of rubbing one another, tails bristling. The act of rallying animates the participants to take risks they would avoid when alone—attack another pack or clan, perhaps.[23] This boldness may loosely reflect what in our species is called the discontinuity effect: groups interacting with other groups are inclined to be more competitive and less cooperative than the individuals who make up the groups would be when interacting one on one.[24] Of course there's a difference between humans and these other animals: people often have symbols—a flag—to rally around. Slavery ants may be an exception to this rule. Sharing a pheromonal flag, the workers rush about in a swarm at their nest before marching off to seize a fresh batch of servants.[25]

Humans follow ritualized patterns more than we may imagine, going far beyond simply mirroring each other's speech and emotions. As children we become clever at duplicating the complex actions of others, typically to no obvious end other than to demonstrate our commitment to a group. Think of a child gaining the acceptance of a hip clique of schoolmates by abiding whatever play rules they concoct. *We dress just so*—okay, check![26] Performing the nonessential bits correctly—getting the ritual right—is a unique way humans assert their identities, to their society included.[27]

Mitkidis finds that those who participate in intense rituals achieve a heightened solidarity that rises above the identity-affirming power of ordinary social markers by making the participants very cohesive indeed, the sort of behavior augmented into outright hazing by fraternities that forge collective unity through hard knocks. When formal expressions of devotion are repeated regularly, people can be driven to so align their fates that they may commit to perilous concerted actions. Cults and gangs instill a sense of unity in much this way; that's especially true for such criminal organizations as the Mafia, which enforce the members' commitment across the generations.

Rituals even more extreme than the winter purification festival cause a radical incarnation of group emotions known as identity fusion. Participants see themselves and their group as truly one and the same, pumping up their commitments to each other while ensuring they stringently uphold group standards.[28] Such "rites of terror" are performed seldom and often by a select few. In nations today they are practiced only in the most intense military programs or during actual combat engagements.[29] Hunter-gatherer and tribal groups commonly exhibited such costly and hard-to-fake rites, which extended to producing painful and irreversible markers like body scarring when their people were embroiled with hostile outsiders on an ongoing basis. Think of their acts as simulating the excruciating pain the players expect during the ordeals to come, readying them for taking action together.

As an ant biologist I'm especially intrigued by the warrior initiation ceremony still performed by the Sateré-Mawé of the northern Amazon. The rite requires young boys to be stung by inch-long, lozenge-shaped horrors some call bullet ants for good reason.[30] The shock of a bullet ant sting once knocked me down even though I had brushed the ant off before it could give me a full injection. That's nothing compared to the experiences of adolescent Sateré-Mawé, stabbed by dozens of bullet ants over five minutes in a level of torment surely matching the agony of many battle wounds.[31] The Sateré-Mawé, not surprisingly, are a fight-prone people.

A people don't require rituals, of course, to be willing to have each other's backs. In dangerous times they may pull together as resolutely as a band of horses does when confronted by a pack of wolves. A commitment to fight for the common good can be a source of joy and inspiration, but it presents dangers, too. Compatriots respond as one even when some have contrary feelings about the group's course of action. For example, they might hold back their opinions because of fear of rejection or of being seen as cowardly, or simply to go along with the thrill of the moment. As independent-minded as people can be as individuals, their differences can be put aside to shore up a group response.[32] Developmental psychologist Bruce Hood remarked of such situations, "Whatever self we believe we have is swamped by others."[33]

Not that everyone will join the displays of unity. Cooperation and trust may be magnified, yet those attitudes aren't always embraced by all the members. Should some openly dissent, the majority, wrapped in the flag, will find it a problem. Championing a personal view can be wildly unpopular or even considered traitorous, a betrayal of loyalty so heinous that in medieval Europe it was deemed a sin worthy of flaying or disembowelment.[34] *We are in this together* turns into something more severe: *You are either with us or against us.*[35] Better to cave in, or at least portray that you had no option if you are called to justify yourself later. Nazi criminals pleaded the Nuremberg defense, arguing that they were just obeying orders. Not only, then, do we abandon our better judgment to our blind faith in our society, but we are absolved of fault for its actions or ours on its behalf.

People who choose to administer electric shocks to others after being told to do so by someone in authority show a dampening of brain activity, suggesting that those acting under orders emotionally distance themselves from the consequences.[36] When we aren't simply listening to a commander but following the communal will and are drunk on group emotions, our sense of culpability may vanish entirely. Our fervor sanctions a path we wouldn't want to be held accountable for pursuing if we had done so alone.[37] All the better, a feeling of alike-ness and interchangeability renders us secure in our anonymity. Whether in a military formation like those at Nuremberg or in the chaos of a battlefield, common soldiers are trained to act as dependably as clones, impossible to tell apart.

If a throng comes together without a plan or clear leader, its actions can be as much an emergent property of the collective as the pooled behavior of an army ant swarm. The mass of humans may give the appearance of achieving objectives through a will of its own despite each individual contributing little. However, it's possible for the response of a crowd of humans to be smartly coordinated and yet stupid—maladaptive, or to people's advantage but normally immoral. A leaderless group is likely to yield sound decisions only under carefully constructed conditions in which a population is polled after each person has been given the leeway to assert his or her individual preference uninfluenced by others. Otherwise, what's often called swarm "intelligence" may more accurately be described as mob rule, with the participants surrendering their individual will and succumbing to the hysteria of the group.[38] Joining the masses gives the normally powerless their crack at awe-inspiring powers, right up to gang violence and genocide.

As a successful and envied ethnicity (one considered cold but competent by hostile outsiders), Jews faced a classic reaction that came to a head under the Nazis. Although they were shown grudging respect and their businesses patronized before the Holocaust, the attacks on them started as troubles escalated—blaming the victim, massively amplified.[39] This pattern has been repeated throughout history. Chinese residents were assaulted for similar reasons during upheavals in Java in 1998. Korean Americans in the Los Angeles riots of 1992 were targets of the sort of collective hysteria that has at times escalated to mass murder.[40] Rationality has little bearing on what transpires when tensions are high.[41] The ease with which ordinary people are roused to harm based on the haziest of justifications is dismaying. After the Rwanda genocide most Hutus admitted that the Tutsi had been fine neighbors. Instrumental in changing this point of view wasn't just the spread of comparisons of Tutsis to bugs: treating the murderousness as normal encouraged previously held but hidden prejudices to surface until, in the words of one of the participants, "when the violence began, it fell upon us like a sudden rain."[42]

Small groups of thoughtful individuals may fare little better. Groupthink can materialize, a desire for fraternity and we-ness eclipsing the goal to achieve sound solutions.[43] People can misrepresent their perceptions of facts to conform to what the group expects.[44]

A tragedy of the human condition is that our group's behaviors can justify the fears and prejudices that outsiders have about us. We can indeed turn out to act like each other, and as a unit. The perceived interchangeability of people most sharply matches reality when tensions are high and we may come to be as much in lockstep as soldiers on parade. Within our society, and equally in theirs, such tension stimulates cooperation, or at least acquiescence to the society's course of action. Subsequently, it reinforces the sensation that our fates are linked—and at odds with those similarly united others.

This feeling of alikeness with others can extend to everyday behavior, too. The Jívaro of Ecuador were gardeners and game hunters. For the Jívaro, killing those who "speak differently" (that is, other tribes), and then using a special kit to shrink their victims' heads, was a must.[45] This viciousness toward foreigners can be contrasted with the relative gentleness of Bushmen societies in recent history. The brutality of the Jívaro and the kindness of Bushmen are stereotypes certainly open to objection, vehemently so when assumed to apply to individual persons. Yet stereotypes can reflect actual group differences in what is called a collective personality, arising at a population level.[46] This has even been shown to be true for animals; for example, honeybee colonies of the same population differ strikingly in aggressiveness.[47] For humans, the collective personality emerges in much the same way as more obvious markers for the society, as a product of repeated social interactions that cause individuals to look and act alike. The British famously are more reserved than Americans, and Americans are more reactive than the French to displays of temper.[48] Differences in national character probably go way back, with the !Kung Bushmen more openly expressing anger than G/wi Bushmen, and Bushmen being on average more timid than other southern Africans.[49] These "personalities writ large" add further to the impression that societies are entities.

In examining societies from different angles, I've skipped over one incontrovertible fact of psychology: most people's well-being rests heavily on their family ties. Having established how human societies might have functioned from their earliest beginnings, and how the human mind responds to people of the same and different societies, we now take a step sideways to explore the relationship between families and societies.

CHAPTER 16

Putting Kin in Their Place

I t makes sense to insist that a society be more than a simple nuclear family—more than merely two parents with a generation of their dependent offspring. But that doesn't preclude the possibility that families and societies might have similar psychological and biological underpinnings. Could a society represent a kind of extended group of kin, either in terms of the actual pedigree of its people, or perhaps psychologically, in the minds of its members? For that matter, what should we consider kin, in the broad sense of an extended family? Do families have the same sort of membership and clear identity that we expect of societies? How integral, extensive, universal, or accurate is our knowledge of our biological kin—and do the logic and emotion of our connections with our families apply to the way we connect with societies?

Families, of course, are central to daily human life, and in ways not seen in other species. Take the general expectation that dad will stick around with his offspring. No dolphin, elephant, chimpanzee, or bonobo grows up knowing a father. And parents with kids are just a part of the picture. Human family relations are labyrinthine compared to the kin connections of other animals. Among people, not only do parents participate in their children's lives—and their grandchildren's—for as long as they live, but we keep track of a whole array of relations on both mom's and dad's sides of the family. We know our siblings and both parents' siblings, and the assorted spouses and offspring of them all. Humans don't just mate for life (or at least try to)—their bonds tie them to networks of kin.[1]

Investigations of those networks by biologists and anthropologists have been an important foundation for the exploration of social behavior, and principally cooperation and altruism. By the 1960s biologists had developed a body of work around the theory of kin selection. The theory holds that species evolve behaviors that favor the propagation of the genes of relatives into the next generation. Ever since, scientists have construed kinship as a core driving force not only in families but also in our connections to a society: If societies are imagined communities, goes one idea, the human mind imagines a society to be an extension of the family. This would in turn lead us to view our society-mates as if they were kin.

Indeed, there is some overlap in how people express their bonds with kin and their bonds with society. Yet evidence from the animal kingdom and human world alike suggests that understanding and keeping track of kin and understanding and keeping track of society members are distinct tasks, generally carried out to navigate different, though at times overlapping, problems and solutions.

KINSHIP AND SOCIETIES IN NATURE

For kin selection to work, individuals must be able to pick their kin out of the crowd or at least favor kin by happenstance. For ants and other social insects, this isn't an issue: usually the society *is* kin. In such species, a colony is one teeming nuclear family, made up of the queen mother and multiple generations of her offspring.

Yet beyond the ant colony the idea of a society as kin becomes cloudy. A savanna elephant core, often mislabeled as a family, can allow complete outsiders to join its ranks, and those outsiders, as far as anyone has determined, are treated as equals thereafter.[2] That a core typically ends up consisting of kin is a by-product of its history. Because siblings grow up together, they are apt to stay together. Gray wolf packs can be family affairs, but they, too, have proven capable of letting in outsiders as permanent members.[3]

Whatever bloodlines do exist can be transitory. This is true even for species like the wolf or meerkat, whose societies can resemble a social insect colony in that they sometimes consist of a family with multiple generations of young that may assist in rearing pups and other labor. A pack or clan can go on for generations, unity uninterrupted. But look closely: breeding males and females come and go, sometimes peacefully, at times replaced by outsiders in a fight. Ten, twenty, or fifty years hence the same society, on the same bit of earth, can be stolen out from under one lineage after another until it's composed of animals with no blood relationship to its founders.

Often in gray wolves and savanna elephants, and invariably in many other mammals from bottlenose dolphins to highland gorillas, societies are composed of multiple lineages of kin.[4] In fact there are societies with no kin connections at all. None of the adults in a horse band are likely to be related. This is a result of how a horse society forms. Maturing horses of both sexes bond not with childhood friends, as elephants do, but with individuals they meet after hoofing it from the society of their birth. Adult horses lose all connections with the kin they grew up with. Kin selection theory would predict that an animal would risk its life for kin. However, the unrelated members of these horse societies stick by each other for the long term, even standing as one to keep wolves away from a foal.[5]

Among the most stable of all mammal societies are roosting communities of the greater spear-nosed bat, which aren't kin at all. Each cave in their habitat harbors multiple societies, roosting at separate sites, consisting of 8 to 40 females that first met as young adults, and one pretty expendable male. Completely unrelated the females may be, yet they treat each other well. The females forage together in a territory they share, and become so committed to the group throughout their 16-plus years of life that when a baby falls from their roost, they protect it from foreign bats until mother comes to the rescue.[6]

Clearly societies offer advantages—in protecting the young, for example—that can keep animals together regardless of whether they are related. While kinship connections may add to the motivation to stay in certain societies, the presence of kin isn't obligatory for a society to carry on fruitfully. A jointly beneficial coexistence is the key to its success.

What of the nonhuman apes? When adolescent female bonobos and chimpanzees transfer to another community, the females are no longer around any of their kin, yet (most notably in bonobos) they bond readily. Conversely, males stay around their kin yet their closest allies tend not to be brothers or sisters.[7] Rather they befriend individuals with compatible dispositions. Sociable or rambunctious chimps stick together, a nifty advantage of being free to wander in a fission-fusion society.[8]

Furthermore, in genetically diverse societies, both human and nonhuman, kinship tends to be impractical to figure out. Consider the relationship of the baby to its mother. No family member should be simpler for a child to recognize. However, there are stumbling blocks even here, as the baby must learn to differentiate its mother from the nannies, grannies, and others who often help with kids—caretakers that exist in many cooperative breeding species with societies. Humans have a solution: a baby listens to its mother in the womb and bonds to the face associated with that voice within three days of birth.[9] A similar shortcut isn't likely for recognizing any other relative, dad included.

Clues may exist, in the form of similarities among family members. For example, a hamster can not only tell other hamsters apart but is able to detect a relative it's never met by comparing the other's scent to its own.[10] In one study, chimpanzees were able to match photographs of unfamiliar baby apes to photos of the mothers of those babies more often than chance would dictate.[11] For all that, resemblances, even among immediate family members, can be spotty. Human fathers are known to overestimate how much they look like their child. While researchers have shown that outsiders can pick out an infant's parents more often than expected by chance, errors abound.[12] Growing up, people come to recognize family members based not on looks or any proof of genetic relatedness, but rather from trusting others to tell them about who is who, and from being together from birth.

Putting aside the difficulties of detecting kin, there is the question of what kinship even means for humans and other animals. People conceptualize categories of kinship relation, like *mother* or *brother*, and likely other vertebrates do too, as has been documented for baboons.[13] Yet what goes on in the head of a baboon interacting with its sister isn't likely to correspond to a person's idea of *sister*. Baboons rely on association for support, more than on relation or resemblance. Ties between the monkeys arise from intimate familiarity rather than actual, unknowable, genealogy.[14] Each female keeps track of her matriline—a personal network of females that are typically such relations as her offspring, sisters, mother, and so on. In practice, however, who a female includes in her matriline hinges on her tastes, so kin may be left out and agreeable nonrelatives added in.[15]

Who makes the cut? Shakespeare put it best when he said "nature teaches beasts to know their friends." Animals often build supportive relationships with those who hung around mother in their infancy. A childhood pal needn't be kin, although the likelihood is good that some are. This might be an older sibling who still clings to mother, but it might be an unrelated playmate, too. Life can also come to revolve around kin through happenstance, as when a male baboon sticks by a past mate in anticipation of further dalliances with her. In the process he ends up befriending her infant, which has a fair chance of being his.[16]

For the most part, then, animals pursue social connections and not biological families as such. Studies of the psychology of relationships suggest people respond to kith and kin much the same, and value them equally.[17] As the saying goes, friends are the family you choose. An interchangeability of the two can be critical for people with small or broken families, or the aged who have lost all the other family members of their generation.[18]

A band society contained many family lines, with friendships driving social choices as much as, if not more than, kinship. A couple customarily raised their kids in the same band that was occupied by one or the other set of grandparents and possibly a sibling or two. Otherwise, their relations—the families of their brothers and sisters, cousins, aunts and uncles—were spread among the society's bands.[19] What kept a band together was compatibility. As with other fission-fusion species, people seek out likeminded others.[20] Among Bushmen, even today insofar as their hunter-gatherer activities continue, "bands have marked personalities: one may consist of people who are by habit quiet and serious, another may be a gay crowd or perhaps people who enjoy a bit of bite in their humor," one anthropologist reported.[21] Not that a band was homogeneous in the personalities of its members; like any social group, people could get stuck with an outcast nobody liked—perhaps an annoying relative of some abashed person in the same camp.

Even so, proximity can be a deciding psychological factor in our evaluation of others as possible close kin. People tend to avoid sex with anyone they were in frequent contact with as they grew up; this seems to act as a heuristic for avoiding incest.[22] Roaming hunter-gathers also tended to find mates outside their childhood social sphere, choosing to marry residents of bands at some distance from their own, where any kin were likely to be far removed.[23]

THE HUMAN RELATIONSHIP IQ

The obvious difference between people and other animals is that once a toddler has a vocabulary, it can be taught not only who is in its family, but also about the family tree relationships between those individuals (even for kin seldom seen).[24] Indeed, the most archaic and widespread words may be "mama" and "papa"—words a baby as young as six months may apply correctly to its parents.[25] Yet the meaning behind the words comes later, starting with the discovery that other kids have different mamas and papas.

What surprises is how poorly infants grasp such relations. A child needs to mature enough to employ language in building complex representations of relationships before it understands who connects with whom, and concepts like *uncle* take years to comprehend. Understanding kinship can be as arduous as memorizing multiplication tables, with the caveat that while multiplication is the same anywhere mathematics is taught, what children learn about kin depends on what their society requires of them. The intricacy of our cognizance of kinship can change over the course of history as well as across cultures. In Modern English, a first cousin may be a father's brother's son or a mother's sister's daughter, a distinction for which Middle English once had separate words. Even within a society, understanding of kinship can vary from person to person depending on the extensiveness and closeness of the family. Those from small or estranged families might have trouble sorting out relationships others take for granted. (*A second cousin? What exactly is that?*) As the biologist David Haig, an expert on family relationships, once quipped to me, "It's a wise child who knows its father, an even wiser one who knows the father's first half cousin."[26]

Given the significance with which scientists imbue relatedness, one wonders how many humans have an aptitude for keeping track of their family trees. How many would pass an IQ test measuring knowledge of kin? The convoluted classic 1940s song "I'm My Own Grandpa" starts with the singer's father marrying the daughter, by another marriage, of the singer's wife; through a Gordian knot of events the crooner arrives at the realization contained in the title. Following its lyrics gives me a headache, and I suspect many other people would fail a relationship IQ test that included this song.

An intimate knowledge of kin relationships doesn't seem to be essential for human life. Much like tribes that have no words for numbers beyond one and two, some societies are less obsessed than others with sorting out these tangles of kinship, at least as relatedness is conveyed by their vocabularies. Residents of parts of eastern Polynesia have just a few terms expressing kinship. The Amazonian Pirahã tribe is even more stripped down, having a single word that applies to both parents, all four grandparents, and all eight great-grandparents. Their one word for "child" is used for grandchildren and great-grandchildren, too, while the word for "siblings" applies as well to the parents' siblings and the children of siblings. The Pirahã tongue apparently lacks recursion—think of "one's mother's father's mother"—a kind of looping that is required to represent distant kin.[27]

Despite this linguistic simplicity, the lack of a name for a category might not keep the Pirahã from grasping the differences among family members. We can't say from the available evidence, but perhaps those Indians still intuit how someone fits in a genealogy based on her age and who bore her without having a specific term to distinguish sister from cousin. The other possibility is that a category like "cousin" is unfathomable to the Pirahã, much as gravity went unnoticed by most people before Newton gave that word a meaning. This alternative seems plausible given the vexation Pirahã experience thinking about, and communicating, other sorts of categories for which they have few terms, such as numbers.[28]

The point is this: the familial connections that are recognized in a given society take years of training to comprehend, yet a three-month-old, too young to speak, deftly detects members of societies or ethnicities. A practical reason for our struggles with kin is that our relatives aren't a distinct group with a clear cutoff—ever more kin can always be added by going yet one more generation back. The child's relative ease with ethnic and racial groups testifies to the importance of broader groups than kin in our evolution. Societies and the differences that mark them, not families beyond their most immediate members, are an indispensable component of the human mental world.

FROM FICTIVE KINSHIP TO EXTENDED FAMILIES

Even for individuals who have an impressive relationship IQ, knowledge of their actual family tree is likely to go only so far. People's remembrance of relatives is generally restricted to those alive during their lifetimes, such as the great-grandparents they met as children. Few people recall much about more distant ancestors (except the one who did something to brag about). Even societies that practice ancestor worship seldom require anyone to master genealogies, but rather to honor their ancestors across the board.

People in most hunter-gatherer bands put even less emphasis on knowing their exact biological connections. What made retaining such information unlikely was a tenuous relationship to the past and its people.[29] Mentioning ancestors was bad luck and so more or less taboo for Bushmen, which would have made figuring out distant blood relations unlikely. Native Australians never spoke of the dead, who were consequently forgotten within a generation. Indeed, an astonishing reason for language change among some Aborigines was that any vocabulary word that perchance sounded like the name of a person had to be avoided after his or her death. The people had to invent a new word.[30]

As far as hunter-gatherers were concerned, culture and other markers trumped genetics. Families in some American Indian tribes could adopt children taken in battle, a practice that beefed up the tribal reserve of warriors. What bound adoptees to a tribe wasn't blood but their adherence to tribal ways learned alongside other children. This shared upbringing meant that both the families and their societies were culturally uniform and genetically diverse.[31] Even when genealogies were deemed important, they could be invented, as group histories can be. Members of Central Asian tribes that claim a clear pedigree turn out to be no more related to those in their own tribe than to the population as a whole.[32]

What did play a role among hunter-gatherers was the terminology associated with kinship. In band societies, fictive kinship, sometimes called cultural kinship, was commonplace, a method of giving each person a symbolic relationship to others. Thus, people used words such as "father" or "uncle" to address fellow society members, and all fathers and uncles were equal.[33] This may explain why, even though Bushmen valued close relatives such as their genetic parents, grandparents, and siblings, their languages had no word for *family*.[34] When Bushmen spoke of kin, they didn't necessarily mean a blood relative. What they often had in mind was anyone sharing a given name, which bespoke this fictive connection. A man couldn't marry a woman who was considered his sister no matter how distantly related. The primary value of fictive kinship was to maintain social networks based around rules concerning everything from monumental issues, like marriage, to such details as who should exchange gifts with whom. We retain a whiff of this view of relationships when we label an unrelated individual as if she were family—right, sister?

One reason not to invest too much significance in distant blood relatives is that, measured genetically, only one's immediate family is a big deal.[35] Who's a relative is, indeed, relative. We are all related at some level. Beyond the very closest family ties, kin merge into the general population in terms of their shared DNA, and quickly. Do the math and you find that first cousins share 12.5 percent of their genes by descent; second cousins drop to a measly 3 percent. Even two people taken from a community at random may have that tiny fraction of genes in common by chance alone. Whatever relationship IQ humans possess is therefore likely concentrated on a few closest kin.

In practice however, who we consider family is influenced by culture. Many Latinos, to give one example, rely heavily on extended families.[36] Even so, few cultures set absolute rules for family membership beyond the immediate family. In describing our family we don't expect to come up with the same set of names as would one of our cousins, although we can anticipate some overlap. Despite that, connections to a broader web of blood relations can have value for survival and reproduction when those kin are obliged to come to each other's assistance. Of course, helpfulness occurs between nonrelatives, too, and can be equally expected. The phrase "treat each other like family" describes how common such cooperation is among members of tight-knit groups such as military units and religious congregations.[37] Neither is it accurate to say that such aid comes simply or dependably even from genetically close family members. Recall how you clashed with a brother or sister, no bombshell given that sibs compete for the attention of harried parents.[38] Dickens, in *Bleak House*, mentions "a melancholy truth that even great men have their poor relations." However, families can rise up to punish any among them who isn't generous. This makes family obligations hard to escape. For most people, regardless of any unpleasantness, kin are kin for life.[39]

If bands of hunter-gatherers thought so little about actual genealogies, where did our current obsession with, and dependence on, extended family relationships come from? Family trees became a concern for hunter-gatherers who abandoned the lifestyle of owning no more than they could carry. Settled peoples who stood to inherit a social position and material items had good reason to know their family trees.[40] Likewise, in industrial societies, extended families have been most nurtured when there is wealth to share.[41] The very size of those societies also puts a premium on people reliably building a broad set of relationships and kin networks represent a consistently available avenue for people to do that. The learned skill to conceive of extended families—and to keep track of, and value, those relationships—has been a recent add-on in human evolution. It requires complex communication and learning and is heavily dependent on what each society expects.

Might people's relationship to their society be a clerical error in the mind, a misfiring of their mental representation of kin? That would be as if people mistook their society for an enormous family stretching, as the patriotic song puts it, from sea to shining sea.[42] Some hunter-gatherers did seem to apply an approximation of this representation of the society as kin. For them fictive kin were constituted to take in the whole society. Each individual could identify every single other member with a kin term.[43] Such universal kinship persists in faint echoes in words like "brotherhood," shorthand for the ethos of treating everyone as if they shared blood.[44] As we have seen, though, kinship truly was metaphoric among hunter-gatherer bands, having little to do with blood—and their societies even less so.

For our species, societies of distantly related others are the norm. No society, even the "motherland" of a small tribe, was ever made up of the offspring of one mother, the way a colony can be for insects. Even in social insects—whose success is often explained in terms of kin selection theory and the tight genetic relatedness of the mother queen to daughter workers—a queen will mate with several males, producing progeny with different fathers. More impressively, an Argentine ant society, for one, is no tight family. A big supercolony has genetically varied queens. Yet no ant prefers her immediate kin or even recognizes which queen is her mother or which individuals are her siblings. As in any other ant species, identification with the colony, and not kinship, is the focus of each worker's industry.[45]

I find it doubtful that human societies are simply amplified versions of family relationships. However, this does not rule out the possibility that family bonds played a role in the earliest incarnations of societies, in humans and other species. The first glimmer of a society would have sprung up among our forebears well before the chimp, bonobo, and human species split off from each other, set in motion perhaps when a proto-ape extended its attachment to its infant to others.[46] It may have been that, since primates don't breed as prolifically as ant queens, the full advantages of group life—for example in staving off belligerent outsiders—couldn't be attained without extending the size of the group beyond one family. Some anthropologists argue that the markers that identify societies or ethnicities (dress, hairstyle, and so on) substitute for similarities between family members.[47] I don't find this convincing given that kin are seldom consistently alike. Still, from the perspective of very deep history, a society might represent a kind of motherland.

All of this said, there is no denying that kinship is a potent force. The obligations to immediate families are as entrenched in the physical operations of the brain as are the commitments to societies, with societies and families involving inherently different aspects of life. Scientists could benefit from a course correction by focusing a bit less on kin and more on the psychological, and biological, basis of societies.

As has become clear, the human psychology around societies is far ranging. In the past several chapters, we have teased apart how members of a society are registered as having essences, rendered equivalent to biological species. Additionally, we investigated how our inexorable, blazingly fast, and clannish evaluation of other people is based on this perception. Our biases extend right down to how human or animal-like we perceive their capacity for emotion and the overall warmth and competence of their "kind." We have seen, too, how these assessments play out at the level of populations. We are disposed to discount the differences between individuals and to perceive the members of other societies—and to a lesser extent those of our own—as both similar and forming a unified whole. Finally, we have considered the ways in which the psychology of families relates to perceptions of our society, granting the influence of both in human affairs. What we have concluded is that even though biological kinship has a solid basis in genes shared by descent, whereas societies are communities conjured up purely by the imagination, the role of societies in human psyche and thought is foundational, and of vital importance.

That's because choices regarding who should be treated as members of a society can be critical to survival, whether those humans happen to be blood relations or not. When people are perceived as foreign, all bets are off. The likelihood of rivalry—and collaboration—between societies will be our upcoming topic.

Peace and Conflict

CHAPTER 17

Is Conflict Necessary?

O n a trek with the research team of primatologist Richard Wrangham in Uganda's Kibale National Park, I had my first encounter with wild chimpanzees. My heart almost jumped out of my body when I heard the hooting and hollering of a dozen acrobatic chimps rummaging for fruit in a fig tree that loomed over lesser trees above me like a shadowy fist. With their thick bodies, the apes were more intimidating than I had expected. They were endearing nonetheless, holding hands, playing tag, and cuddling. It was part frat party, part Zen gathering. I was surprised to find myself at peace, as if I were among friends.

The reading material I had brought along, including Wrangham's *Demonic Males: Apes and the Origins of Human Violence* and writings by Jane Goodall, undermined my euphoria. I could only imagine the shock Goodall had felt in 1974 when, after years of relative calm at Tanzania's Gombe National Park, a bloodbath began. One community of chimps incrementally destroyed another in what became a very lopsided four-year war. The chimpanzee violence evoked the worst facets of human behavior, in which the members of a society, reacting en masse to others whom in many instances they don't even know, cast aside their qualms about violence to attack outsiders.

The capacity for such violence is a thread that joins humans to chimpanzees and other species. Voltaire wrote, "It is lamentable that to be a good patriot one must become the enemy of the rest of mankind." Chimps, classified in the genus *Pan*, seem always to be willing to battle the rest of Pankind.[1] Still, though it contains more than a grain of truth, Voltaire's blunt conclusion goes too far—at least for people. Humans are flexible in our range of options for obtaining resources and bringing down oppressors, including aggression, tolerance, and collaboration between groups. How human societies choose between these options is the subject of the current section. In focusing first on murder and mayhem, this chapter, more so than the last few, returns to nature to see what insights can be gleaned about human behavior. In our morbid captivation with the dark side of our past, what we really want to know is whether human societies are fated for violence.

Before Goodall witnessed Gombe's killing spree, documented conflicts among chimps involved animals *within* a society. Males were known to fight, at times to the death, over social position; and females, it was subsequently discovered, were capable of killing the infant of a rival. It wasn't simply that no one noticed violence between chimp societies. Most thought societies didn't exist; they weren't aware that chimps lived within strict territorial borders, let alone that they protected their space ruthlessly.[2] You may experience a sense of déjà vu: the primatologists' initial ignorance of chimpanzee communities recalls the entomologists' original view of the supposedly peaceful Argentine ants, before the ants were found slaughtering each other along territory borders. Because society membership is not always visible in daily life, societies, profoundly important as they are, can be easy to miss.

ACTS OF EXTRAORDINARY BRUTALITY

Aggression between societies is unlike anything normally seen between fellow society members. Within a chimpanzee community, aggression mainly involves fights between individuals, or sometimes several allies ganging up on a victim. Fights between groups within a community, involving many individuals on both sides, haven't been reported. A party of chimps, for example, may be cautious upon drawing near another party of their own community but are never outright hostile. Foreign societies are the targets of group violence.

Among chimpanzees, such violence is unleashed in raids on the neighbors. Their fission-fusion social structure makes the animals vulnerable to attack. Raids stay away from any parties that, by the sound of them, are large enough to put up resistance and instead single out whatever target, male or female, they happen to find alone. These attacks seem to be the goal of the raiders. The apes don't appear to be hungry and don't stop to feed. To keep the enemy from staging such stealthy assaults and getting out unscathed, patrols keep vigil along the borderlines, at times moving quietly, other times with bravado. Both patrols and raids are almost entirely composed of males, the hypercompetitive sex focused on territoriality.[3]

The raids come out of the blue. By killing off foreigners, raiders can over time weaken and sometimes eventually eliminate the targeted society, as occurred at Gombe. The long-term consequence is that the aggressors expand the society's realm into the adjacent territory, improving the community's access to food to raise young and attracting more females—even adding a surviving female or two from the losing group.[4]

There are animals for whom conflicts between societies are tests of strength that as a rule fall short of violence. Ringtailed lemur troops face off with females cuffing and lunging and giving a ruckus of cries, while the males flutter their tails to waft an intimidating scent. Meerkat clans leap face to face in a war dance, their tails erect. Even in these species, though, matters escalate if the two sides are equally matched and neither backs down. Losers are wounded or killed and sometimes forfeit their property. A few other species are more like the chimpanzee in their unrestrained violence. Fights between spotted hyena clans and naked mole rat colonies can be bloodbaths. Closest to the chimp in attack strategy is the New World spider monkey, another fission-fusion species. The males join up to raid the neighbors, taking the unusual measure for a treetop animal of tiptoeing single file on the ground as quietly as chimps on a raid.[5] Yet it's the gray wolf that most closely matches the chimpanzee for pure viciousness. Wolves regularly kill members of foreign packs, often in the course of brazen invasions of another territory in search of game.[6]

Wolves don't dispatch members of outside packs with the quick chomp to the neck they employ to take down elk. While visiting the wolf researchers based at Yellowstone, I learned that a pack had just killed an old female and her companion from another pack. Both died from abdomen and chest bites apparently inflicted over many hours. As for the Gombe violence, "There were gang attacks of extraordinary brutality," Goodall would recall. "They did things to their fellow chimps that they would never do within a community but which they do when trying to kill a prey animal."[7] That's an understatement: the viciousness of chimps and wolves toward foreign animals can exceed what they employ to bring down prey or kill rivals within their own community.

Violence is far less common in humans than chimpanzees, which experience aggression every day.[8] That said, conflicts between human societies reach extremes of depravity and probably always have. Evidence of early wholesale slaughter among hunter-gatherers comes from the ancient cemetery site of Jebel Sahaba, northern Sudan, where 58 men, women, and children were buried approximately 13,000 to 14,000 years ago after each was impaled with 15 to 30 spears or arrows, more than necessary to kill a person, suggesting a community was brutally razed.[9] Accounts also exist of Aborigines slaying each other in outright battles, one involving 300 individuals. An early European traveler recounted both sexes "fighting furiously and indiscriminately, covered with blood . . . without intermission for two hours." In the end, the winners tracked the losers to a camp, where they beat them to death. The account continued: "The bodies of the dead they mutilated in a shocking manner, cutting the arms and legs off, with flints, and shells, and tomahawks."[10] Warriors across history have made trophies of their victims' body parts, from shrunken heads to scalps to genitalia, often to strengthen the essence of their own people by capturing the life force of an alien group.[11] The pincushion-worthy surfeit of arrows and grotesque mutilations bring to mind the wolf killings at Yellowstone and chimps mauling outsiders more hideously than if they were prey animals. An otherwise psychopathic quality of violence is normalized when denigration turns to demonization. What would at any other time be condemned as heinous becomes reason for celebration.

As it is in other animals, group identity is key to understanding when and why aggression turns, literally, to overkill. Neighboring bands of roaming hunter-gatherers more than likely belonged to the same society and would have shown no animus.[12] Of course, not everyone got along, and conflicts between individuals could turn nasty, but whole bands didn't hold grudges or regard the other bands of their society as hostile. Group aggression was generally directed at other societies. The extent and form of this violence has long been a contentious question in anthropology. What's certain is that the nomadic hunter-gatherers shied from high-risk engagements.[13] Their situation was comparable to that of ants with small colonies, which similarly have no permanent structures and few possessions to protect: when outsiders threatened, it made sense to simply move away.[14] The nomads carried out more dangerous actions only in times of crushing competition and conflict, as we saw when =Au//ei Bushmen turned to battling aggressive neighbors in the nineteenth century. Massacres such as the one at Jebel Sahaba would have been rare for people in bands—there was probably a settlement of hunter-gatherers at that site. For human nomads as for equally fission-fusion chimpanzees and spider monkeys, furtive strikes were the preferred choice.

Raids were often justified as retribution against perceived wrong-doing—maybe witchcraft or territorial incursion by the other side.[15] Even if the raiders could identify the specific person who had carried out the offense that spurred their attack, they usually targeted whom-ever they could easily lay their hands on. Indifference as to the choice of victim was a convenient outcome of seeing outsiders (in the case of nomads, probably ones the aggressors knew) as identical and inter-changeable. It also enabled the interlopers to get in and out quickly and unscathed. Once a people are reduced to a category, they are all equally targets. The biblical prescription of an eye for an eye does not distinguish between eyes; injustices inflicted by just one or a few for-eigners are viewed as righted with attacks on anyone of their "kind." Where this social substitution backfires is that the victims are viewed by their compatriots as unique and blameless persons. The suffering inflicted on those victims is felt as harm to all; thus retaliation is inev-itable.[16] No animal we know of directs aggressive reprisals at foreign groups in this way.

Given that hunter-gatherer bands had few material goods, what were the payoffs of these engagements, other than the immediate sating of bloodlust? In human societies, as in chimpanzee societies, the aggressors were predominantly males who wielded control of a private domain that (framed from a biologist's perspective) produced resources essential to raising offspring. One valued resource was the women who birthed the children, and women could be seized during raids. In addition the territory itself could be an asset worth taking. Descriptions of hunter-gatherers annexing territories are scarce, how-ever.[17] A detail provided by the anthropologist Ernest Burch about the Iñupiaq Eskimo gives a clue as to why: "The great majority of people thought that their own estates were the best places to live, and they could hold forth at length on what was special about them."[18] Per-haps a grass-is-always-greener perspective seldom made sense in tradi-tional societies where people grew up wise to every nook and cranny of the home turf. Coveting a neighbor's land would be illogical unless it offered a huge improvement over a region already intimately un-derstood. Nonetheless, raids could continue until a rival group was abolished, and no swath of ground would have gone unclaimed for long. And as tiny as band societies seem to us now, their relative sizes must have been critical to their success. Bigger groups could have displaced smaller ones even when they chose not to make a show of their might—the motto "peace through strength" taken to its prehistoric fountainhead.

VIOLENCE AND IDENTITY

Another payoff of aggression lies in reinforcing unity. "Are wars . . . anything but the means whereby a nation is nourished, whereby it is strengthened, whereby it is buttressed?" the Marquis de Sade wrote after the French Revolution, bringing to mind the Nazi rallies at Nuremberg.[19] When the members' rancor is directed at foreigners, their identification with their society heightens; their sense of common purpose and shared fate, of the people rising as one, is invigorated.[20] In the Yankee North of the United States, it took the Civil War to forge that sort of oneness. "Before the war our patriotism was a firework, a salute, a serenade for holidays and summer evenings," Ralph Waldo Emerson mused. "Now the deaths of thousands and the determination of millions of men and women show that it is real."[21]

William Graham Sumner, the sociologist who coined the term "ethnocentrism," wrote in a much-discussed passage a century ago: "The exigencies of war with outsiders are what make peace inside, lest internal discord should weaken the we-group for war."[22] To Sumner, external war and internal peace play out a horrid game of interdependence. Any competition and conflict with outsiders redirect people's attention from their competition and conflicts with each other and toward their identity as a group.[23]

Whether or not violence toward outsiders is obligatory to keep a society intact, it is well understood that contrasting ourselves to outsiders, and primarily those we cast as enemies, helps place our society front and center in daily life. We are drawn together by the urge for self-protection. The Israeli psychologist Daniel Bar-Tal tells his readers that every society picks out groups to "serve as symbols of malice, evil or wickedness," whose threat, even if real, tends to be played up.[24] In recent years Russia, North Korea, and Iran have taken on that role in the United States. Should an adversary be lacking, people may flounder until a new one is found—or invented. We rise up as one against terrorists, asylum seekers, illegal aliens taking jobs, or members of our own society thought to have false beliefs, shifting our ire from one to another with ease. When this bad blood is deeply embedded in a group's self-identity, it becomes too precious to relinquish. Such intractable positions are held by many Israelis and Palestinians; each side has come to function with exceptional solidarity and staunch commitment to recognizing their differences.

Making matters worse, our flawed capacity for assessing risk means that our responses to outgroups are frequently overreactive. Part of the problem is the way people cherry-pick information, a proclivity that doesn't bode well for national or ethnic relations. We are more likely to remember a foreigner doing our society harm than one performing an equal measure of good.[25] News of terrorists triggers our prejudices even though dying from an act of terrorism is much less likely than a fatal slip in a bathtub. Our touchiness makes sense when you consider that the human mind would have evolved to discern anything motivated to hurt the people who came together in the small groups of earlier epochs. Our hypersensitivity to such malicious hazards—a holdover from the distant past—can set humans too hastily on the warpath.[26]

Then again, maybe anxiety about physical harm isn't the sole cause of our trepidation. In understanding our fears, we should not underestimate the role of identity. That includes our grasp of the identities of disliked or dreaded outsiders. Humans are susceptible to latching on to negative stereotypes even when they have no basis in reality (perhaps explaining the belief once commonly expressed among tribes that the feared people of the next tribe over were cannibals).[27] Intense emotions can be bolstered by potent aspects of human markers, especially objects of symbolic power and how they are treated.[28] And so it is that if foreigners show deference to our flag, we register them as trustworthy and are warm in return. Conversely, any perception that a symbol is mistreated enrages: consider the clamor that arises when our leader is beaten in effigy. Our attachment to emblems of identity suggests that the reason we panic about terrorism and not about tubs is not that terrorists are likely to endanger us as persons but that they can harm objects of symbolic importance to our society. Think of the Twin Towers and the Pentagon. Fears of a repeat of 9/11 have put group emotions on high alert.

MOVING AWAY FROM VIOLENCE:
LESSONS FROM NATURE

The question of questions remains as to whether Alfred Lord Tennyson's characterization of nature, "red in tooth and claw," applies to societies. Does structuring the world into groups inherently lead to alienation between people? This is indeed the case for ants, which show no alternatives to conflict between the societies of their species, invariably battling or scrambling for resources before neighboring colonies snap them up. One might expect it to hold true for people too, for a simple reason: there's only so much room for the societies of this world, just as there's only so much space for individuals. If societies thrive from the competitive edge they afford their members against outsiders, it may seldom pay to willingly relinquish anything to outsiders unless the act entails some obvious reciprocal gain. Hence the Latin proverb *Homo homini lupus*: man is wolf to man.

Animals can let down their guard around outsiders under some circumstances. Something approaching harmony has even been documented in the pugnacious gray wolf. Wolves, like spotted hyenas, can pass through foreign territories to pursue migrating herds. Admittedly it's hard to tell if this is politeness by the owners or sneaky trespass. More convincing is the behavior of wolves at Algonquin Provincial Park in Canada, where 40 years ago the packs initiated a tradition that's been seen nowhere else. Instead of occupying their territories year round, the wolves follow the exodus of deer each winter to a small plot of land near the park called the deer yard. Here, the deer become concentrated enough to feed all the wolves. The packs don't exactly turn convivial, yet at tenfold their usual density, they live in peace. In one instance two packs joined for a time without incident; three packs once ate, with no fuss, from the same carcass over the course of a day. Such cautious indifference can be considered a minimalist form of partnership. It's also a testament to the flexibility of an ordinarily xenophobic species that we assume operates on instinct.

Among the species celebrated for their friendliness, the savanna elephants and bottlenose dolphins stand out. In the elephants, the affection is typically most intense in cores that share a history, having split off from each other in the past. Members of bonded cores perform greeting ceremonies, rumbling and trumpeting while flapping their ears and spinning in place. The members of one core know those of the other, and certain individuals are tight friends. The dolphins in Florida enter each other's communities and interact in a friendly way. But not all is honey and roses even for these species. The battle scars of male bottlenose dolphins have been attributed to border skirmishes. Savanna elephants steer clear of cores they don't know or don't like by eavesdropping on the others' infrasonic calls. Often a big core, or one with a strong matriarch, will displace a weak one from a tree or water hole. These power plays can affect the health and reproduction of the individual members and the survival of the societies themselves.[29]

Most renowned for peaceful relations between societies is the bonobo, with communities openly intermixing. Each community has a home range, yet rather than being rigidly defensive about these spaces like the chimpanzee, bonobos pass through to make social calls, the kids tagging along. Bonobos even prefer to give food to strangers over community members, a remarkable indicator of how enthusiastically they build foreign relations.[30] That said, bonobos are hardly cavalier about their social visits. While callers don't sneak in like raiding chimps, territorial borders exist and are crossed with due caution. The residents may respond to their appearance with frantic chases and screams, bites, and scratches. Everyone usually calms down, but should a social mixer not be in the cards, the visitors go home. Moreover, certain communities are never seen together. Much the way two individuals can be on bad terms, some societies apparently have their irreconcilable differences. Under these circumstances the bonobos do stick to their side of the territorial lines.[31]

Such caveats aside, why are bonobo societies generally so permissive? The rarity of violence has been attributed to the general abundance of rich sources of food in their habitats.[32] If that's the case and bonobo communities are only fair-weather friends, good relations could be rescinded just as the truce among Algonquin wolf packs would collapse if the winter deer population didn't fill the stomachs of all of them. During intervals of conflict, the merits of being in a strong bonobo community would become obvious. Luckily hard times seem to be few and far between in the part of the Congo that all bonobos call home. In any case, bonobos are fully capable of putting up a fight. "Veterinarians sometimes have been called in following altercations to stitch back on a scrotum or penis," anthropologist Sarah Hrdy dryly reports of captive bonobos.[33] In addition to the occasional assault, they may kill in the wild. In one case several bonobos ganged up on a male of their own community. Researchers suspected a murder, though no body was found.[34]

Even in some of the least violent species, the most the animals manage in terms of foreign relations is to ignore or avoid outsiders, as the wolves in the deer yard do. Sperm whales live among foreign clans but stay out of their way—they are so massive that any clash could be life threatening. The geladas are masters of disregarding outsiders: their troops intermingle in a show of indifference, aside from the caution the alpha males display around bachelors out for trouble. This may tell us only that among these primates, which eat abundant grasses, there is little or no competition over food.[35]

Species usually eager to fight outsiders can have their moments of tolerance, too. When a lion pride splits up and each half takes over part of the original territory, they appear to give the others breathing room to settle in. Even so, within a year or two the former pride mates are just as hostile toward one another as they are to complete strangers.[36] Baboon and mountain gorilla troops intermix and let the youngsters play together when males have no sexually receptive females to come to blows over. Prairie dogs allow for a détente in communal foraging grounds outside their territories on land too poor to dig burrows, and therefore apparently not worth defending.

And what about the chimpanzees—can the territorial males reverse the demonic nature that Wrangham attributes to them and coexist with neighbors? When it comes to accommodating foreign communities, this species may be at the bottom of the list. The best that can be said is that some populations raid and kill foreigners less often. Their relative restraint has more to do with lack of opportunity than peacemongering, however: the chimps in those areas stick to big parties and consequently are seldom vulnerable to attack.[37]

Given the genetic proximity of both chimpanzees and bonobos to humans, it seems reasonable to conclude that our base level of distrust of foreigners, and appetite for doing them harm, is part of our shared heritage with the chimp, while the ability to put aside our qualms and bond, therefore, is a gift we have in common with our bonobo relations. Our two cousin species become contrarian angels standing on our shoulders, whispering their good and bad advice. Fortunately the human species has cut down on the impulsive sort of violence common in chimpanzees, our greater emotional control and toleration of each other an adaptation we share with bonobos.[38] Yet the message from nature is hardly optimistic. Time and again we find that animal societies can be on good terms, or at least not harmful to each other, when conditions suit such behavior: when there's little competition for property or mates, or when fighting over them is not worth the trouble. Such ideal situations are seldom lasting. In Algonquin wolves, for example, violence returns as conditions go downhill, after the deer spread out again for the summer and become more difficult to hunt.[39] Once the packs are compelled to compete over food again, the wolves reestablish their nastier habits.

Considering the state of affairs we see across species when competition is intense, and the human response to competition and clashes of identity with foreign powers, it seems fair to assume that equanimity with the neighbors will be a perennial struggle. Plato is said to have claimed that only the dead have seen the end of war, and no doubt that's true. War can unite people, as William Sumner described, and as troubling as that seems, a society unable to draw on this selective focus would be defenseless in times of peril.

Nevertheless, this much is certain: societies don't need to be antagonistic toward outsiders any more than they require their members to cooperate within their borders. Inimical societies spend most of their days not fighting, even when their peace is uneasy. They may not have a high regard for the other people, how those people live, or their symbols, yet they aren't forever burning each other's flags, either. At times that's all two societies can achieve—the forbearance to set aside disputes and go about their business like the wolves in the deer yard. Such arrangements have worked out during periods of intense conflict, if only briefly, an example being the Christmas truce of 1914, when the German and Allied troops felt free to enter the no-man's-land along the Western Front and sing carols and drink together for the day. Even seeing foreigners as beneath us doesn't need to equate to hostility, as is attested to by the macaque. I have mentioned that these monkeys promptly connect an outsider with vermin (*horrors, a spider!*); the usual outcome of this bias is that the troops give each other space rather than go on the attack.[40] The negative stereotyping and species-like distinctions between societies that come so naturally to humans don't force a violent response out of us, and perhaps this monkey-like lack of intolerance, so to speak, was a steppingstone to the first rudimentary human alliances.

While a common adversary can unquestionably motivate people to rally around their society, war is a tactical choice, and we can have enemies we feel no compulsion to kill. Indeed, the fact that international alliances occur at all is proof positive that being loyal to, and favoring, one's own society is a psychologically independent act from thinking poorly of outsiders. One can exist without the other.[41] So long as their mutual antipathy is not deeply set, the cooperation between human societies can surpass anything observed in nature. Up next, the conditions under which such warm relationships thrive, and what that tells us about humanity.

CHAPTER 18

Playing Well with Others

L ike many explorers, I've had the experience of trading fishhooks for something I needed—in my case, a dugout canoe
made by a tribe in the rainforest near the Pacific coast of
Colombia. Nothing seemed remarkable about closing the deal, even
if it did require downing a fermented beverage that wasn't exactly
to my taste. Yet however ordinary that transaction may have seemed,
given everything we know about strife between societies in the natural
world, this willingness on both sides not only to forgo hostilities, but
to see a foreigner as other than toxic, is pretty special.

As you may have surmised based on the verdict of the previous
chapter, any positive contacts between animal societies are problematic, and actual collaborations between them are rare or absent. Oftentimes their positive interactions look more one-directional. Rather
than representing one side of a trade or agreement, the bonobo's
generosity in giving food to strangers is better thought of as an olive
branch. The apes don't appear to expect anything from the foreign
animals in return other than, perhaps, their tolerance—let alone to
team up with them to get anything done as a group. Savanna elephant
and bottlenose dolphin societies similarly show no evidence of moving
beyond camaraderie to partnership. The arrangement closest to an alliance exists in sperm whales, which team up with other unit societies
to catch squid more effectively than either unit could alone.[1]

Human societies can accomplish a greater payoff by working together rather than against each other, mitigating any competition
over resources, a feat that other animals seldom achieve. The societies

of our species can, for example, turn scarcity into plenty by extracting more from the environment with outside help (just as two whale units do when hunting squid). In the face of the very real risks posed by foreigners, it's worth asking how and why humans evolved these enterprises. And once they did, the question becomes how they were able to balance the demands of being in an alliance considering the pressures for their societies to remain separate, and distinctive. Evidence from hunter-gatherers guides us in addressing these issues.

THE VARIETY OF ALLIANCES

One of the most remarkable collaborations to be found among hunter-gatherers anywhere was the one between the eel-collecting Aborigines of the Mount Eccles region of Australia. At least five groups speaking different dialects of Gunditjmara, and quite likely other peoples of the region as well, constructed the expansive waterworks that made it possible for all of them to catch the fish. Their relationships weren't without bloodshed. Sometimes wars broke out. But compared to the battle-prone Indians of the Pacific Northwest, similarly dependent on fish runs, the Australian tribes treated each other almost collegially, and with a clear mutual payoff: everyone's success relied on their labor to maintain the waterways. Eel harvesting was an international effort.[2]

Some hunter-gatherer alliances involved the movement of personnel to achieve a mutual goal, usually defense. Although the behavior is unknown in other animals, combining forces against a common enemy has been a justification for binding human societies, much as it can be for binding the members of a single society when outsiders threaten. The American Indians collectively called the Iroquois, hunters with rudimentary farming who occupied what is now western New York State, hammered out a confederacy before the arrival of Europeans, sometime between 1450 and 1600.[3] Member tribes were self-sufficient and independent; however, a grand council met when necessary to direct their relations and coordinate defense against outsiders, Europeans eventually among them.[4] Beyond that, the tribes collaborated in an ever-changing manner to suit their interests, as tribal and hunter-gatherer peoples were able to do elsewhere in North America and around the world.[5]

For humans, the creation of plenty by way of intergroup relations is often accomplished through trade.[6] The possibility of this arrangement is universal. The Arawak presumed that Columbus, though from a society far outside their experience, would be willing to swap goods. Unfortunately, in seeking mutual gain they found their own ultimate demise, slain by forced labor, musket fire, and diseases until they vanished as an independent entity. Lewis and Clark passed through the territories of Native Americans who were normally hostile to outsiders and might have had no compunction about killing them—except the Europeans presented themselves as prospective trade partners. To sustain these associations over the long term requires a delicate interplay among the parties. Ideally, they must see each other as independent but on a par in strength and importance. Either that, or the stronger group can dictate the terms of the arrangement to the disadvantage of the others.

The Arawaks' fate reminds us that contact with foreigners is a fraught business. Humans have always turned first to those of their own society, whose shared identity makes social and economic interchanges easier to navigate. There's greater risk of duplicity or misunderstanding in negotiating with those whose ways and values, and, of course, language, are foreign. Judging from research on interactions between races in modern societies, getting past the differences may consume enough mental attention to make mistakes likely.[7] Adding to the anxiety is the possibility of rejection or reprisals should things go sideways.[8] Self-sufficiency is desirable. Everything, then, from finding a mate to procuring meals to keeping enemies at bay, tends to be handled internally whenever possible.[9]

Nor do societies have to be open to trading and cultural exchanges. Some Bushmen cooperated with neighbors readily, some did not.[10] Think of the dividing line between societies as a bottleneck to regulate the movement of people, information, raw materials, and prepared goods. All these assets move reasonably unimpeded within a society, but in a regulated, often sharply curtailed flow among societies. Once in place the bottleneck can be widened or narrowed depending, respectively, on the payoff or loss brought about by contact. When a foreign society was disliked, its influence felt as a blight on the local culture, the bottleneck could tighten. Even so, acceptance of desirable goods and innovations, whatever the source, was unstoppable, which could account for the wide spread of newly invented stone tools in the early archaeological record.

KEEPING THINGS MOVING

It's hard to say when the first collaborations existed between human societies. Not only are hunter-gatherers known to have traded with farmers for centuries, they have long sought out commodities from each other's societies as well. Unfortunately, alliances, and even trading, don't leave clear evidence. Finding a tool chipped from rock that could only have come from a site many kilometers away may be used as evidence of trade between societies.[11] Yet the implement could have arrived there by other means. Rather than being acquired in a transaction with outsiders, possibly the same intrepid person carried it the whole distance, or it was exchanged between members of one society whose territory covered that area.

Or perhaps its migration occurred over a series of steps from one society to the next, in what's called a transfer chain. Illustrative are ancient pieces of Chinese pottery that switched hands until they reached villages far off in the interior of Borneo.[12] Even then trade doesn't always explain the movements. *Coenobita compressus*, a type of hermit crab, shifts to a larger shell each time it outgrows its home. Oftentimes the shell it chooses had been previously jettisoned by another crab, much as one person may carry off a bauble dropped by another. The shells move under crab power 2,410 meters a year on average, one fusspot crab to the next. Scaled to human dimensions, that's equivalent to each shell voyaging a kilometer every single day.[13]

Nuts buried by rodents are an example of a transfer chain, in this case propagated by theft. Squirrels steal acorns that others have buried for later eating. In this way, a nut bounces from one spot to another for kilometers until it sprouts or one of the thieves eats it.[14] Some ants and honeybees rob other colonies, grabbing food from foreign workers and dashing off.[15] These examples remind us that stealing is a time-honored alternative to trading as a means of moving objects over the landscape, one that predates humankind.

While we don't know whether early human societies depended more on commerce or plunder, brute force would seldom get people what they craved on a reliable basis. This reflects that human necessities are more complicated and diverse than those of squirrels or honeybees. Once people started to rely on goods like arrowheads and body paint, locating adequate food and water was no longer their sole concern. Not all the merchandise a society fancied was likely to be available on the real estate they controlled. While people met whatever material needs they could locally, building relationships with foreigners open to long-term commerce became, in many instances, not a luxury or token gesture of friendship, like a bonobo's gift of tidbits to an outsider, but a requirement.[16]

Still, commerce may have begun with something very like the bonobo's easy and amiable contacts with outsiders. Even before that, perhaps, mere mutual tolerance was sufficient to enable the flow of goods in both directions. Band societies often acquiesced to outsiders entering their land to harvest a desired resource themselves. Not that it was ever possible to completely stop trespassers. While a blackbird able to scope out every corner of its small domain from a central perch can see and try to block all incursions into its private space, the dominions of the societies of humans and many other animals are too expansive for that. A meerkat clan will go so far as to brazenly sleep in one of the burrows of another clan when its owners are elsewhere. At best the residents can confront the interlopers they discover, repulsing them most zealously from the interior of the territory, the geographic center that in many other species contains the den.

But hunter-gatherers were canny at reading the environment. American Indians and Bushmen not only detected days-old footprints, they could identify the sex, age, and often the exact person who made them.[17] Intruders were likely to be found out if only after the fact. That made it prudent to avoid retribution by seeking approval to enter the land of foreign groups. In any case outsiders often couldn't simply creep in and find what they wanted—they depended on the locals' up-to-date advice on when and where to go. Thus guarding the land was usually unnecessary as well as impractical.[18] Permission to access resources could be sought to solve an urgent predicament. Accounts of hunter-gatherers are full of stories of water holes drying out and of game migrating from one territory into another. In any case, the tolerance for visits came with conditions. Both sides were expected to return the favor on a continuing basis, and a reliance on such reciprocity kept people on their best behavior.

Depending on the cost and benefits of defending or granting access to resources, hunter-gatherers differed in their possessiveness toward their realm.[19] Some took the take-no-prisoners approach of chimps, others selectively protected resources as baboons do, while yet others were as open to outsiders as bonobos—with the possibility of negotiating all kinds of concessions in between. But rarely was there ambiguity about who owned what. Precious assets, among them materials venerated for their symbolic importance (a pigment used in a ceremony, perhaps), could escalate rivalries. In general, though, the low cost of negotiating an easement added to the reasons not to seize foreign territories. This has always fostered contact, and familiarity, between human societies.

Those with a superabundance of a valued resource could afford to be generous. Bogong moths make annual flights up Australia's Snowy Mountains in such numbers that when the time was right, foreigners visiting for the season were given spots on the slope where they caught and consumed around a kilogram of insects per person each day. The tradition lasted a thousand years. That weight in moths has the fat content of thirty Big Macs. By season's end the formerly scrawny people headed back to their home territories plump and content—and one imagines with the expectation that they owed the locals something in return.[20]

THE FIRST MARKETPLACES

Regardless of whether tolerance of outsiders entering one's territory to harvest goods was the forerunner of trade, trade in the usual sense began when parties swapped goods face to face. The first simple markets may have been arranged at territory borders to respect each side's private spaces while allowing spot checks to ensure that exchanges were equitable.

But what was fair? Dealings among people from the same band society customarily took the form of casual exchanges.[21] There was enough trust that exact parity wasn't expected, much the way Christians exchange holiday gifts. Should one side come up short, they could make it up next time. Trade between societies was different, being less predictable, with more haggling and oversight and risk of relations turning sour.

The easygoing attitude toward exchanges within societies faded as settlements grew. When traders usually offered very different goods and services and were strangers or barely knew each other, specific values had to be affixed to their offerings. As a result, interactions within societies became more on par with commerce between societies. Settlements of the Chumash hunter-gatherers of California used beads as a form of currency, setting value to goods in the modern sense.

For recent hunter-gatherers, transfer chains fueled primarily by trade accounted for wide connections between band societies.[22] Such items as medicinal plants, grinding stones, and ocher moved from one Australian group to the next, sometimes across the breadth of the continent. Their value grew with distance, as was true for American Indian goods, too.[23] Pearl shell may have seemed magical stuff when it arrived far inland for use in ornaments. Some items weren't used as originally intended. Boomerangs ceased being made in northern

Australia centuries ago. Southerners who continued carving them were eventually able to swap the boomerangs for other goods when a fad swept across the north to employ these missiles not as weapons, but as percussion instruments for musical events.[24]

In addition to raw materials and worked-up products, band societies traded ideas. Anything from a trendy word to an improved technique for making a tool could be copied across great distances. The circumcision performed during initiation ceremonies on Aboriginal boys was probably learned from Indonesian traders in the 1700s. The procedure gained traction over a wide expanse of Australia, some taking the practice to extremes by slicing the penis its full length.[25] The Aborigines also copied each other's songs and dances. Best documented was an example first reported in 1897. The Molonga ritual of the Workaia people entailed several long nights of fantastical performances, the key characters in elaborate costume. Over the next 25 years, the Molonga spread across a 1,500-kilometer span of central Australia, even though only the Workaia understood its words.[26]

For societies to interact without a hitch, some awareness of a social connection was a boost. Hunter-gatherers often arranged marriages between their societies for the purpose of courting allies; spouses usually had opportunities to visit home, giving them the equivalent of dual citizenship, something unheard of in other animals.[27] Understanding each other was key. It helped that as a consequence of the history of intergroup relations, many hunter-gatherers spoke the languages of their neighbors. Both Australians and the Great Plains Indians also widely shared sign languages used in diplomacy. Some of the gestures were visible from afar so that negotiators could communicate beyond the distance a spear could travel.[28] The same gestures had a second function: warriors on a raid could signal to each other to soundlessly coordinate an attack.

TRADING AND CULTURAL DIFFERENCES

Just as individuals interact more comfortably the more they have in common, the same is true of societies: similarities ease the way to friendship.[29] For example, similar languages and compatible cultures simplified the partnership of the Iroquois. Archaeologists speak of the interplay between societies as an "interaction sphere," in which comparable values and other aspects of identity facilitate the traffic of goods.[30] The very act of trading further enhances the resemblance between the societies. This is decidedly true when the transferred goods are more than simply raw materials, such as when the groups trade new ways of making or doing things or the manufactured items themselves.

Yet societies have to stay sufficiently distinct to preserve their members' sense of worth and meaning, or so the findings of psychologists would suggest. Herein lies the balancing act that has influenced the course of so much history. Commonalities are a plus—to a point. Too much interchange can be seen as a threat a people's unique identity. Perhaps compounding the problem, being alike can also backfire should societies find themselves desiring, and coming to blows over, the same scarce items.

Earlier I introduced the theory of optimal distinctiveness, in which individuals strive to be enough like other members of their society to earn their respect yet at the same time seek to be different enough to feel special. A reasonable hypothesis is that in building relationships with their neighbors, societies similarly gravitate to this middle ground—an intense bonding that arises from the solace brought on by their similarities, and pride at their distinctiveness. To be a robust society, or a well-adjusted person, is to be the same *and* different. Even the most look-alike societies must preserve hallmark differences close to people's hearts.

Reducing overlap among societies could correspondingly diminish competition; so goes a theory that has been proposed to explain, for example, distinctions in diet between tribes living in close proximity in the Amazon.[31] And a desire for cherished differences could be fulfilled productively by the emergence of distinct economic roles between societies. After all, parties offering identical wares have little reason to trade. A society could exchange a surfeit of tools made by its members for items they found too arduous to produce themselves. Given the versatility necessary in a hunter-gatherer band, whose members honed generalized skills that differed only by sex and age, specialization among people may have originated at a societal level before working its way down to the individuals within societies.

While such differences in aptitude were not ubiquitous between hunter-gatherer societies, the evidence suggests they often occurred. "Each locality tended to make certain objects with a skill or flair which was admired in other localities," Geoffrey Blainey, an Australian historian, writes of the Aborigines. Different groups crafted spears, shields, bowls, grinding stones, jewelry, and so on. Indeed, Blainey adds, "much of the specialization had existed for generations, and its origin was even the theme of tribal myths."[32] In the Pacific Northwest, the blankets woven by the Chilkat Tlingit and the adze blades of certain other tribes were exchanged (or stolen) along the length of the coast. Numerous other examples of reciprocal relationships have been documented between small tribal societies dependent on domesticated food. For instance, the Fur agriculturists of the Sudan supplied millet to various cattle-herding tribes in return for milk and beef.[33]

By achieving an optimal level of distinctiveness, societies would, firstly, become interdependent over long spans of time, their peoples finding all the more reason to avoid acrimony and seek each other out for their mutual benefit. Though to my point here, each society would preserve its boundaries regardless of how much its people were exposed to outsiders or how much they swapped with them or otherwise relied on them. That was true whether—to consider two examples given from Australia—they were barely scraping by in the Western Desert or were well-fed fishermen managing the waterways full of eels around Mount Eccles. No nation was ever ruined by trade, Benjamin Franklin said, and this is so not just economically but socially.[34] The Mandan and Hidatsa of the North American Great Plains maintained clear identities even as their cultural centers transformed into trade hubs, a development that forced other tribes to learn their languages.[35] As for the Iroquois, the confederation had to be loose enough that its tribes didn't cede autonomy or land. In fact, there was little contact between the common people of different Iroquois tribes and their separation persisted, and arguably strengthened, in spite of their interdependence.

Dissimilar peoples find ways to profit from each other, too. In fact, the path can be expedited the more radical their distinctions. Australian Aborigines welcomed the Indonesian fishermen who visited the northern coastline in the eighteenth century, while Bushmen traded goods with the Bantu herders who have lived among them for two millennia.[36] Pygmies and their farmer neighbors took things a step further by codifying a relationship that bolstered the survival of both in a forest habitat where filling one's belly can be a struggle, whether by tilling poor soil or hunting scarce jungle prey. Subsisting as hunter-gatherers much of the time, every Pygmy group became associated with a village. There, each Pygmy kept a lifelong relationship with one of its farmers, working his fields for part of the year and providing him with bush meat and honey in return for crops and other goods. Connections between Pygmy and farmer are so ingrained and ancient that some farmers believe the Pygmies first introduced them to the forest.[37]

One proposed role of markers is that they keep people from imitating outsiders when foreign ways could inflict damage.[38] This I doubt very much. Of course, an outside influence can be detrimental, as when harmful drugs cross a border. But even if neighbors differ sharply, both tend to adopt from the other what suits them, without the interchange bringing about a calamity. Indeed, people can turn the mismatch into a boon. Certainly the problem of communication with a dissimilar other could present an imposing handicap at first. Still, their peoples would likely have vastly different material needs; even if one considered the other its inferior, they might not be rivals for the same things. In fact, their ways of life and skill sets, such as those of the Pygmies and the farmers, could be complementary.

The similarities and differences between societies, great or small, must factor into people's views about outsiders—for example, their warmth and competence. That would in turn influence the extent to which the outsiders are seen as a formidable, potentially hostile entity or as persons who can be negotiated with in good faith. Once early humans refined their options for playing well with others, what emerged from these assessments were interactions between societies that could be as varied, fine-tuned, and adjustable over time as they are today. The constituent tribes of the Iroquois had been violent toward each other prior to their union; indeed, the peace between them was brought about only through war. As one expert on them grimly noted, "At times the best way to make somebody stop fighting is to fight him till he does stop."[39] After the Iroquois had negotiated an armistice, however, the tribes farther afield took their turn at feeling nervous. Harmony between societies can ironically bring about violence across a region, by establishing a more dangerous opponent for those left out of the entente.[40] One enemy replaces another.

The neural wiring of the amygdala remains in place to fire its age-old cautionary reflexes of fight or flight. Overcoming those base urges to establish mutual trust between societies prejudiced against each other is a thorny business, a core problem of diplomacy. Even for groups on excellent terms, sublimated biases ensure the playing field is never perfectly level, as each side jockeys for a better deal. Our collective identities embolden us to be self-serving and cutthroat, undermining good relations and creating fertile ground for the birth of enemies when times are hard.[41] Rivalries between groups do not create ethnocentrism; however, the competition draws out its more repugnant facets.[42]

How do we avoid conflict when resources and opportunities dry up? In recent centuries, even with mass atrocities factored in, the probability of dying from an act of aggression between societies has declined globally. Arguably peace is fostered by the increased contact between countries. Nations also draw ever more on talents and resources from beyond their borders.[43] Ideally, this interconnectedness and interdependence should see nations through periods of shortfall, at times when what tranquility exists between the societies of other animals tends to collapse. Avoiding violence when the potential social and material gains from war are high takes more than good intentions, however: it requires cultivating, and recognizing, the greater payoff, over the long term, of peace over conflict—even among hated adversaries. Whenever that minimum can't be met, every nation must be committed to act against those who refuse to abide by the rules that safeguard the international order. This is a lofty goal. Hopefully it's an achievable objective in light of the perils of modern warfare. In no other species do societies coordinate to preserve the peace.

The volatility of relationships between societies finds its parallel within a society, where bonds among people are never static. The members' identities undergo trajectories of change over the very long term that can be predicted in broad outline, a fickleness that bears on the rise and fall of societies.

The Life and Death
of Societies

CHAPTER 19

The Lifecycle of Societies

"We do not even know how to determine approximately the moment when a society is born and when it dies," the eminent French scholar Émile Durkheim lamented well over a century ago.[1] Despite the obvious practical and academic importance of the major questions regarding societies—how they are founded, how they develop, and how new societies replace them—no definitive answer has revealed itself since Durkheim's assertion in 1895. Durkheim pointed out that even the biologists of his time had resolved little about the life and death of societies. Although those lifecycles have by now been well studied for certain groups of organisms, the topic has indeed been largely ignored by the natural sciences as a subject of general inquiry right up to the present day. As for sociologists and historians, their reflex tends to be to treat the birth or dissolution of any society—be it early Egypt or the former Czechoslovakia—as an affair exclusive to its time and place.

Of course, the devil is in the details. Nevertheless, the rise and fall of societies across the natural world indicates that our social groups have been engineered by evolution to come and go, much as the bodies of individual organisms are. This ebb and flow of societies is linked to how members perceive the identities of others. The behavior of animals and the ways in which human beings formulate our identities in changing social environments are closely tied to societal loss and rebuilding. Related to this deep question of identity is the critical topic of trauma and whether it is a sad necessity in the lifecycles of societies.

The dynamics of society genesis and transformation play out in a manner unique to each species, constituting the basic historical chronicle for its kind. That narrative depends on the rules by which the members of the species interact and identify each other, and the resources available at a certain time. Yet a theme emerges: Life requires the constant fulfilling of needs—food, sanctuary, mates. When those needs aren't met, escalating physical and social stressors spur the decay of societies. More often than not, the troubles will be most severe when a society has overgrown what its environment can support. Even though a large society can overrun smaller neighbors, its swelling population aggravates competition among its own members, with the added burden on every individual to keep track of who is who—should the societies of a species have opportunities to grow to a size where that is a problem. The members' proficiency at managing relationships and coordinating activities declines.[2] This provokes a shift in allegiance to a subset of the society membership—splinter groups in which everyone fares better.

The separation of subgroups into independent societies is de rigueur in vertebrates. Several lionesses, for example, will leave a pride that's grown too big to feed all its members. Should an aggressive male join a pride, females rearing cubs fathered by another male may depart, splitting the pride to avoid infanticide by the newcomer. Lions in an oversized pride are forced to make a fresh start with those they know and get along with the best, as is typical of species that rely on individual recognition.[3] Such a severing of relationships differs sharply from the phenomenon I've been calling fission: the temporary and casual separation of a society's members that regularly occurs in fission-fusion animals—species such as lions, chimps, and humans where individuals are free to separate and rejoin each other. When a society divides, there's little chance of piecing it back together.[4]

A FRESH START IN THE CHIMPANZEE AND BONOBO

How new societies are born remains the most serious gap in our knowledge of our fellow great apes, the chimpanzee and bonobo. This seminal event is rare: vertebrate societies generally arise once in decades, if not centuries. The infrequency presents a problem. Data is scarce, for one thing. Worse, it makes it easy to shrug off incidents critical to the creation or destruction of societies, to ignore them as anomalies simply because they are uncommon. Such an occurrence might be a new arrival from outside the society or the death of a key animal in it. Either change can threaten the stability of the group.

An excellent example is the brutal conflicts among the Gombe chimpanzees documented by Jane Goodall in the early 1970s, puzzling at the time but which turned out to shed light on how societies split apart. Primatologists now realize what set off that spate of especially virulent nastiness: what had been one society had, under the watchful eyes of Goodall and her assistants, divided into two. The cleavage was the endpoint of a protracted process. A first indication that something was awry came in 1970, when some chimps were clearly associating more with each other than with the rest of the community to create two subgroups, which I will call factions. There's no evidence the factions had already been present, even loosely, when Goodall first came to Gombe ten years before. In any case by 1971 they had firmed up, one habitually occupying the north and the other the south part of the territory.[5]

Initially, members of those splinter groups mixed amicably when they did get together. The dominant males in the factions would furiously charge each other when they met, but that was nothing unusual, as contenders for alpha status within a community often challenge rivals. However, in 1972 the factions parted ways, establishing independent societies that no longer mingled. Goodall, recognizing that the chimpanzees had separated into two societies with distinct membership, named the communities Kasakela and Kahama. Violence began after that split, with the Kasakela apes raiding the weaker Kahama to the south, in the end obliterating the Kahama community and taking back much of its territory.[6]

The two-step process at Gombe—the emergence of internal factions followed by division—seems to be omnipresent among primates that live in societies, having now been documented for the troops of at least a couple dozen monkey species.[7] Why it takes place I can only conjecture. Just as people do, other vertebrates seek out allies and mates, dodge or fight enemies, and ignore others. Among chimpanzees or bonobos on the move by fission-fusion, each individual can pick whichever of the shifting parties fulfills its best interests at the time. Normally an ape cultivates relationships widely, creating social opportunities wherever in the territory it travels. Such behavior helps keep the whole community interlinked. But with stress on the upsurge, perhaps because there are too many chimps, factions must take shape as the members focus their attention toward a more manageable set of generally agreeable individuals. At first those apes will remain part of the original community, the factions mixing without mishaps, given the many social links remaining between them. Because of their greater time apart, however, whatever allies each animal had on the "other side" will eventually be phased out of its life with the firmness we would expect of a person dropping a friend who joins a cult. Months or years after the factions initially form, they sever relations—among chimpanzees all remaining ties between them are cut (with the rare exception, brought up in Chapter 4, of friendships between females of different communities, which they keep secret).[8] A single community has generated independent entities, a pair of societies just as incompatible as any two colonies of ants.

Primatologists admittedly know few specifics about how the story unfolds in the chimpanzee. The only division anyone has watched is the one at Gombe. Likewise, there is a single record of a bonobo community division. The course of events pretty much paralleled that at Gombe. Unlike Gombe, the bonobo factions were already in place when the study began, so we can't say how or why they formed. The two subgroups were stable over the entire nine years of study prior to the breakup, aside from two females switching sides, and one male temporarily doing so. As time passed the factions had noisy quarrels. After they did divide, the communities kept apart for a year, at which point they became amicable, as is commonplace for the separate societies in this species.[9]

Divisions must be triggered when societies approach their uppermost populations. Chimpanzee communities seldom reach much beyond 120 individuals, and those of the bonobo are a bit smaller. At that size a community can be thought of as mature since it is large enough to dominate its neighbors, but such a size comes with difficulties. The relations within the society become strained, as they do among lions in an oversized pride who no longer know each other all that well. This might make it seem that maturity is the point at which societies divide, which must often be true. Yet at least at Gombe that wasn't the case: the community there broke up when it numbered only about 30 adults. Clearly the pressures that sever a society can build at any time. At Gombe, the researchers probably set off the division when they supplied bananas to attract the apes for study—a seemingly good idea with unintended consequences. Chimpanzees mostly avoid competing with others in the community by dispersing. The tactic works out because most of their foods have a patchy distribution. But once all the Gombe animals were scrambling for the same concentrated food source, conflicts escalated when the individuals that would later form Kasakela, and wipe out Kahama, took control of the baits. The antagonism between those factions then went from bad to worse when the fruit was removed and they all faced a shortfall.

A struggle for dominance also could have pushed the division along. Gombe's factions crystallized in the months after Leakey, the dominant male, died, leaving a power vacuum to be filled. The runner up, Humphrey, refused to concede the ascendency of Charlie and Charlie's brother Hugh. One can imagine the dispute between the males compelled everyone else to choose a side, the preferred option being the subgroup offering superior social stability, defense capabilities, food, and mates. Alternatively each of the apes may have simply chosen the portion of the territory that one or the pair of other alpha males happened to favor. Humphrey preferred the northern half of the territorial range where the Kasakela community eventually came to be. Fights over alpha status have fragmented societies in species from mountain gorillas to horses and wolves. A baboon troop can split if the females pick different male favorites or rebel against a tyrannical female.[10]

Social antagonism isn't always a factor when a society divides. Across the social insects, including the majority of ants, queens leave their birth nests to found new nests on their own, without clashing. Honeybee and army ant societies can't function without big populations, and instead form colonies by division, but the mechanism is different from the procedure in other animals and requires no aggression. The workers separate into two groups, half sticking with the original queen and the others following a new one, her daughter. Everything goes smoothly despite this distinction in royal allegiance.[11] And there are cordial divisions even among the vertebrates. An oversized elephant core often grows uncoordinated with the death of its matriarch, the instability causing the members to cluster around different females nearest to her in seniority; these factions increasingly separate until they become independent—at times, though not always, on friendly terms. Sperm whales, too, form new societies without much strain, as when an outsized unit of more than 15 adults has trouble carrying out activities. The unit diverges into subgroups that wander ever farther away until they stay apart, without social tensions. But otherwise among the vertebrates some acrimony is pretty much guaranteed.

We have a lot yet to learn, and fingers crossed another division of chimpanzees is on the way that can be subject to a more detailed documentation. One community in Uganda has peaked at 200 members, the largest ever recorded. Factions formed within it at least 18 years back. That the community has persisted suggests the apes can sort out their social preferences over a very long time before achieving independence.[12] I imagine each faction as a recipe on simmer, the individual members adjusting to the overall mélange before congealing into a functional society that can stand alone.

OTHER WAYS TO FOUND A SOCIETY

Division isn't the only means to generate societies, probably even in our own species. In some other mammals, a solitary individual or mated pair can establish a society in a style paralleling that of most ants and termites, with their dispersing queens and males. Compared to a division, this course puts the individual at risk. The security of sticking tight to a society is a benefit that the members seldom forgo except under conditions of great opportunity (free space opens up that can support a lone animal) or peril (as when an animal or a group of them is aggressively pushed out by competitors). Among the mammals, only naked mole rats go solo as part of a regular cycle of society formation. Naked and defenseless they may be, but mole rats of both sexes, fattened for the trial ahead, make a dangerous sojourn aboveground, wandering from their birth nest to dig a starter chamber for a new colony. There the lone animal waits for a mate, or several mates, to discover it.[13] Occasionally a pair of prairie dogs, and possibly of spotted hyenas, set up shop on a bit of land unoccupied by others of their species, although division is the usual course in both animals. As a last example, a pregnant gray wolf may go off on her own yet has it tough hunting and warding off enemies unassisted. Such a lone wolf is rarely lone for long, though life is hardly less dicey if a male joins her, considering that a pair of wolves is no match for a full pack.

In a pinch a chimpanzee can squeak by on its own as well. At one site in Guinea, males sometimes abandon their birth community, a behavior typical of females. A male deserter doesn't have the female's option of joining another community, presumably because its resident males would kill him. The apes aren't cramped for space in Guinea, however. Should the male be able to find a safe haven between community territories, he will stay put and try to mate with any female passerby seeking to immigrate. Whether such a female will remain with him to start a community from scratch is unknown, though their chance of success must be slim.[14]

Putting aside for now the topic of the importance to humans of being in a society over the long term, the "obligatory interdependence" some claim for our kind is a touch exaggerated.[15] Apart from our child-hood reliance on our elders, going it alone, or as a couple or fam-ily, is occasionally expedient. I've mentioned the Western Shoshone breaking up seasonally into families, but their bands would get back together again each year. Few have survived in unbroken isolation the way 24-year-old hiker Chris McCandless attempted to do in Alaska in 1992, with the tragic results recounted in Jon Krakauer's book *Into the Wild*. The perilousness of living alone puts the chance of a hermit cou-ple begetting a whole society from scratch near zero.[16] William Peasley's 1983 account *The Last of the Nomads* tells the gripping tale of Yatungka and Warri, Mandildjara hunter-gatherers who struck out alone in Aus-tralia because their relationship was not recognized under tribal law. The pair was rescued several years later following their near-death in a drought.[17] With good weather they may have had grandchildren by now. Even so as a germ of a society their descendants would have been dangerously inbred. Overall, then, going solo is a last resort. This op-tion depends on numbers: ant colonies cast off prospective queens and the males to mate them by the hundreds, so colonies reproduce even as almost all queens die. No vertebrate breeds so prolifically.

Where a couple fails, a small group might stand a chance, in hu-mans and other species. The exodus of a few animals from a society is called "budding" if they succeed in forming an independent group.[18] Ideally they need not go far. Several wolves or lions might carve off a corner of their former society's territory, taking advantage of their access to an area they already know intimately. Should such a little en-tourage travel farther, its gains can be extraordinary if, by serendipity, the journey takes it to unoccupied lands flowing with milk and honey, or the equivalents thereof for the species. The supreme example of such a coup is an Argentine ant invasion, where what originally had been a handful of colonists mushrooms into supercolonies of billions. Some prehistoric migrations of humans would have been of this ilk. As with all invasive species, early peoples had the most success when they located a spot with few to no competitors. Some North American tribes started this way, as when the Athabaskans of the subarctic moved to what is now Mexico and the American southwest to become the forebears of the Apache and Navajo over a half millennium ago. More dramatic still was the relocation of hardy souls to truly far-off lands; whoever sailed on the first raft from Asia to Australia is a case in point. For such people, cut off entirely from their former society-mates, all of the terrain was theirs to claim. For each fortunate boatload that survived the journey, horrific numbers must have perished.

Still other modes of forming a society exist in nature, but they seem to be largely closed to our species. A meerkat clan or painted dog pack often begins when a few males from one clan or pack join several females from another. This group-dating approach bestows the relative safety of a goodly population on the fledgling society from the get-go.[19] A band of horses often originates when rambling individuals from different sources link up in what might be considered a mini version of a melting pot. The closest approximation of this among humans occurs when people from decimated groups come together to form a community, as happened with some American Indians and with escaped African slaves, known as Maroons, who erected societies scattered throughout the New World.[20]

BREAKING HUMAN SOCIETIES

Division, then, seems to be the usual road to birthing societies in humans and most other vertebrates. Dividing entails clear advantages across species—after all, both sides generally start off populous.[21] The division of a human society is not likely to resemble the stress-free automated undertaking typical in honeybees, however. Groups of disgruntled mutineers don't rise up within insect societies, but people, after all, are typical quarrelsome vertebrates. Enough information exists that we can identify factors that would have caused societies made up of bands of hunter-gatherers to break down, and evaluate how those factors may play into the dissolution of settled societies, including the nations of today.

Divisions wouldn't have been pushed along in nomadic hunter-gatherers by local social troubles, such as family disputes or social overstimulation for people who had little privacy.[22] Given how fluidly people could move to bands located elsewhere in the territory, such conflicts could be resolved without the pain of a societal rupture.[23] Tellingly, a dysfunctional band that split up left everyone's sense of who they were—their identity—unaffected. Life went on after people sorted out with whom in their society they were more comfortable. A societal division would more typically be the outcome, instead, of rifts between broader groups of people, across many bands. It's here that the two-step course of action observed in our primate cousins and other mammals would have played out in our species: factions would have emerged to which members became increasingly attached, followed, often years later, by a severing of relationships.

What led to such factions is the question, since many of the factors that drive their occurrence in other vertebrates appear to matter little in human societies. Still, they are worth considering in turn. Shortages of food, water, mates, or safe havens, significant in divisions in other species, ultimately helped precipitate the fall of many human societies. Even so, resource shortfalls aren't necessary to the process. Because band societies were leaderless, their fate would not likely have hinged on the actions of specific individuals, in the sense described for the Gombe chimpanzees, which favored different contenders for dominant status. In any case, a division couldn't be coerced; people in bands would rise up against the idea.[24] Difficulties in coordinating activities could have presented occasional problems, but there was usually little need for those living far apart in a band society to cooperate in any numbers anyway. Also, because band-dwelling people put little emphasis on blood kin outside their immediate families, weakening ties of biological kinship across expanding populations would not have mattered either. After all, fictive kin, people one could call father, aunt, and so on, would still be everywhere in the society. However, staying connected with allies might have become more difficult as societies expanded beyond a thousand people, but probably not by much. And because of the use of shared markers large (rituals, language) and small (mannerisms, gestures), the presence of strangers—or leastways individuals who were ever more unfamiliar—would no longer have been an issue the way it could be for chimps or bonobos in a growing community.

Indeed, by sharing markers across their society, people add a twist to the mammalian pattern of separating into factions before dividing—and this shift from individual recognition to anonymous societies would have made a difference in the proximate means by which societies unraveled. The crucial role markers play in societal breakdowns might not be immediately obvious. After all, shared markers have the power to mitigate the tensions between individuals that drive other primates to sever their connections. A recurrent theme of history is that when humans identify with each other strongly, they not just endure but join together and flourish under all but the most brutal conditions.[25] Starve or persecute them, pack them together or spread them apart, yet rest assured: the ties that bind people most tenaciously, beyond those to their immediate families, will be their identification with a society. Where relationships would have permanently eroded in a monkey troop or prairie dog coterie, markers give humans the resilience to stay faithful to other members. In point of fact, given the advantage of a big population in out-competing other societies, we might predict that once our ancestors used markers to distinguish insider from outsider, human societies could swell without bounds. After all, when the markers of a social insect are reliable, no more effort has to be expended to bond to an astronomically huge society than to a tiny one—Argentine ants continue to share an identity with the other ants in their supercolony even after it has spread across continents, obliterating all contenders from the territory.

However, there's a distinction to be made between the fixed molecular brew that defines an ant society and the indescribably diverse markers that bind a human one. Despite the social sturdiness markers give humans, over time the stability they confer cannot be trusted. Our markers aren't set in stone, but are subject to alterations, yielding social class distinctions, regional variations, and more. Even though population size isn't an issue for *Homo sapiens* in the same way it is for chimps, disruptive differences in markers are certain to materialize in our species when many individuals interact seldom—the usual situation when early humans spread out in bands. The more changes in a society's markers to accumulate without the members adjusting to them, the more factious that society is likely to be. Eventually—among bands of people, sooner rather than later, as I will illustrate next—every society reaches a breaking point.

CHAPTER 20

The Dynamic "Us"

The Walbiri people, living in the deserts north and west of Alice Springs, Australia, and famed for their tribal dances and art, were recorded by anthropologists in the 1950s as believing they had existed, with the same religious ties to the land.[1] Yet for them as for all of us, societal stability is an illusion. We suffer from a cultural amnesia, a selective memory that lets us imagine our people's undergirding essence as bedrock, etched out by its valued labels of identity. In truth, markers are in flux. The number of stars on the US flag increased from 13 to 50 without undermining the citizens' connection to their nation—indeed, the multiplication of stars has been a point of pride. Even ways of doing things that seem essential to the life of a society, such as keeping slaves, mutate or die out. What matters in the long run isn't the specific markers we prize at the moment, but that whatever markers are in vogue ensure a society's separation from outsiders—and this means societies are open to transformation.[2] Ever since the ratcheting of human cultures accelerated in prehistory, societies have been subject to steady enhancement, reinterpretation, and outright reconstruction. The way things are properly done changes with the times without destabilizing a society or eroding the discontinuities between that society and others. When this social elasticity fails, however, we find ourselves on unstable ground.

IMPROVEMENTS AND INNOVATIONS

Over the long term, then, the boundaries between societies trump the markers that define them. Still, the members act to minimize changes affecting what they regard as their society's preeminent qualities. For preliterate peoples many ingredients of identity, from spiritual beliefs to dances, were preserved through the ages with surprising accuracy. Repetition and ritualization served to "comprise almost unbreakable 'codes' for the uninitiated," according to anthropologists at the University of Connecticut, all but guaranteeing that minutiae stuck.[3] You don't have to turn to hunter-gatherer ceremonies and storytelling to be convinced of this point: early Greeks orally transmitted the *Iliad* and the *Odyssey* before inventing an alphabet. For those aspects of culture that require this kind of perseverance to learn, humans generally rise to the occasion—call it being mature, since the rites of passage central to most societies signify the taking on of the behaviors and responsibilities of adulthood. Despite this continuity of tradition, however, hunter-gatherers had no immutable standards about how to act, and certainly no means of enforcing their markers across the centuries. Early humans did not live in a vacuum, static and unreactive, even if the archaeological evidence indicates they underwent change at a near imperceptible pace.

Skills imperative for survival stayed most firmly in place, the clearest evidence of this being the consistency of stone tool types over long stretches of time. That didn't stop people from going so far as to reconfigure their livelihood when they had to, despite the fact that some stuck obstinately with their ways regardless of the dire consequences. Greenland's Vikings, for example, were apparently tethered to the homeland by sporadic trade and pressured by the church to keep their agricultural customs. A few of their communities may have starved after woeful attempts to raise imported livestock rather than take up the local Inuit practices of hunting whales and seals.[4]

Nevertheless, a willingness to pursue opportunities is a trademark of being human. The Pumé are a superb exemplar of such adaptability. On the savannas of Venezuela, where farming would be hard, Pumé in hunter-gatherer bands feast on lizards, armadillos, and wild plants, while along the rivers, Pumé villages raise gardens of manioc and plantains. Among the Pumé these differences mean little. All of them carry out the same all-night Toja ritual, share the same language, and think of each other as Pumé.[5]

The flexibility of human identities means that differences in sub sistence aren't the cornerstone of differentiating our societies the way they are for animal species, with their niches or ecological roles. Yes, societies can opt for distinct approaches to nourish themselves, which can reduce competition. Coastal societies might reasonably rely on fishing, for example, while their neighbors hunt game—and the people could see these choices as part of what defined them. Yet societies occupying the same habitat can eat the same food and make the same tools, the only outward distinctions between them being matters of arbitrary variation in myths or dress.

Not all changes in identity come about on purpose. People en act the traditional ways as best they can, but faulty recall across the generations can alter their behavior inadvertently and at times detri mentally, such as when, as I instanced previously, the once ocean-savvy Tasmanians forgot how to fish. Written records slow but don't stop the loss, even for the things people treasure most. Poor recollection and new frames of mind can affect perceptions of well-documented events when we envision the past in allegorical terms. Members of preliterate societies, dependent on each other's memories, lived out a game of telephone, in which sentences spoken from person to person grew garbled and unrecognizable—except for them the garbling could ex tend across everything they do.

Shifts in speech itself evidence this drift most vividly. Even in our age of globalized interchange, a rich diversity of languages and the dialects within them persist. The speech approximating that of the Midwestern states is often pointed to as standard for American En glish. After listening to this accent on television and radio for gener ations and adopting bits of it, however, English-speaking populations the world over have kept their own distinctive speech patterns and trajectories of language change. Midwesterners themselves continue to deviate from "their" standard: a shift in vowel sounds around the Great Lakes began in the 1960s, typified by the lengthening of the *a* sound in certain words so that "trap," for instance, has come to sound like *tryep*.[6] Linguists delight in such variations, recording them within all societies from !Kung Bushmen bands to British royalty.

Anything that can serve as a marker of identity, be it language or cuisine or gesture, is constantly being reshaped in this manner. Some shifts arguably are a consequence of the boredom people feel constantly doing things the same way.[7] Novelties can be introduced bottom up, say with the influx of goods and ideas by trade or theft, or are an outcome of the diffusion of trends among the populace. The essayist Louis Menand summarized the evidence: "We get attracted to things that we see other people are attracted to, and we like things more the longer we like them."[8] Hunter-gatherers did not adopt fads with the regularity that we take for granted, from changing hemlines to the popularity of cell phone apps, nor did their cultures boast the shifting subcultures of modern life. Yet they painted their skin and played music in ways that varied, if only subtly. People no doubt tentatively espoused, and in time came to like more and more, novel social choices.

Complete novelties found their place, too. Imagine a hunter-gatherer coming up with something radical. If the innovation was invaluable it would spread no matter its source. Otherwise the reaction an individual had to it would depend on the innovator. People prefer to follow those who match their values, but they may cut a quirky close companion some slack. While hunter-gatherer bands had no clear leaders, novelties could be introduced top down by anyone with a modicum of influence. Such a role model could nudge everyone's choices in new directions, likely by arousing the subconscious impulse to mirror the behavior of admired others.[9] For example, people could follow the advice of someone with spiritual skills, as this story told by an anthropologist who observed the Andaman Islanders lays bare:

> Long-established customs may be altered overnight as the result of a "revelation" by some seer, only to have the new customs overthrown themselves in the course of time by the next "revelation." I had an experience of this in the Onges when Enagaghe, a well-known seer, announced one day that he had received an order from the spirits about the way in which hunting trophies were to be displayed. No longer were pig jawbones to be impaled one behind the other on poles hung horizontally along the sloping roof of the hut a little above the hunter's bed.[10]

Today much cultural variation emanates from teens, who lead the way in challenging "proper" behavior. Even though their choices can't extend outside a society's permissible wiggle room without a backlash, as time passes, everyone from hippie to skinhead affects the mainstream; counterbalancing the shift are older generations, who hold back the generational changes until their influence wanes. Such battles between young and old seem timeless, but whether they were waged in hunter-gatherer societies isn't clear: most accounts of people living in bands focus on how children learned traditions rather than bucked or invented them. Kids are kids, though, and defiance appears integral to how a human child grows to be an independent being. As boys and girls are generally, the children of hunter-gatherers would have been open to new experiences, from messing about with hairstyles to exploring uncharted terrain.[11] When chic ideas, methods, or products emerged in the distant past, it seems more likely than not a youth introduced them.

THE BIRTH OF AN OUTGROUP

That the details of how band societies split up are incomplete is to be expected, given that catching one of them at the moment of a division is almost impossible—in fact has never been done. The frequency with which languages are born might be taken as a rough guideline to how long societies endure. Languages drift apart over time much as genetic sequences drift between species in what's known as a molecular clock, and measurements of this linguistic drift suggest they divide on average about every 500 years.[12] Not all societies develop what linguists judge to be their own language, however—some show only a difference in dialect. Hence a half millennium might overestimate the life span of societies. Still, the few estimates of the longevity of band societies suggest the figure isn't far off.[13] Nor is this longevity unique: chimpanzee communities achieve a similar age.[14]

Yet while the final sundering might occur once in approximately five centuries, the lead-up to that moment takes long enough to leave behind plenty of evidence as to what went on. By combining this fragmentary information with what we understand about how human groups split apart generally, I aim for a general picture of what the lifecycle of societies looked like throughout most of human existence.

Among bands the game-of-telephone effect was compounded by an inaccurate knowledge of what went on elsewhere. Variations crept in most prolifically when society members were seldom in contact. In an anonymous society, individuals living far apart might not need to know each other, yet if their markers were to remain the same, they did need to be aware of what distant members were doing. Certain factors could amplify the spatial variation between widely separated people in a band society. On a territorial border, obviously, exposure to outside ideas and goods was greater. Outlying bands with the most contact with foreign societies and the least exposure to other parts of their own society would come to differ from their compatriots else-where.[15] Making things even more complicated, different stretches of the border would abut different neighbors. Hence bands faced sharply divergent problems and opportunities from place to place that compounded their differences in identity with others of their society.[16] As an outcome, people at the territorial margins grew marginalized.[17] Factions were born among these outliers.

A human attribute that helps keep a society together in the face of potentially disruptive diversity is that people can be blind to such differences even when confronted by them. Philosopher Ross Poole articulated this perfectly: "What is important is not so much that everyone imagines the same nation, but that they imagine that they imagine the same nation."[18] Even when discrepancies were no-ticed—in hunter-gatherers, perhaps during a reunion of bands for a feast—people would have sought to avoid openly expressing dissen-sion around matters of identity if confrontation was likely, much as we do today.[19] Nevertheless, at a certain point, even differences that once might have appeared random and inconsequential could come to be seen as important, and too uncomfortable to ignore. The re-unions were times of animated gossip. Among the subjects must have been any oddball behavior, especially when people saw members with whom they had little or no acquaintance do something unexpected. Humans tend to project more negative motivations on strangers, and of course the larger a society is—even a society of hunter-gatherers—the more such poorly known others there are.[20] With minds swayed one way or another and no leader to compel anyone to conform, the stage would be set for ever more distinct and independent factions to emerge. Call them outgroups in the making.

In the same way that a chimpanzee's fate can be determined by which emerging faction it joins and thereby where it ends up living, and with whom, a person's fate could rest on his or her faction of choice—possibly more than on his or her choice of spouse. In making their decision, however, hunter-gatherers would often have had little to go on, especially since they were unlikely to have any idea what lay ahead. Early humans, like animals experiencing a society breakup, would seldom have gone through a division before. They could not fully comprehend their shifting circumstances or accurately envision an ideal outcome. To make matters worse, studies of decision making tell us that people can be remarkably uncertain about their best in-terests even when a lot is at stake—for example, often they go along with an idea they think is popular but which few actually believe.[21]

As regards to issues of identity, the cart can come before the horse. Whether a choice is bad or good may only be sorted out after people are forced to take a stand on the subject.[22]

Despite all that, it's more than likely most hunter-gatherers' choice of faction would be predictable. Humans find contentment in being with familiar others—for a band society, namely, those of their band and, probably, other bands nearby. People's connection to a certain general area within the territory—their "home turf"—might in itself be a binder. Even the factions at Gombe formed around chimpanzees with a fondness for the same bit of earth. What made human factions very likely to correspond to where people spent most of their time was that newfangled markers would spread from their point of origin; because of this diffusion, hunter-gatherer factions, like many cultural distinctions today, would be regional.

Different factions didn't have to be adversarial, and certainly not right away. Like the Gombe chimpanzees that at first still socialized, the people remained identified with the original interconnected society. In recent times the Walbiri had four subgroups on good terms, each with its take on dreamings and rituals, while the Comanche fell into three factions with their own minor dialects, dances, and military associations.[23] Given that human minds treat societies as species, I imagine we handle this variety within a society much as we do differences within an animal species, for example dog breeds that we (and, as it turns out, the dogs also) see as variations on a single theme; likewise we register members of other factions within our society as versions of our own "kind."[24]

Troubles mounted when the factions became a source of irritation. "The master defense against accurate social perception and change is always and in every society the tremendous conviction of rightness about any behavior form which exists," psychologist John Dollard averred.[25] Once again it's important to recognize that it was the members themselves who decided what behavior was proper or offensive. Just about any difference could set off this "rightness" response and initiate the process of firming up the factions.

One imagines the division might be precipitated by either an accumulation of many peculiarities or that one especially bothersome difference; in his children's story *The Sneetches*, Dr. Seuss described a world where those who had stars on their bellies refused to associate with those who did not. Among the little changes that could turn into something of belly-star importance, language would be a chief contender, a truth made plain by the Tower of Babel story.[26]

Hunter-gatherer societies could develop several regional dialects as they grew.[27] In the 1970s a linguist reported that the Dyirbalnan of Australia who live in the northern part of their territory not only have their own dialect but call themselves by a different name, suggesting a division wasn't far off.[28]

Doubtless the black sheep effect could be a factor, where people grow hostile toward any member whose outrageous displays seem like an affront to conceptions of what their society is. Psychologists study black sheep as one or a few persons: rebellious teen turned criminal, maybe. But what if, as American sociologist Charles Cooley said, "the one who seems to be out of step with the procession is really keeping time to another music"?[29] Such an outlier may not be treated as a black sheep among the company he or she keeps. People weed out some social variants and absorb others into their permissible diversity. What they allow may not match throughout the society, however. Among the likeminded, a person others may see as deviant can thrive, his or her choices emulated by those tuned in to the beat of a different drummer. The eating of the dead by the Ypety Ache, an act steeped in spiritual meaning yet misunderstood and feared by other groups of Ache, may have been a deal-breaker back when those groups parted ways.[30]

Barriers in geography could bring about enough divergences that a society could cleave without any response to deviant behavior. Humans who became utterly isolated, such as the first Aboriginal arrivals to Australia, were cut loose to transform in any direction that suited the place.[31] Elsewhere, topographical features led people to split up despite some contact with their former society-mates. Another group of Ache divided in the 1930s when a highway was constructed down the middle of their territory. Afraid of the foreigners who traveled that route, the hunter-gatherers gave the road a wide berth. With the social ties between them reduced almost to naught, the Yvytyruzu Ache drifted from the Northern Ache until both groups thought of themselves as independent people.[32]

THE ULTIMATE SEVERING

Describing the fall of the Roman Empire, the nineteenth-century American senator Edward Everett wrote that the society had broken "into hostile atoms, whose only movement was that of mutual repulsion."[33] We can assume that for hunter-gatherers, too, the division would have been set in motion when each faction came to see the other as a hostile atom, its identity intolerable, its actions falling beyond the society's boundaries of acceptable behavior, if only because of disagreements as to where those boundaries lay. If a society gives the world meaning by conveying a story about how life should work, what had been one story would have split in two.

No one has specifically modeled how a society might divide, but research by social psychologist Fabio Sani, much of it with colleague Steve Reicher, on schisms within different sorts of groups suggests factors that might come into play for entire societies, too.[34] The Church of England split up after 1994 when the members who saw the ordination of women as contrary to the true nature of the church went their own way by devising other denominations. In a second example from the same time, the Italian communist party entered the mainstream and took a new name, causing a minority faction to splinter off into a new party that stuck to its principles and retained the party's original symbols. In both situations the members who felt that the refinements enhanced their identity seized on the changes as mandatory. For them, those changes strengthened the group. The other members, though, interpreted the refinements as harmful deviations from the intangible essence that defined them, threatening their solidarity. The belief that their identity would be subverted drove a wedge between them and supporters of the change.

We might expect that most divisions have been prompted, now and in the past, by alterations in identity on both sides. Sani's studies suggest there can be imbalances, however. The more conservative faction, the one espousing the least transformation—in hunter-gatherer times perhaps the people in the core of the territory who were buffered from outside influence and retained most of the society's oldest attributes and original name—might have attracted those individuals we speak of today as nationalists, who abhor change. In the view of such people, dangerous behavior, which might have originated with a single black sheep, would have spread. What had been one estranged person worthy of chastising, or worse, would have metastasized into a faction acting inappropriately, even maliciously, as a unit. Such a faction can elicit a strong commitment from its members because of how likeminded they are. Factions express a narrower set of behavior than present across the whole society, making their solidarity easy to come by. However, the more radical side—occupying the fringes of the territory—would have seen things the same way, and felt just as unified about their own viewpoint. Because they would have seen the changes they promoted as essential to strengthening the same society, for them the dissidents would have been those conservatives refusing the improvement.

When factions emerge around differences they find disagreeable, we can expect little effort by either side to see things from the other side's perspective. Psychological studies show that this lessening of attention to the other side manifests even for factions that are just getting their start and so are still trivially different.[35] As a result, communication breaks down and factions strengthen. To individuate the others becomes not merely hard, but destructive. Doing so might give us cause to question our beliefs when we are confident our motives are pure and they are in the wrong—evil, even. From the vantage point of both sides, then, the threat that sundered the society isn't a shortage of food or of places to live, even when the dire straits brought on by such misfortunes may factor into its demise. Rather it is defects in the collective identity that had once tied them together.[36]

We have seen that people conceive of societies as if they were separate species. In fact the metamorphosis of societies as they splinter off from each other has been labeled "pseudospeciation," in effect turning what had been one species of human into two.[37] In this splitting, markers serve a role similar to genes.[38] As it does for biologists chronicling the origin of species, the pervasiveness of change makes figuring out who divided from whom in prehistory a challenge. The trials of anthropologists are made all the greater by the readiness with which societies trade, borrow, and steal.[39] Darwin's final words about living things in *Origin of Species* apply equally to societies: "Endless forms most beautiful and most wonderful have been, and are being, evolved."[40] A more detailed consideration of the psychology behind the separation of human societies will confirm that this beauty hasn't come easy.

CHAPTER 21

Inventing Foreigners and the Death of Societies

The disbanding of a society is a time of reinvention. Any reading of history suggests society breakups mirror the breakup of a marriage. When one can't turn back from a split, years of repressed opinions come pouring out that may express the opposite of what had been professed a month, if not the day, before. As pressures to conform to social norms shift, diminish, or vanish entirely, people on both sides gain the latitude to explore ways of interacting that had been out of favor or considered heretical. Previously unacceptable acts can leap to the forefront, helping each group distance itself from those who are now *other*, reimagined as outsiders, such that they come to appear ever more foreign.

The evidence indicates that many of the modifications of daughter societies—their character displacement, to borrow again a term from biology—occur in the initial years after they go their separate ways. Their newfound freedom of expression may be a reason why. That's when language—and no doubt many other, less studied, aspects of identity—undergoes the fastest rate of change, before settling into a relative stasis thereafter.[1] Indeed, distinctions between societies, often enough, are an outcome not of their ignorance of each other due to geographical separation, but of their awareness of and interaction with each other. This would be conspicuously true after societies split up. The opportunities for independent thought and invention afforded by a newly minted society, leading to a convergence of perceptions around themes the members can celebrate as their own, can make its formative years a golden age. For example, the Declaration of Independence and US Constitution remain the reference points that Americans turn to for guidance when questions about the nation's governance arise. Based on what is known about modifications in identity, I believe this would have been the case over the course of our evolution as it is now.

Yet there would be a deeper psychological impetus for a reworking of identity to bloom right after a division. The sense of being adrift, their fates severed from the meaning and purpose the larger society once provided, would heighten the urgency of the people's search for a strong identity, and essence, that stands apart.[2] Moreover, their identification with each other must actually *matter*. Certain groups, such as people experiencing homelessness or those who are obese, may be marginalized but don't create societies with identities of their own. Neither do sick or disabled chimps or elephants, even when others treat them as outcasts. These outliers fail to bond since they do not see others with their condition in a good light. They lack what psychologists describe as positive distinctiveness.[3]

Hence the insights of psychologists suggest that the members of a start-up society will toil to distinguish themselves favorably. To achieve this, they improvise cherished attributes or express old ones in a special way. The process is analogous to the development of traits that biologists studying the divergence of species call isolating mechanisms. Whatever commonalities remain with the other society can be denied or ignored. Like divorcees not on speaking terms, the societies can break off contact, which would mean any shared history would be eschewed or forgotten.[4] Regardless, no matter how alike the newbie societies might seem to outside eyes, reunification would quickly be impossible.

DIVISIONS AND THE PERCEPTION OF US AND THEM

One remarkable aspect of a society division is that the relationships between the former comrades have to be recast at the level of individuals—every one of them.

A split must yield an unequivocal grasp of who belongs where, if only so each offshoot maintains order and independence from the start. The traumatic nature of this recasting of identity in chimpanzees is what made the Kasakela raids on the Kahama community all the more horrifying. The slayed animals were not just known to them; many had been among their friends. Best buddies Hugo and Goliath found themselves on opposite sides of the chimpanzee schism, but continued grooming each other as the factions drifted apart (with Goliath on the losing side). Hugo took no part in his killing, yet another male, Figan, did, even though Goliath had been "one of his childhood heroes," Jane Goodall recounts.[5]

What doomed Goliath was the shift that occurred in how the chimpanzees discerned their erstwhile compatriots—a change in how they categorized each other. The apes' deftness with categories reminds us that, as one primatologist put it, "chimpanzees, like humans, separate the world into 'us' versus 'them.'"[6] Here's Goodall's further take on the subject:

> [The chimpanzees'] sense of group identity is strong and they clearly know who "belongs" and who does not. . . . And this is not simple "fear of strangers"—members of the Kahama community were familiar to the Kasakela aggressors, yet they were attacked brutally. By separating themselves, it was as though they forfeited their "right" to be treated as group members. Moreover, some patterns of attack directed against non-group individuals have never been seen during fights between members of the same community—the twisting of limbs, the tearing off of strips of skin, the drinking of blood.

"The victims," Goodall concludes, "have thus been, to all intents and purposes, 'dechimpized.'"[7] As dehumanization likely can be for people, the ability to switch off an awareness of belonging to "our kind" becomes a mechanism by which the members of a new society firm up the irrevocable separation from their former society-mates.[8] This change of perception can be a gradual process, preceding the division itself. In the division of a macaque troop, for example, fights shifted from conflicts between individuals in the different factions early on to showdowns between the factions acting en masse as the breakup approached. It was as if the monkeys no longer treated the others as separate beings but as a collective.[9]

What happened at Gombe to ultimately cause one group to hive off from the other—the tipping point, after all those years as factions that tolerated each other, when the chimpanzees at last severed any remaining ties? Somehow they all recast their former fellow members as dechimpized others. Did the scientists miss a row that caused each and every ape to change its outlook on the members of the other faction?

That some sort of deciding incident could play a role in the division seems plausible for monkeys that stay in constant contact with

everyone in their troop and therefore seldom miss anything of consequence. This thoroughness of knowledge is impossible for chimpanzees, spread out as they are. Not every ape would have borne witness
to any pivotal event, and chimps can discern little from others about
what transpired. This draws our attention to a key difference between
chimp and bonobo divisions and those of people: in deciding such
matters as *Who do I hang with?* or *Are the others now foreign?*, our ape relations at best act solely on what little information they are able to glean
from community members that perchance are close by. While the
dynamics of how divisions unfold might have evolved before speech,
humans can ascertain what's taken place elsewhere, and express and
bolster opinions as to who must be banished from the society and who
belongs.

Whatever caused the final sundering of the Gombe community,
one thing seems certain: dechimpization could not have come about
by each animal negotiating a change in relationship with every other
one, individually. Perhaps by the time of the division, the apes were
prepared to revise their treatment of those in the other faction all at
once. Voilà! Past companions became foreigners in a single transference of identity, and a society was born. For aggressive animals like
chimps and gray wolves, such a boundary shift distinguishing who
is in from who isn't would render past relationships of little consequence. In one fell swoop, the once beloved Goliath became something alien—even dangerous.

Of course, chimpanzees, wolves, and most other mammals do
not have markers to which they affix their identity. What our species
evolved to do (if we give this crude hypothesis any credence) is connect this mass transference of identity to the unique attributes that the
members come to associate with their faction. This shift in perspective
creates stress lines along which people clash while they are still part
of the same society. We can expect that once humans see a group of
compatriots as coming together around behaviors they find truly un-
forgivable, even appalling, those people become irredeemably *other*.
Prejudices now come into sharp relief, as their markers, viewed unfavorably, turn into the chief thing registered about them.

Other species with anonymous societies behave much the same way,
though some use a far more formally organized and painless strategy
than humans. In a dividing colony of the honeybee, the two offshoot
hives will share the same identifying marker at first—the same odor.
As far as anyone knows the only reason the offshoots don't recombine
is that one of them flies to a location far from the original nest. While

the matter hasn't been studied, a reasonable guess is that once the two colonies settle at their respective spots, differences in diet, or more likely in the genetics of each queen's offspring, cause their "national" scents to drift apart. As a result, each colony, belatedly, achieves an identity of its own.

A review of civil conflicts shows that shifts in how humans respond to others in their own society can turn, disconcertingly, on a dime, ranging all the way to outrageous dehumanization and outright brutality. The best-documented examples concern modern ethnicities rather than emerging factions, and show the problem at its extreme. Sleuthing by the Polish American historian Jan Gross confirms that everyday citizens can sever relationships as utterly and gruesomely as any chimpanzee. Gross reconstructed the story of Jedwabne, a Polish town that butchered more than 1,500 of its Jewish residents over the course of one day in 1941.[10] While I doubt violence of this intensity was common in hunter-gatherer societies, a breakaway faction could well be looked at as inferior. Depending on a people's response to its members' newly adopted ways, this could even manifest with disgust from its inception. Forged on a sword of social disintegration, this outlook could be tough to alter. Rapidly, the groups would split firmly apart.

The very sense of being spurned can be a psychological blow. Indeed, rejection even by those we have rejected ourselves can bring on pain or depression. It hurts to be excluded by a group we dislike; among the studies that attest to this is one titled "The KKK Won't Let Me Play."[11] An anticipated outcome is that people who identify strongly with a faction bond all the more tightly the worse they are treated.[12] Those joining separatist movements in Quebec, Wales, Scotland, and Catalonia unify with profound indignation around anything they perceive as unfair, such as overtaxation and repression of civil rights, sharpening the fault lines along which their societies could cleave.[13]

For the most part, close ties tend to weather a societal division. Having grown up together, family members often think similarly and are likely to choose the same faction. This may explain why hunter-gatherers were on average related to others in their society to a slightly greater degree after it splintered, a pattern also present following divisions in other primates.[14] The trend is sharper still in tribal settlements. People in them depend on extended families to share goods and bequeath property, causing most divisions to fall neatly along family lines.[15] Of course those kin and allies that do take opposing sides in a social schism find their relationships severely tested. Anyone who shifts allegiance to another faction runs the risk of being disowned and replaced with friends whose loyalty is aligned.[16] History is rife with tales of brother killing brother. That was at no time more poignantly documented than during the American Civil War, known as a war of brothers for how family and town loyalties were sundered, damaging relationships for generations afterward.[17]

I don't know what the tipping point might have been to cause early hunter-gatherers to finalize a split, but recorded human history and evidence from other primates suggest that a complete lack of enmity would have been rare. This remains true even after the advent of language, when our ancestors could in theory have annulled the relationship by level-headed negotiation. Fights broke out in the events leading up to the division of the Gombe community; however, these altercations were of a fairly normal sort for chimpanzees. Small consolation, that: the killing started once the community came apart.

Yet for people the comparison to divorce, with the common presumption of irreconcilability, may not be the best analogy for picturing the aftermath of a division. As with some other mammals, outright violence need not play into the division or mar the relationship between the societies thereafter. Savannah elephants (and it seems bonobos, too) can face turmoil and uncertainty as divisions play out, and endure a similar global recasting of membership. However, they often (although not always) rebuild ties later. With those species, as with ours, a division might be more usefully likened to conflicts of teens with parents: a fraught growth phase that must be passed through for both parties to achieve independence. In all but the most tumultuous circumstances there may be a fair prospect of later making amends, even if the two sides are now firmly separate.

THE MAGIC NUMBER

For hundreds of millennia the volatility of human identities is what made the breakup of societies certain at a small size. Indeed, that size was so predictable that some anthropologists have proclaimed 500 a "magic number." As a ballpark average for all parts of the globe, this was the number of people living in a band society.[18] Much as 120 seems to be the size beyond which a chimpanzee community is likely to become unstable, it's reasonable to think that 500 represented an approximate upper limit on the population of a stable society for most of the prehistory of *Homo sapiens*.[19]

One can infer a practical reason for a society to contain at least 500: by some calculations, a population of this magnitude gives humans the opportunity to select a spouse who isn't a close relative.[20] This may explain why people, unlike the many mammal species that live in societies of a few dozen, rarely show the restless and risky drive to join a foreign society. Thanks to the abundance of mates, most people throughout history have had the option to stay with their birth society for life. But what set the divisions in motion at this size rather than at some larger one, which would have given our ancestors an even wider choice in spouses, as well as the defensive advantages of a more imposing group? That number doesn't seem to reflect the checks and balances of societies living enmeshed in nature, as ecological factors like predators and food availability differ remarkably between the jungles and tundras where hunter-gatherers lived. The territories hunter-gatherers occupied differed in total area in those different ecosystems, with Arctic people ranging farther, but the populations of the societies were much the same everywhere.

That band societies hit a low population ceiling may have been a function of the psychology governing the expression of human individuality. Maintaining a balance was essential: members had to feel similar enough to each other to share a sense of community yet different enough to think of themselves as unique. I argued in Chapter 10 that people had little motivation to set themselves apart from others when everyone in the society lived in a few bands, hence the conspicuous paucity of cliques among those hunter-gatherers. But once their population swelled, those same people would hunger for the differences offered by connections with narrower groups. This increased drive toward diversification of identity would promote the emergence of factions that ended up causing discord among the bands and severing their relationships. The situation would be different for societies that settled down, eventually in colossal populations. Unlike people in bands, most settled people found opportunities to connect to social groups that were not factions but were rather widely acceptable and necessary for the society to function—jobs, professional associations, social clubs, and niches in the social hierarchy or among extended kin.

I have left open the question of what is special about the anthropologist's magic 500. My suspicion is that it's at about this population that the human wherewithal to have even the crudest knowledge of everyone in the society begins to be sufficiently strained for individuals to lean heavily on markers as a crutch in their interactions. In groups larger than 500, humans would start to feel truly anonymous, something that has no effect on ants but that undercuts people's desire to matter as individuals. This loss of self-worth would make people eager to enhance their distinctiveness by accepting what novelties come their way. Without the tight organization and oversight of a major settled community, such novelties would spur a division. What I can't say is whether this attribute of social identity evolved in the Paleolithic to keep band societies at a size that (for reasons yet to be fully understood) was ideal for life at that time. Alternatively, it could have been an adaptation to enhance people's individual social interactions, in which case perhaps the 500 number was an accidental outcome of this psychological feature. More likely both were important.

HOW SOCIETIES DIE

Societies come and go. Ant and termite colonies ordinarily die with their queen. Every generation of colony starts afresh with a new foundress. Some mammal societies reach almost as decisive an ending. Should the breeding pair in a gray wolf or painted dog pack or a meerkat clan die with no viable successor, their society is doomed. Unlike the insects, however, the other members live on and with luck disperse to join a pack or clan elsewhere.

Divisions ensure that most vertebrate societies don't come to such an absolute dead end. Under ideal conditions a society can propagate like an amoeba that keeps dividing. The biologist Craig Packer has watched several lion prides divide through a dozen generations over five decades of research in Africa.

Closest to an immortal society are the supercolonies of Argentine ants and a few other ant species, whose members keep the colony identification as they spread globally. Even for them, though, new supercolonies presumably come along at some point. For this to transpire, the identity of the ants from the same supercolony must change to become incompatible. Because ants never experiment with the behavioral choices that incite people to part ways, their divergence in identity would be underlaid by genetics. Perhaps a queen with a mutation affecting her "national" scent avoids being killed by her birth colony by lucking onto an isolated spot. There she establishes a nest with her unique identity odor, which could—however improbable it might be—grow into a supercolony.

Among the vertebrates, the bottlenose dolphin communities along the Florida coast may match supercolonies in permanence, albeit certainly not (given their individual recognition societies) in scale. A community occupying the same corner of Sarasota Bay has stayed steady at about 120 animals for over four decades of study, its territory handed down through the generations—inheritance of real estate being a prime advantage of society membership in many species. Genetic evidence suggests that this community has been there for centuries.[21]

Yet despite the potentially endless repetition of divisions, the demise of societies must be inevitable at some point, if not for the bottlenose dolphin then for everyone else. Even an amoeba, which divides endlessly, faces an uphill battle when its diet is curtailed. Eking by at the point of starvation, one of the two amoebas dies after each division, leaving its sister to split again. This genetically programmed response keeps the overall number of amoebas near the carrying capacity—the highest number the environment can sustain.[22]

That sharp death rate brings to mind the tragedy that befell Gombe's chimps when one community wiped out the other, and for good reason. Malthusian realities apply to both, as follows: Populations of even the slowest-growing species reach the carrying capacity in a few generations. Therefore, barring environmental change, about the same number of chimpanzee communities will have existed on any suitable patch of habitat throughout the millennia. Before the chimps (or lions or humans) attain this peak in population, the products of each division can find space to settle without much of a fight. Yet once societies are cheek-to-jowl, conflict will likely loom large not only between neighbor societies but also internally as the members brawl and scramble for resources, straining their relationships until they crumble. Should a large society fissure under these overcrowded conditions, its members will be no better off. The animals will continue to struggle over the resources that had been accessible to all of them when they were part of one society. Division in itself doesn't ease population stress.

Still, dividing a society can provide a savagely practical way around the logjam. When a community of chimpanzees, all competing with one another, is riven in half, the daughter society that knocks out the other—as Kasakela did Kahama—seizes everything for itself while coincidentally reducing the level of conflict among its own members. It's the culling game plan of a starving amoeba implemented by brute force. With societies at or near carrying capacity, as societies generally will be, the way forward may indeed be cruel. If your former societymates, now on their own, didn't bring you down, other societies could. One doesn't need to be a mathematician to understand that over time most societies die out.

The merciless calculus of survivorship tells us that, in the ultimate sense of evolutionary causes, competition lies at the root of why societies divide. Nevertheless, given how much their markers shifted from place to place, division of band societies must have been unavoidable at some point, regardless of population pressure. The question becomes what occurred when humans filled the landscape to capacity, so that the offshoot societies, butting up against competitive neighbors, had no place to go or space to grow. People tend to unite around a threat when competition with outsiders is their overriding concern. A hemmed-in society's unified front would impede the emergence of conflicting points of view, presumably including the factions whose conflicts are a prerequisite of division. Possibly a society in such a bind could maintain a single coherent, if slowly changing, identity even if barraged by outside influences.[23] Still, I suspect this situation would delay the society's division though not stop it. Indeed, should social friction impel a division when the daughter societies are packed tight, the loss of the more vulnerable of those daughters may be all but guaranteed. Arguably, this is what took place in the forest patch at Gombe when Kahama chimpanzees had little room to retreat. Such an end must have befallen many human societies in prehistory, often from attrition brought about by raids, their few survivors, usually women, perhaps joining the winning group.

"The past is a foreign country," or so the writer L. P. Hartley claimed.[24] The archetypal rules by which humans alter their identity thus ensure such a built-in obsolescence. Even the societies that prevail transmute with time to forms their original founders would reject as insufferably foreign. Given time, a society alters as inevitably as does a biological species that evolves until the living generations wouldn't recognize their predecessors if they met them, and paleontologists deem it worthy of a new species name.

So it was in countless repetition before the dawn of history, societies severing from each other, some wiped out yet even the winners in time changing beyond recognition to divide again. Every sundering would have been a source of heartbreak and agony, carried out for reasons that were momentous at that moment but ultimately forgotten. These divisions have been a part of the rhythm of life as primal as love or personal mortality, yet unfolding over a long crawl of generations beyond our ability to grasp. Win or lose, a society, like its individual members, is fleeting.

How many societies has the cycle birthed? If we can take estimates of the total number of languages to have ever existed as a rough indication, human societies have numbered into the hundreds of thousands.[25] Since not every society has a distinct language, it's conservative to assume a million-plus societies have come and gone, each made up of men and women confident of their society's significance, abiding existence, and success over its predecessors, of which yours is one.

Thus the sundering and death of societies is certain. In most vertebrate species, to review, the process has nothing to do with the markers that make people relaxed around each other. Chimpanzees have socially learned traditions, yet don't discriminate against those who act strangely. Furthermore there's no reason to think they come up with customs that some members take on and others spurn, bringing about a severing of a community. For humans, however, being part of a society comes with obligations of acting properly and abiding by its rules and expectations. Still, those rules and expectations are a moving target. Ever since humans first employed markers for societies, those societies have broken down around factions of people who take alternative views about their identity. Our psychology orchestrates the change by turning what once was familiar into the foreign.

In charting the birth of the societies of our hunter-gatherer forebears, I've said little that is specific to the rise and fall of nations, which will be one of the subjects of the section to follow. Before getting to this question I must address the social pathways that made such immense societies possible. As it turns out, what drove the success of civilizations was anything but dovishness. Societies that turned to conquering each other—in the course of time incorporating ethnicities and races—enter the picture now. The residual effects of their struggles, some ancient, some ongoing, linger in every corner of the earth's surface today.

SECTION VIII

Tribes to Nations

CHAPTER 22

Turning a Village into a
Conquering Society

T en thousand years in the past—a single heartbeat ago in the
drawn-out prehistory of our species—the last ice age began to
draw to a close. As the climate warmed, some hunter-gatherers
turned to farming, a shift known to archaeologists as the Neolithic
Revolution. This metamorphosis took place independently in six parts
of the world: first and most prominently in Mesopotamia, a region of
the Middle East, 11,000 years ago; followed by the area China now
occupies, 9,000 years ago; in highland New Guinea, 7,000 years ago, if
not earlier; in central Mexico and roughly contemporaneously in the
Andes centered around Peru between 5,000 and 4,000 years ago; and
in the eastern United States, between 4,000 and 3,000 years ago.[1]

From these humble beginnings, enormous kingdoms emerged in
four of the areas: China, the Middle East (I include India here, which
was sustained by crops brought in from that region), Mexico (starting with the Maya and later the Aztecs), and the Andes (culminating
in the Inca). Nothing seems grander and more deserving of detailed
treatment than the forging of a civilization, with its cities and elaborate architecture and culture. Yet after the growth of societies beyond
the size of hunter-gatherers living in bands, starting with people settling down into tribes, our story accelerates, for the simple reason that
the prerequisites for turning those tribes into civilizations are simpler
than they may appear.

We don't see anything that approximates a contemporary nation in scale or intricacy among other mammals; in nature only certain social insects develop what in effect are civilizations of their own. Clearly the evolution of anonymous societies was crucial for civilizations to come into existence and spread. But anonymity is insufficient to explain either why we humans are able to develop large societies, or how we maintain them. A number of extra factors must come into play—some, like a sufficient food supply, are quite obvious. Others, such as having the means to work out social problems and to give the members ample avenues to distinguish themselves, are much less so. Achieving anything close to what we call a civilization took something further still: violence and the machinations of power.

History books overflow with the splendor of nations, their clashes and partnerships, the strivings of their colorful characters, of governments that march forward or fall to the wayside. We take the details personally, often because these are the stories of our people, and they matter to us. For all that, though, the differences between nations and most of the societies that preceded them are matters of degree, rather than of kind.

I say *most* of the societies that preceded them because one transition was a paramount shift as societies grew in size and complexity. Societies began to absorb each other. From there the road to nations as we know them today, opened up by the Neolithic Revolution, was short indeed. For these conquering societies to take root, the several elements that needed to be in place began with the most basic of resources.

FOOD AND SPACE

Having more people requires more food. This self-evident truth makes it easy to assume that a sufficient supply of food is enough to spur the growth of societies. It is not. Consider the monkeys clamoring around the marketplaces of New Delhi. Urban macaques live off the fruits (and meat and vegetables) of agriculture, stolen from street vendors. While the relative plenty does indeed support a higher overall population of monkeys, urban troops nonetheless remain much the same size as those of their country bumpkin and deep-forest brethren.[2] There are simply more troops, packed tightly as to leave no unoccupied space. Much the same goes for the Argentine ants still living in Argentina, where colonies, penned in by many hostile neighboring colonies, cannot grow very large, however much they have to eat. Those in California, of course, have escaped that constraint.

Plainly, humankind has not ended up like urban monkeys. The Nile Valley wasn't home to thousands of mini-Egypts, but instead to the grand one that birthed Ramses II. That said, across wide expanses of the globe humans actually did respond to the availability of dependable food, whether from farming or abundant wild resources, by propagating many little societies instead of creating one large society. In the 1930s, for example, when outsiders first hiked into the New Guinea Highlands, people already occupied it in the hundreds of thousands. A walk of just a few kilometers was all it took to reach the territory of another settled tribe. Each tribe committed to the food supply generated in their area, which typically included plants and animals domesticated on that island. Explorers found the same pattern in the Amazon basin and elsewhere. Although most of these horticultural tribes had less flamboyant cultures than the hunter-gatherers of the Pacific Northwest, they also lived in villages, or at least had a central lodge as a retreat (although a few tribes, especially in parts of Africa and Asia, were nomads that herded domesticated animals). In the past, a tiny minority of these tribal groups would prove to be points of transition to the grand societies of today. Understanding how tribes were organized, and the special features that permitted these few to increase in size and complexity, is the next part of our journey.

VILLAGE SOCIETIES

Life in a tribe could be one big soap opera. As among the settled hunter-gatherers, there was no shortage of occasions for petty squabbling and violence. Conflicts included disputes over matters that might spoil a family reunion, such as what's for supper, but also accusations of sorcery, fights over spouses, and arguments about the distribution of responsibility.[3] These disagreements could precipitate the breakup of a village, the people at times becoming so disgruntled they moved far off to avoid each other as much as possible. Many villagers would experience such a social cataclysm one or more times in the course of their lives. Prehistoric villages in the American Southwest, for example, typically lasted from 15 to 70 years.[4] An example of village fissioning is that practiced by the Hutterites. Arising in the sixteenth century in what is now Germany, this extant communal sect came into being relatively recently by the standards of the groups we normally refer to as tribes. After several centuries of migration, Hutterites emigrated in 1874 from Russia to the American West, where they live in colonies of up to 175, each of which operates a farm. Social stresses rise as a colony grows, until finally the members arrange for the colony to split, an adjustment made on average every 14 years. Although these transitions are handled in a more orderly fashion than the fissioning of preliterate villages, the dynamics are much the same.[5]

For tribes to stay together at all, solutions were necessary to ameliorate, or at least manage, conflict. In this, most tribes dependent on horticulture devised strategies similar to those of hunter-gatherers who settled down. One approach to emerge repeatedly was to reduce competitiveness between people by adding dimensions to the social distinctions acceptable in their society.[6] Among these were differences in their work and status. Thus, even if some tribes started out as egalitarian as hunter-gatherers in bands, such an outlook seldom lasted. Also contributing to the members' differences were opportunities to be a part of social groups. Across the tribes worldwide the differentiation arguably reached its apogee in New Guinea. The most complicated were (and still are) the Enga, a regional population that manages a social existence that can seem to have a Rube Goldberg–like intricacy. The thousand-plus individuals within each Enga tribe celebrate their history as a people. However, every tribal member is born into a clan and subclan that claims its own garden space, and at times there can be bickering, or outright fighting, between clans. Nevertheless each tribe has remained intact for the long term.[7]

Some sort of centralized management for dealing with social issues is also all but assured in human settlements, even if rudimentary in the simplest of them. As I described earlier for settled hunter-gatherers, people ensconced at one place had more patience for displays of authority than hunter-gatherers did in bands—yet often just a bit more patience; each village tended to have a headman, but his significance came to the fore during conflicts, and even then he spent most of his time convincing people rather than leading them.

Still, even tribes such as the Southeast Asian highlanders described by the anthropologist James Scott in his book *The Art of Not Being Governed* didn't lack governance. The title alludes to tribal efforts to avoid being devoured by the powerful civilizations spreading like amoebas from the lowlands. The mountain-dwellers had chiefs who could be tyrannical.[8] Elsewhere the topic of who should have power could itself be a subject of contention. Supervising social matters didn't always require a certain person at the helm. The Nyangatom of southern Sudan and Ethiopia are pastoralists distributed across many villages, each of which moves several times a year to spots suitable for the herds (they can almost be portrayed as hunter-gatherers who happen not to hunt since they drive domesticated cattle rather than stalk wild ones). The Nyangatom have a few specialists, such as men skilled at castrating cattle or making symbolic scars on the chests of warriors, and yet they maintain peace without a permanent ruler; instead, every male has a crack at leadership of the tribe, shared with others of his age cohort.[9]

As for tribal peoples that live in settlements—whether they survive by hunting and gathering, gardening, or gasoline fueled agriculture—social friction has limited their populations at one place to something on the order of a hundred or several hundred individuals, up to a few thousand for New Guinea highlanders, who spaced their homes out in a way I imagine reduced strife. A Yanomami village in the South American rainforest, where by contrast everyone hammocked almost on top of each other in an oval shelter, was often smaller, with as few as 30 but up to 300 members.[10]

In some situations, one such village was all there was to a society. Often, however, there was a grouping above the village. The Hutterites, the Yanomami, and the Korowai of New Guinea, renowned for their tree houses, are all examples of tribes comprised of more than one village. The structural and functional equivalent of a nomadic hunter-gatherer society was such a regional cluster of settled people—call it a tribal or village society.[11]

As anthropologists have often widely underappreciated band societies in favor of studying bands, the individual village has similarly been a centerpiece of anthropological research, rather than all the villages collectively. This bias in attention is due, firstly, to the autonomy of the villages. No outsider, even from another village of the tribe, told the village residents what to do, any more than one band of hunter-gatherers had any say over another. But another reason researchers have focused on the village as the level of study is that the relationships between villages can be dramatic: villages were often famed for their conflicts with each other, including, among the Yanomami, revenge killings.

Just the same, the tribes were important to their people. Yanomami villages had deadly feuds—reminiscent of the Hatfields fighting the McCoys, though in the case of the Yanomami, multiple families were involved. Even so the villages continually renegotiated their relationships. Fights alternated with reconciliations involving marriages, feasts or trade. Everyone had friends in other Yanomami villages, and, as with members of hunter-gatherer bands, people could relocate to another village—although villagers, committed to tending a garden (and in New Guinea, domesticated pigs), accomplished such a move less readily than did the nomads who foraged unencumbered for wild foods. Indeed, much as a villager could change village, whole villages could fuse. These dynamics worked as they did for bands of hunter-gatherers, with the fissioning and fusing of villages driven by social relationships.[12] The biggest distinction between village societies and band societies is simply that villages changed their locations (usually to freshly cleared gardening spots) and split and combined less often.[13]

Seen this way, a village and a band were not much different. As with the people in a band, the residents of a village seldom had to think about the larger society of which they were a part, but it was there for them the rare times they needed it. Their mutual identity came to the fore when a difficulty or opportunity arose at a societal scale. The villages of Jívaro in present-day Ecuador, famed for their shrunken heads, joined forces when raiding foreign tribes. In 1599, in the largest assault of this kind, 20,000 Jívaro from many villages freed their territory from foreign rule in a coordinated massacre of 30,000 Spaniards.[14] Such village societies also had words that encompassed their whole society, just as hunter-gatherers did. Yanomami, for example, is the name the people give themselves, while Yanomami tapa is their way of indicating all their villages. As we saw with many band societies' names for themselves, the monikers Yanomami and Jívaro mean "human."

Collective markers of identity defined these tribal clusters of villages as a society just as they bonded other societies across the globe. The Yanomami are a case in point. The common dress, housing, rituals, and other shared traits of the members enabled the fission and fusion of villages—similarities that the people recognized. When the people of a village came to an impasse and went their own ways, the parting was just like that of a band of hunter-gatherers. Even if those individuals felt animosity toward each other, in the end they nonetheless had the same language and way of living, and remained part of the same society.[15] Differences did accumulate, however, across the villages of a tribe. For settled tribes as much as for hunter-gatherer bands, that's where things went south permanently. The Yanomami today seem to be diverging into several tribes, each with a population of a few thousand, as a result of changing identities. Indeed, some linguists distinguish five, still very similar, Yanomami languages. The Yanomami themselves are aware of the divergences, and they mock the odd Yanomami from distant villages.[16]

Tribes like the nomadic Nyangatom and the settled Enga and Yanomami managed to stay intact at overall populations often in the many thousands—larger than was normal for a band society. At the same time these observations about the intolerance of the Yanomami toward differences among their own people point to one reason why very few tribes took the path to building an empire. Being hemmed in by neighboring tribes wasn't the only obstruction. A tribe came up against the same problem hunter-gatherers faced: its own people's identities started to clash.

Yet should a tribe have managed to retain a consistent identity, it still would never yield a sprawling civilization solely from the growth of its population. This would have been true even under the most ideal conditions—a favorable birth rate afforded by ample food and space, able leadership, and plenty of social differentiation. That these attributes were insufficient is demonstrated by the fact that all large human societies, on close inspection, contain not the descendants of a homogeneous stock of people, but populations of diverse heritage and identity. The failure of hunter-gatherer and tribal societies to adjust to variation in markers stands in stark contrast to the resounding success of nations in this regard. Indeed, to understand the birth of civilizations we must grasp how they came to be made up of a mixture of citizens, eventually comprising today's ethnicities and races.

SOCIETIES DON'T FREELY MERGE

One possible explanation for the heterogeneity of civilizations is that growing a society entails societies voluntarily merging. The evidence indicates this isn't so. Across the animals the merger of societies is so rare as to be nearly nonexistent even in species where societies compete little.[17] Bonobos and chimps exemplify this: the only "mergers" of their communities strain the word's meaning. Primatologist Frans de Waal tells me that bonobos who are strangers to each other can forge a community from scratch with little fuss. Bonobos' aptitude for befriending strangers must simplify their adjustment to each other. Such arranged societies are artifacts of zoo confinement, however. In nature, bonobo communities, even if on good terms, keep their memberships apart. Integrating captive chimpanzees into one community is by comparison a nightmare, requiring months of careful introductions, with lots of bloody skirmishes along the way. The only reason the xenophobic chimps adjust to each other is that, with their original communities lost to them, they, like captive bonobos, are refugees, with no choice. Similarly the handful of permanent mergers recorded for wild monkeys occurred after the animals' troops were decimated down to a few. The survivors, like apes in a zoo, were a kind of refugee, forgoing their past societies to join another as a matter of survival. Their link-ups don't come close to the mass unions we see between entire human groups.[18]

Under ordinary conditions the same holds true among social insects: combining mature colonies isn't a part of their pin-brained ethos.[19] To my knowledge, permanent mergers of healthy societies take place only in African savanna elephants, and then rarely, and just between two cores that once belonged to a single core that divided in the past. By reconstituting their original membership, sometimes years after splitting up, the pachyderms apparently affirm they never forget.[20] Otherwise, once a society has gelled so that its members identify with each other and stay above a sufficient head count to persevere, that society remains distinct from all others.

The same is true of humans: once society members have their identity set in place, a merger freely made with another society is highly unlikely. I've seen no evidence, for example, that goodly numbers of foreigners were absorbed into hunter-gatherer bands. This is true even in cases of close cultural similarity between the outsiders and the society they might conceivably join. So while bands of the same hunter-gatherer society could fuse, as could villages of a tribe, societies stayed firmly apart. The hurtle of merging societies is compounded in humans by the unlikelihood that each can adjust to the foreign identities of the other. The only examples of mergers again bring to mind a fugitive situation: coalescent societies come about when people are too few to subsist on their own. Coalescence was the fate of uprooted American Indians from the 1540s through the eighteenth century, especially in the Southeast, after years of dying from fighting Europeans and their diseases—among the famous examples are the Seminoles and Creeks. Once combined, these refugee populations often took the name and much of the lifestyle of the dominant tribe in their mix, with a few allowances for the social markers of everyone else.[21]

Even taking to a foreign lifestyle doesn't lead to a merger. For example, the Fur people of Darfur live on arid land that normally supports little livestock. A lucky family with surplus cattle can keep their livestock fed only by pulling up stakes to join what are known as the Baggara. But this is no change in identity. The Baggara aren't a tribe; rather, the term is an Arabic word that applies to the herding lifestyle, which many tribes in Darfur pursue. So although a Fur family might become herders among those other tribes, and even be respected enough by the other herders to gain acceptance as allies, they remain distinct. Even a Fur who marries into another tribe among the Baggara lacks the upbringing to be mistaken for a native-born member of that tribe.[22]

Despite the human capacity for generating foreign partnerships, full mergers of societies are likewise never an outcome of alliances. Indeed, psychologists find that societies heavily dependent on each other are inclined to set themselves apart.[23] The Iroquois confederacy was crucial in fighting mutual enemies—originally other Indians, then Europeans—with the tribes tasked with defending different borders of their combined realm. However, the independence of its six tribes was never in doubt.[24] Coalitions such as theirs could be a source of pride, yet that didn't diminish the importance of their original societies.

Of this, then, we can be reasonably sure: societies, from the conjoined bands of hunter-gatherers to great empires, never freely relinquish their sovereignties to build a still greater society.[25] Aggressive acquisitions of both people and their lands, and not willing mergers, brought different societies into one fold. Greek philosopher Heraclitus was right to proclaim war to be the father of all things. In locations from the Middle East to Japan, China to Peru, the only way a society created a civilization was to combine an explosion in its population with the expansion of its realm by force or domination.[26]

ACCEPTING AN OUTSIDER

The absorption of outsiders into human societies would not have begun with aggression. It started—harmlessly enough considering that both sides in the arrangement could benefit—with the acceptance of the occasional outsider as a member, as is necessary for finding sex partners in many species. While band societies tended to be populous enough for members to choose a mate from among themselves, some transfers would have taken place to seal a partnership between groups and to minimize inbreeding over long spans of time.[27]

Bringing in such a foreigner would not have been an easy proposition even for early humans. The mate (often a woman) or adopted refugee or outcast would have borne the effort of adapting. A few of the newcomer's exotic behaviors might have been encouraged, as when a society gained from his or her skills—a better option than trading for a tool you can't make was getting ahold of the toolmaker! Nevertheless, given how people nurture and protect their identities in the face of contact with outsiders, a newcomer would have had little effect on the society's conduct.[28] Any new arrival inept at or averse to changing his or her ways faced a hard life and possible rejection. Still, the foreign-born could alter his or her identity to suit the new society only so far. Conversion of membership can never be absolute. Even when people try hard to fit in, their inner essences remain alien, unalterable.[29] After years among the Yanomami, anthropologist Napoleon Chagnon wrote,

> More and more of them began to regard me as less of a foreigner or a sub-human person and I became more and more like a real person to them, part of their society. Eventually they began telling me, almost as though it were an admission on their part: "You are almost a human being, you are almost a Yanomamö."[30]

Giving credence to the Yanomami perception of Chagnon is the fact that no newcomer to any human society could ever perfect all the markers of a foreign people. Telltale signs would divulge a person's origins even if he or she managed the changes that mattered most to the society members well enough to find a place among them.

Even so, adding an outsider or two in this manner wasn't sufficient for anything remotely close to an entire ethnic *group* to materialize. An unpleasant truth is that ethnicities made their nascent appearance by way of an industry that didn't make wide headway except among people who had settled down: slavery.

TAKING IN SLAVES

Slavery is an almost exclusively human activity. Of course there are the slavemaker ants overviewed earlier in the book. Among other vertebrates the only behavior at all like enslavement occurs among langur monkeys. A female that has never birthed an infant can steal one from another troop to raise it (though the odds of a practice baby surviving are dismal).[31] In West Africa, raiding male chimps sometimes goad a foreign female to their territory for sex rather than kill her, but she slips off for home at the first opportunity, the very same day.[32]

Seizing a foreigner and keeping him or her around permanently was seldom an obvious option for people in bands. Escape was too easy. Even so, raiding parties could take any surviving women, who had little alternative but to marry the winners. Slavery was regularly practiced by a few band and small tribal societies, such as among the

Indians of the Great Plains, who did not just take captives but traded them as commodities.[33] Although such captives could run away, they may have found their former identity so defiled, they could never return home. We find an example in an 11-year-old Spanish girl captured by the Yanomami in 1937. After 24 years, Helena Valero ran off only to find her half-indigenous children shunned by the Spanish community since, as she bitterly recounted to anthropologist Ettore Biocca, "she was an Indian and her children were Indians."[34] A woman seized by the Comanche in 1785 refused rescue despite being the daughter of the Chihuahua governor. She sent back word that she would be miserable going home with a face inked with tribal tattoos, the sort of indelible marker that all but ensures a lifetime commitment to a society—and turned that Mexican woman into a foreigner among the people of her birth.[35] In both instances the captive was European, yet tribal people caught by another tribe faced the same problem.

Bondage grew in significance as settlements became organized to hold captives, although not all settled people kept slaves. Even the Northwest Indians had been living in settlements for centuries before taking to slavery with a vengeance. Often those tribes made sure slaves had little opportunity for escape by kidnapping them on expeditions to villages so remote that getting home was well-nigh impossible.[36]

Slavery took the inequality in the relationship of newcomer to receiving society to the absolute extreme, granting the latter total dominance over a prisoner. Captives retained their status as aliens and were dissuaded, or prohibited, from identifying with the society. Not surprisingly, perhaps, given the importance of markers in human life, some sort of brand could be forced on seized men and women to ensure their identification as slaves. Tattoos as well as actual branding with a hot iron were prevalent in the Americas and in medieval Europe. Shaving heads was widespread, too: with hairstyle a matter of pride as an indicator of identity, the loss was an intended psychological blow. Adding insult to injury, many societies put slaves through fiendish initiations and banned them from using their birth names, practices that severed any hope a slave might have had of reestablishing former connections while making clear to everyone his or her meager status and loss of meaningful identity.[37]

Once slaves had been rendered damaged goods, permanently unable to go home, slaveholders could take advantage of the slaves' insights about the societies of their birth by bringing them along to capture more slaves.[38] Slaves were valuable, too, for their language skills in negotiating truces or trade deals. One of history's best-known captives was Sacagawea, who was kidnapped from the Shoshone by the Hidatsa at the turn of the nineteenth century and later served as a guide for Lewis and Clark.[39]

The payoff of having slaves was enormous. A hostage taken during a brief assault could yield a lifetime of labor at a cost to the captors no different from what they expended on beasts of burden—food and shelter—without the time and expense of raising him or her from birth. The fact is, North American Indians lacked work animals, so the slaves of Pacific Northwest tribes were as economically crucial as horses or oxen were for many Old World societies. Indeed, history is rife with explicit comparisons of slave to beast. More than anything else, such comparisons starkly reveal the antiquity of the human penchant to consider only one's own people fully human and to assign varying lesser degrees of humanity to outsiders. Even egalitarian hunter-gatherers could think of foreigners as less than human; slavery put the concept into daily practice, while giving those nonhumans value as goods. Writing of the Pacific Northwest Indians' view of other tribes, one scholar tells us that "free populations of the coast can be seen as analogous to uncaught salmon and uncut trees. And just as the fisherman turned the salmon into food and the woodworker trees into shelter, the predatory warrior converted freemen into wealth."[40]

A slave's status as an animal, extreme as it might be, was a straightforward extension of the imbalances in prestige that often emerged between people in settled societies. Shorn of identity, the lowest of the low, slaves fell neatly at the base of the pecking order generated by human minds. This seemingly preordained hierarchy allowed their abysmal rank to register as natural and just to people from before Aristotle's time and on through the millennia. Indeed, one reason to take slaves from afar was that the novelty of the captives' appearance made it easier to see them as distinct and thus inferior.

While members of the elite owned most slaves, their existence was a boon for low-status society members, too, who were freed from the view of themselves as occupying a society's bottom rung and from doing the degrading labor that came with it. This suggests another reason hunter-gatherer bands seldom took captives: bondage rarely made sense when everyone carried out equivalent tasks with leisure time to spare. Supervising slaves would have simply added to their workload. Yet not every slave was treated badly or did menial or hazardous labor such as hauling waste or quarrying stone.[41] Slaves performed best in prime condition, and a leader's slaves had to be worthy of his status. Regardless of their roles, however, slaves were identity-affirming reference points. For example, the historian Theda Perdue explains that among the Cherokee, slaves "functioned as deviants," an invaluable service given that "the members of a society frequently establish their identity not by proclaiming what they are, or the norm, but by carefully defining what they are not, or deviance."[42]

Once a society was dependent on slaves, it typically had an ongoing commitment to seize more, because slaves rarely bred enough slaves. Males were routinely castrated to make them easy to handle, and stress could depress reproduction for both sexes. Just as slavemaking ants must undertake repeated raids, at times on the same nests, to replenish slaves that have no queens of their own to reproduce, slaveholding humans had to conduct further raids, often on the same "inferior" societies, to keep up slave numbers.

THE CONQUERING SOCIETIES

The very existence of slaves required a society's social boundaries to expand to allow for their numbers and strangeness, a radical achievement. Yet in its initial form among the hunter-gatherer and most of the tribal societies in which it was practiced, enslavement meant adding only a few souls now and then. Despite the fact that these enslaved were greatly outnumbered and could be treated as mere animals, they were a harbinger of the diversity within societies to come. Indeed, by their mere presence slaves made the inclusion of outsiders in substantial numbers a comprehensible notion. However, the question remains how societies came to engulf entire populations and consider them as fellow members.

What started the process along was a change in the motivation for warmongering. When people set down roots around a rich source of food, wild or domesticated, they often had to defend themselves from greedy neighbors. Pacific Northwest tribes had substantial homes and valuable fish runs and stockpiles that could be seized; they also possessed the manpower to protect them or to steal booty from others.[43] Traces of fortifications are ample evidence that ancient villages worldwide needed to defend against the menace of outsiders.[44] The mere existence of a settlement would likely have amplified the fear and distrust, since the concentration of individuals at one place can create an appearance of potentially dangerous unity—a logical reaction for outsiders to have, as individuals in close proximity can act in concert quickly.[45]

The density of goods, which necessitated a defensive posture, led to the expansion of societies in space as well as population. Although taking someone's land, let alone the people on it, would rarely have occurred to nomadic hunter-gatherers, tribal societies found that the spoils of a fight could be multiplied manyfold when a pugnacious group annexed a productive territory and let its residents live. In fact, once people began to live in anything larger than band societies that could fall apart after repeated raids, completely destroying a foreign group was seldom either the goal or the outcome of clashes in the past.[46] In this light, much of the history of the past century, such as the Nazi vision of a world without Jews or the dedication of ISIS to obliterating Shia Muslims, looks all the more aberrant.[47] The Bible claims every man, woman, and child of the Canaanites of Sodom and Gomorrah were killed, but as genetic studies have shown, they are the ancestors of the modern-day Lebanese.[48]

Wiping people out makes no sense given the incentives to profit from a conquered population. Much as enslavement can prove economically more lucrative than killing people, subjugation applied that calculus to taking in entire societies, from which tribute or labor could be extracted on a continuing basis. No other animal shows anything approximating this colonial conquest. It's absent among ants, for which surrender followed by subjugation is impossible. In dispatching the spoils of war, across all the ant species there are only two options: to take slaves or to wipe out the losers, in which case cannibalism is common, as it has been for humans.[49]

Hunter-gatherers in bands couldn't pull off conquests, yet villages could. Not that all tribal societies took to appropriating territories. Indians across the Pacific Northwest ranged far to seize slaves but rarely if ever took control of a foreign people and its land. By contrast a tribe with a culture bent on expansion and dominion most often fought with the occupants of territories adjacent to their own, both because a hostile neighbor presented more of a threat than one far off, and because the people on nearby properties were easier to control. The victors who used this tactic are called a chiefdom; their leader, a chief.

Chiefdoms were never more than a minority of societies. Still, European explorers came upon chiefdoms by the hundreds, some with many thousands of subjects. Much of eastern North America, for instance, was claimed by chiefdoms renowned for maize agriculture and earthwork mounds. Not every chiefdom relied on agriculture, however. For example, the Calusa of Florida were a chiefdom of settled hunter-gatherers.

Chiefdoms were a turning point as pivotal to the advance of societies as was the evolution of markers of identity. No civilizations would exist without the pattern set in motion by chiefdoms after the

Neolithic Revolution: the subjugation of foreign societies rather than routing, ruining, enslaving, or murdering them.

To conquer others, a village required an effective supervisor. While tribal leadership, as noted earlier, was usually weak, the occasional person could gain prominence. Such a figure, coined a Big Man by anthropologists (as with chiefs, most were male), achieved a following usually after showing himself to be a superb warrior. In New Guinea, where tribes were at odds constantly, there were (and still are) many Big Men. Depending on the threat to their people, these headmen could shift in influence—and then disappear, as I related for the =Au//ei Bushmen, who transitioned almost directly from bands to Big Men societies and back within historical memory. When there was a perceived risk of neighbors striking first, Big Men could turn to the brute-force techniques employed by many alpha male chimpanzees, but to much more far-reaching effect.[50] In that situation a Big Man could come to seem indispensable for coordinating the many, who would have acquiesced to being a united hive for the sake of strengthening the "we-group" for war, as sociologist William Sumner put it.

A Big Man could become a chief by taking control of other villages. That didn't always happen by seizing enemies. At times he strong-armed a permanent union from what had originally been a fair alliance between friendly autonomous villages, put together to serve the needs of the moment in combating a common adversary. Such a predatory individual could then commandeer the whole region as a base to further expand his domain.[51] Vigorous chiefdoms came to take over what had once been many independent villages, and eventually entire other chiefdoms, to attain populations in the tens of thousands or more.

Few chiefdoms lasted long. For one to persist, its chief had to put a stop to insurrections over the long haul. Like a Big Man, a weak chief had to continue to earn his people's respect, and their faith in him rarely lasted and seldom automatically extended to his children. A chief's best bet was to act on people's dread of being attacked by keeping the battles going. Ultimately, though, a chiefdom had to persist through times of peace, which required that a chief's position, and that of his chosen heirs, be set firmly in place. Inherited status exists among certain animals: a female spotted hyena or baboon assumes the social position of her mother. As for humans, supporting leaders over the long term may be aided by a psychological disposition to see the status quo as legitimate. Displays of power were a basic part of the job, and the extravagant apparel of royalty traces back to headdresses of early chiefs. Today the most downtrodden are prone to believe the high standing of others as warranted, and to assume that important in-dividuals are smart and competent.[52] This perhaps innate belief could have evolved to protect people from the impulse to try to overthrow the powerful and thereby put themselves in danger, which would ex-plain why people concentrated in settlements have always been sus-ceptible to dictators, autocrats, and notions of divine right. Belief that only a deity has power over a leader ensures his ascendency.

Maintaining control of many people, especially when they belong to multiple groups, has always been a laborious business.[53] For an expanding chiefdom to continue functioning, the defeated, though they might be dehumanized to an extent, couldn't be as denigrated as slaves. Their former identity was not entirely lost to them. Many remained on their land with family and community, a situation that allowed their populations—unlike those of most slaves—to increase. Yet life in a chiefdom could be tough. The residents of independent villages, like members of a band, had little reason to exert themselves more than was necessary to get by. Once conquered, however, the van-quished may have been a step up from slaves, but they were still often regarded much like resources to be exploited. Consolidation into a larger society meant that market relations had spread beyond local village campfires, and in theory the goods taken from the subjugated bolstered the economy to the benefit of all. Nevertheless, the spoils disproportionately went to the chief's people or were funneled into further conquests. Compounding the rapacity of chiefdoms was the demand for resources by an expanding part of the society, from priests to artists, disconnected from food production altogether.

In a band and village society, the constituent villages and bands had been fully able to act on their own and did so most of the time. With chiefdoms, such loose connections between the populations of a society became a thing of the past. Chiefdoms were thus a formative step in a consolidation of societies into a unit—a litmus test for what we think of as a sturdy nation today. To endure for generations, soci-eties starting with chiefdoms needed to achieve what is impossible in other species: a sustained tolerance, if not melding, of groups that had been formerly distinct. Perhaps counterintuitively, then, this forma-tion-of-the-whole was strongest not in societies whose members were most alike, but in those where people of varied origins came to coexist and depend on each other. This was notably true for the state societ-ies that would rise from the most lavishly successful chiefdoms, whose political organization and stability, and influence in welding together people with different backgrounds, we will consider next.

CHAPTER 23

Building and Breaking a Nation

F ive and a half millennia ago, Uruk, a handful of interconnected towns located east of the Euphrates River, in what is now Iraq, swelled in population and complexity. The largest of them housed thousands of people, supported by a range of goods and services the likes of which had not been seen before. There were streets, temples, and workshops. The numerous tablets with cuneiform writing recovered in the area suggest that many aspects of life were carefully watched over.[1] Uruk is among the first examples of a society that transformed so starkly that what began as a chiefdom had taken on a new means of organization: what academics call a state society (or nation, as I will sometimes refer to modern states, following everyday, informal usage). Although some of the earliest states were no bigger than hamlets by modern standards, they were nevertheless the sort of society we pledge allegiance to today.

States shared a number of important attributes from their initial appearance. Crucially, in a state society the leader escaped many of the encumbrances that burdened chiefs. A chief had a limited power base, and could be relatively easily overthrown. The fatal flaw in a chiefdom was the chief's inability to delegate authority. When the chiefdom grew large, the former chiefs of its subjugated villages could be allowed to retain their position, but the paramount chief had to oversee each of them personally. This kind of flimsy oversight, based largely on the leader's dominance or powers of persuasion, became impractical once the combined territory took more than a day to traverse.[2]

All that changed with the emergence of states. Heads of state not only asserted the sole right to enact their will, they could back the claim up with the support of a formal infrastructure. In a state, division of labor and hierarchies of control made their way into institutions dedicated to governance. It was, then, with the proud birth of bureaucracy that societies came to improve their cohesion and rule great expanses. When one state conquered another, each of the former state's territories was typically converted into a province, its capital refashioned into an administrative center.[3] Government agents, each master of a particular job, were distributed as needed. This system of oversight meant that societies could be ruled more coercively than before, even if lag times in communication from the capital to the outer reaches continued to be a handicap for early states. In fact, with a strong enough infrastructure, a state could survive the overthrow of its leaders or reign in their worst impulses.

States are distinct from chiefdoms in a few other particulars as well. For one, laws are truly laid down: while people carried out vigilante justice in societies with weak leaders, in a state those with authority mete out punishments. In states, too, notions of private property are fully realized, including luxury goods sought after by an upper class. Indeed, while people in chiefdoms could gain prestige and sometimes showed differences in social class, in states inequalities reach extremes. Differential access to power and resources can be either earned or inherited, with some people working for others. Finally, states extract tributes, taxes, or labor from citizens in a more formal way than chiefdoms; in return, they offer infrastructure and services that ensure the members are more dependent on their society than ever before.

ORGANIZATION AND IDENTITY IN STATE SOCIETIES

States around the world bear similarities not just in such defining features as their administration of power, but also in the organization of those infrastructures and services. Like any society, a state is a problem-solving organization, and its big problems often require complex solutions.[4] On this point, we discern in states many of the patterns we have already found among the social insects. When a society, human or ant, becomes sizable, the demands on it to provide for and protect its members grow intricate and diverse. Consequently, so must the means by which these obligations are met. Methods must be found for the transport of supplies, troops, and other personnel when and where goods and services are required. Failing to meet basic needs could be cataclysmic. Hence, while there's more than one way to configure a state worthy of the term "civilization," with its impressive urban centers, the range of possibilities is actually quite constrained.[5] When states and their cities expand, land use becomes more structured, institutions from intellectual academies to police forces grow more elaborate, and the range of job opportunities skyrockets.

There are also improved economics of scale. For instance, it can be easier to feed and house each member, and these lowered costs can lead to a resource surplus that ants invest in warfare and humans in the military, though our species can divert the surplus as well into the sciences, the arts, and such nonessential projects as the Taj Mahal, the pyramids, and the Hubble telescope, which require ant-like levels of coordination and industry.[6]

In fact, the similarities between world civilizations, even those with completely separate histories, are remarkable, if not downright eerie. Historian and novelist Ronald Wright put it this way:

> What took place in the early 1500s was truly exceptional, something that had never happened before and never will again. Two cultural experiments, running in isolation for 15,000 years or more, at last came face to face. . . . When Cortés landed in Mexico he found roads, canals, cities, palaces, schools, law courts, markets, irrigation works, kings, priests, temples, peasants, artisans, armies, astronomers, merchants, sports, theatre, art, music, and books. High civilization, differing in detail but alike in essentials, had evolved independently on both sides of the earth.[7]

Many of these innovations not only made it possible to feed and house large populations; they also contributed to keeping societies intact by influencing how the people thought of others. Keeping up an affiliation across a few hunter-gatherer bands was improbable enough; maintaining a common identity grew daunting as societies expanded—at times to the breadth of a continent—and tribes turned into nations, with all their varied subjects. Much of the problem was, and is, a matter of connectivity. To slow disruptive shifts in identity, the population must be interconnected: the more, and more up-to-date, knowledge that citizens have about each other, the better. People can either stop a change or adjust to it, yet only if there is an efficient exchange of information within the society.[8]

One factor that allowed human interactions to skyrocket was our species' evolved flexibility with regard to personal space. As described earlier, this was important even for the first human settlements. This flexibility has reached its zenith today, however. The populations of Manila and Dhaka work out to roughly one person for every 20 square meters, a million times the density of some hunter-gatherers. Our comfort with the proximity of others depends on upbringing. Yet barring a disability such as agoraphobia or monophobia (a fear of crowds and of being alone, respectively), people existing cheek by jowl exhibit few pathologies.[9]

Proximity is just one, very low-tech, way for people to stay in tune with the identity of others. Even still, it is not essential. After all, part of the population of any state must live in rural areas to cultivate the crops. Societies developed other methods of maintaining contact throughout their realms. The domestication of the horse in Eurasia, the invention of writing by Mesopotamians and of ocean-going ships by Phoenicians, the long-distance roads of the Inca and Romans, the printing press in Europe—all such innovations promoted the stability and expansion of societies. In addition to facilitating the transport of goods and the extended control of central authority, such innovations improved the spread of information—and notably, information about identity. This was true not just for states. Once they had horses, tribal nomads such as the Tartars and hunter-gatherers like the Shoshone kept their identities intact over much greater distances than would have been possible when their ancestors moved only on foot, though the complexity became far greater as states took in more, and more dissimilar, groups.

Jump ahead to ancient Rome. At its height, all but the most remote parts of the empire were firmly linked by an identity that encompassed vocabulary; fashion from apparel to accessories to hairstyles; crafts such as pottery, floor mosaics, and stucco walls; casual customs, formal traditions, and religious practices; cuisine; and home designs and improvements such as plumbing and central heating. Identification with Rome carried over as well to public works like the layout of towns, roads, and aqueducts. This is not to say the provinces conformed to a uniform standard of Romanization. Rome, like all societies, made allowances for diversity, and across its range people expressed their identification with the empire with geographically local flourishes that reflected their ancestry—and class-based ones, as well, with the affluent opting for the most and most expensive emblems of Roman identity.[10] This sweeping accommodation of markers required efficient communication beyond anything that was possible in a tribal society.

Whatever form leadership took—the rotating arrangement, leadership by committee, or single-person rule—a leader helped shape the social fabric, and part of that mission was shoring up people's identities. Sometimes the influential just set down what was broadly agreed to be acceptable behavior, sometimes their own eccentricities became fashionable, and sometimes leaders could compel behavior of their choosing, setting standards for anything from speech to dress. Given the vulnerability of their positions, the first and most innocuous of these roles, serving as the Voice of the People, was the bread and butter for the leaders of tribal societies and chiefdoms. By setting a firm example, an effective leader could secure his position by contributing to the citizens' sense of shared identity and fate, helping to keep the people's bonds strong even when they were numerous. But once a population was firmly under a leader's thumb, his or her authority tended to amplify: kings seldom felt obliged to show the generosity of a headman at a potlatch. In many historical instances, the influence of leaders reflected a firm control of the roadways, printing presses, and other means by which information flowed across their societies.

By the time a society manifested the organization of a state, the role of religion had typically changed in ways that further enforced people's identification. Hunter-gatherers admired those imbued with healing or spiritual skills, yet their animist philosophies asked little of their followers.[11] Most tribes and chiefdoms weren't much different in this regard, but the invisibility that a state's large population afforded to its members required they be more strictly supervised. The notion of omnipotent gods provided a mechanism to influence people's behavior even behind closed doors, through fear of divine punishment.[12]

So long as their governance isn't overly despotic, the benefits states offer can be profound. Not only can the hive-like level of interaction within them strengthen people's collective identity; the rapid-fire exchange of information brings on the opposite of the Tasmanian effect, in which sparse and weakly connected populations forget the innovations of their forebears. Once multitudes interact, fresh perspectives don't just catch on, they are caught in the mill of constant social change. The cultural ratcheting that began in earnest 50,000 years ago has accelerated to the point that no one grows old in a society that hasn't altered markedly since he or she was born. Such enhancements make identifying with a society more of a moving target than it once was.[13] What's more, the mass connections proliferating in states are matched by mass ignorance. In contrast to the near comprehensive cultural smarts of hunter-gatherers living in bands, not even the leader of a state knows more than a fraction of what is required to keep the society functioning. People today often find it necessary to track rapidly changing social trends in deciding what they themselves should do.

The safeguarding and control of large populations require a heightened level of organization that can be traced back to the first state societies. States assembled armies to put down rebellions and carry out further territorial invasions using offensive strategies that changed as societies grew. Cautious raids made sense for hunter-gatherer and tribal peoples, since they could afford to lose few fighters and the aim was to inflict damage or kill rather than subjugate. Big Men and chiefs often had to personally lead warriors in order to motivate and maintain their following. Even then each combatant was prone to act alone, with plans breaking down in the heat of the moment with little accountability.

By contrast, the ruler of a state remained safely ensconced in the capital, where he or she could direct attacks and, if the battle was won, oversee the takeover of the territory and its survivors. Military tasks could be delegated to specialists. For a state to sustain a large population and fuel these war efforts, farming, usually of energy-rich grains such as wheat, rice, and corn, had to be carried out at massive scale. A state might overrun its competitors not just because of its greater supply of fighters, but also through its superior tactics, weaponry, and communications. Notable too was a state's tight control over its regiments, composed of citizens drafted for war. The army's strong collective identity, instilled through training and a submersion in patriotic symbols, cemented the soldiers' resolve to "stiffen the sinews, summon up the blood," as Shakespeare put it. Disciplined training ensured much greater reliability, and uniformity, among the troops. This uniformity and the sheer scale of wars guaranteed the impersonal nature of the enterprise. All traces of individuality were stifled. Combat between large states took on ant-like qualities in giant anonymous societies. Social substitutability—the sense that retribution for wrongdoing can be inflicted on anyone in the guilty group—was amplified from the time when raids of hunter-gatherers killed whatever foreigner they came upon, often someone they knew: armies faced soldiers who were interchangeable, and indistinguishable, strangers. Negative stereotypes about the enemy would have clouded any perception of them as individuals, as often happens with foreigners. There was also the fact that the enemy attacked in numbers and dressed in uniforms that made it next to impossible—and completely unnecessary—to tell them apart.

THE MARCH OF CIVILIZATIONS

"War made the state, and the state made war," sociologist Charles Tilly aptly observed.[14] There have been no truly pacifist states. Whether we speak of a chiefdom or our country, peace masks generations of power plays and almost always swordplay. Any society larger than a handful of villages is composed of once-independent groups. Minoa, a Bronze Age civilization on the island of Crete, was renowned for a tranquil culture of merchants and artisans.[15] Yet even the population of Minoa, serene in its heyday, must have come together, in this case before historical records, by force. The same is true of the modern-day people of such long-peaceful states as Luxembourg and Iceland, when their history is traced back far enough. Just as chiefdoms swallowed tribes and then each other, the pattern of the expansion of the nations and empires that ensued stayed the same. Throughout recorded history conquering was followed by consolidation and control, repeated ad infinitum.

The birth of states in conflict, and their obligatory inclusion of people from varied sources, have a simple explanation: by the time states emerged, essentially no unoccupied habitable land remained. Some group—whether roaming hunter-gatherers, a tribe, a chiefdom, or a state—was already there and willing to do whatever it could to retain its independence. Any expanding society had to push aside, conquer, or destroy other peoples whose territories would have been part of an unbroken quilt on a map. Not that every battle was directed at the control of land—much of the time the plan was to bring back booty and slaves. All the same, the most avaricious and successful states pushed their territorial borders ever outward. Only a few hunter-gatherers would have been left to eke out an existence in places skipped over because the land was too barren for agriculture.

Superiority in battle wasn't enough, however, for a society to make the climb from chiefdom to large state. The few titanic civilizations that emerged generally did so in settings where the societies they took over were packed in a tight space. Under this condition, described by the anthropologist Robert Carneiro as circumscribed, conquest paid off royally. "War appears when mobility is not an option" is how the anthropologist Robert Kelly framed it.[16] Tribes farming fertile sites encircled by inhospitable regions experienced such circumscription, locking them in a fight from which in time just one would emerge.[17] Think of the Nile Valley hemmed in by deserts, where ancient Egypt took hold; or the Hawaiian and Polynesian island chains, specks in an overwhelming ocean, where giant chiefdoms, some containing 100,000 persons, claimed their domain.[18]

Though far from guaranteed by a circumscribed setting, civilizations were more likely to emerge in such places than elsewhere.[19] Where there was no circumscription, a chiefdom or state could reach a modest size, then be unable to pursue further expansion as surrounding societies shuffled their locations to escape takeover. That was the situation in New Guinea, where an entire tribe, for example of Enga, could move as one unit to keep from being cornered—reminiscent of the flight response a small ant colony can employ to avoid conflicts. To settle in a new place, the fleeing tribe would have to negotiate each shift through alliances with its new neighbors.[20] Because people develop emotional ties to territories, these migrations signaled extreme duress. Yet apparently as a result of this fleet-footedness, the island of New Guinea never supported even a small state society until the country called Papua New Guinea was founded at the eastern end of the island in 1975. Even then it took years for many of PNG's citizens to learn that such a thing as a "country" existed. To the majority of the people, their tribes continue to matter most.

When an aggressively expanding society couldn't be escaped, often victory was pulled off without bloodshed. Neighbors could be cowed by displays of prowess heightened by uniforms and flags that blurred the marshaled troops into a united "bristling monster," as one battle account described.[21] "It was an explicit policy of the Incas, in expanding their empire, to try persuasion before resorting to force of arms," Garcilaso de la Vega, an early seventeenth-century chronicler of South American history, explained.[22] But showmanship no doubt set the tone of the negotiations, and it may have been safer to acquiesce to subjugation than to be incessantly terrorized by hordes of well-armed outsiders.

FRAGMENTATION, SIMPLIFICATION, AND CYCLES

It bears reiterating that nothing about the transition to states was preordained. Nevertheless, the emergence of large, complex states—civilizations—can have the appearance of inevitability, if only because their expansion slashed the number of small societies. No wonder states have been by far the rarest form of society, but are now ubiquitous.

Conquering chiefdoms and states sprang up the world over after the advent of agriculture. Most were a flash in the pan, this transience a reminder that no level of complexity secures a society's continued existence. The epigraph of this book quotes Plato, who wrote that a state must only be allowed to increase to a size consistent with unity. Should the learned rulers of such a state heed Plato's advice, however, its unity would still be transient. States undergo a cycle, never an enduring stasis or inexorable ascent.[23]

In *Collapse*, Jared Diamond explores how such factors as environmental degradation and competition can hasten a society's demise.[24] But what Diamond calls collapses are a few extreme instances of what is actually the ever-changing nature of societies. Most significantly, such a collapse—when the word is used to indicate the abrupt termination of a society rather than a bleak economic downturn—is more accurately a fracture. Though band societies and tribes divide, the modus operandi for breaking apart chiefdoms and states has been a complex yet often predictable sort of splintering. These breaks were sure to happen, in the end, regardless of resource abundance. More specifically, the manner by which chiefdoms and states absorb outsiders makes them likely to sunder along the demarcations on a map that conform roughly to their past military occupations, even if political and topographic factors and the vagaries of war can have an impact on where the dividing lines ultimately fall.[25] For example, when a Maya civilization shattered, its families didn't just slip into the jungle, and certainly not across a wide area or over the protracted time the word "collapse" might suggest. Generally, a king lost control. The top tier of the society vanished. Former sacred and symbolic objects and public works were often desecrated.[26] Outlying districts, subjugated in a distant era, would thereby break free under leaders who no longer had anyone to dictate what they could do. The fracturing of the final Maya kingdom, Mayapan, decades prior to the arrival of the Spanish, left 16 small state societies for the conquistadors to find.[27]

Conquering and fractioning alternated through time. Had the Spaniards made it to Mexico a century later, at a different point in this loop, they may well have been met by another Maya empire like Mayapan, reborn, or perhaps—if a recent breakdown had been especially harsh—by isolated farming villages. In the former case the palaces, temples, and art would have stunned them far beyond anything they actually did see. These public works fall to ruin each time control of labor and essential supplies dry up. As a result, the smaller societies left behind after a breakdown experience a cultural simplification. Though often still in possession of their palaces, their more rural populations would no longer have the resources and manpower to maintain them in their previous splendor; in extreme situations people could see-saw between agriculture and hunting and gathering.[28]

The Roman Empire foundered, in the end, into smaller states as well. These stayed intact by delegating leadership responsibilities to lords in what became a feudal system, a form of social organization in many ways functionally equivalent to a chiefdom.[29] At that point societies, in the sense of people's identification with a greater group, would seem to have lost their sway across much of Europe. Yet careful scholarship demonstrates that the local lords crippled, but didn't erase, the sense of belonging peasants felt with the region beyond the fiefs. That nations quickly sprung up during and after the Middle Ages reveals how people can reconnect to feelings of solidarity that stretch far back in time.[30]

Cycles of conquest and dissolution are evident everywhere. Historical records are replete with chiefdoms and states that gained, then lost, members and property after expanding too far to hold their territory. This latter phenomenon, which Yale historian Paul Kennedy calls "imperial overstretch," though it's not at all specific to empires, is precipitated by the way that overambitious societies can be worn down by invasions and civil wars.[31] In such affairs, economics mattered. Control of distant peoples could be an expensive logistical ordeal. There would be a fine line between perceiving an outlying population as a benefit or a nuisance, either because it competed with its conquerors for resources or because it didn't contribute enough to justify the conqueror's investment, or both. Meanwhile the province seldom recouped enough rewards to offset the resources taken from it, and the imbalance fomented discontent. Overburdened families could creep into the hinterlands, living as refugees beyond the reach of the government. Yet countless times through history those people who chose to stick to their land eventually won their independence, whether through the conqueror's willing retreat or in a struggle.

Imperial overstretch could have been motivated by the fact that the most exotic and desirable goods often came from societies that were radically different and immensely far away. At some point the practical option shifted from conquest to commerce, a calculus that led to construction of the Silk Road and the trade networks of modernity.[32]

Other patterns emerge at grand scales. Mesopotamia and Mesoamerica were like Europe: collections of states within a geographic region that, due to both trade and their historical connections from past conquests, shared much in common. Both individual societies and these regional expressions of them could go extinct. Sumer, Akkadia, Babylonia, Assyria, and other states around the Tigris-Euphrates river system came and went. Still, they all exhibited common threads we call Mesopotamian, with overlapping artistic styles and polytheistic religions. Then, in the first century AD, after three millennia of brilliance, Mesopotamian cultural traditions, political organizations, and languages all but vanished. They were brought down by the invasion of nomads arriving from the south and west, leaving only trace influences on the modern populations of the Middle East.[33]

All this was a wild ride played out in super slow motion. Each rise and fall left its imprint on the population of a region, yielding a fresh medley of cultural ingredients. A conquered people re-freed could revert to the ways of their forebears, but not entirely. The greater society of which they had been a part left its mark, as they would have learned the language or taken to the beliefs of those who had once been in power. Across a geographical expanse, then (for example, over the full extent of Mesoamerica occupied by the many and successive Mayan civilizations), people retained evidence of their past unions in the increasing similarities of their identities. Such an overlap would have made their conquering and control of each other easier the next time round.

The imprint of societies past may explain certain curiosities of world geography. Distantly related peoples of the South American rainforest, for example, have similar crafts, traditions, and languages not easily explained by trade. Their hybrid cultures could be a signature of powerful chiefdoms, now vanished, that overpowered and refashioned the identities of many. Archaeologist Anna Roosevelt, great-granddaughter of Theodore, has argued that the Amazon Basin once had urban centers. The evidence for these cities, such as prehistoric canals I have seen in Suriname, lies buried beneath a green carpet.[34]

THE ADVANCE AND RETREAT OF NATIONS

For the last century, at least, most wars have been civil, often brutal attempts to break societies rather than expand their reach.[35] The factors already spelled out for divisions of early human societies continue to play a role. Identity remains pivotal in both building and fragmenting nations, with a key difference. Schisms in a band society emerged from the ground up across a territory because of distinctions brought about by faltering communication between people who had been pretty much the same when the society formed. Factions arose from scratch among the society's members in a way that would have been hard to foresee when the tribe was young. For nations nowadays, however, many factions are symptomatic of groups whose dissimilarities extend into the remote past—indeed, that predated the state. Populations struggling for independence, such as the Scots or Catalonians, aren't concerned only with political and economic interests, but have a substantial psychological driver: they picture themselves as different going back to the mists of time.

It is fascinating to consider what this means in psychological terms. The response of early humans to a schism in their ranks would likely have emerged very slowly, until a crisis point was reached among members whose forefathers, at the time of that society's birth, would have seen each other as equals. By contrast, the citizens of a state can react to many uprisings by zeroing in immediately on the timeworn social differences of the recalcitrant outliers, a choice that at times can bring centuries-old conflicts to renewed prominence. In all likelihood, derogatory stereotypical beliefs would surface, ideas that could be traced back, in a rough form at least, to a time when those people belonged to separate societies.

Separatist groups today draw on, and indeed manipulate to intense emotional effect, symbols with a meaningful connection to their past to keep their people united in the face of hardships or to catalyze their secession.[36] The breakup of Yugoslavia yielded a half-dozen republics that recast the anthems, flags, and holidays previously valued by their people; they also added new ones on the spot. These revolutions hark back to how hunter-gatherers swiftly diverged to establish differences after the trauma of a division.[37] But laying claim to time-honored markers is a quick-and-dirty way to revitalize a people's freshly minted independence.

As happened in the dismemberment of early state societies, such as the Maya, the populations of the resultant countries often must get by on less than they did as part of the more inclusive nation. Unless the original regime had been extremely repressive, this makes partition a step back economically. Today, seceding nations receiving little outside assistance rebound slowly—among the products of former Yugoslavia, Bosnia and Kosovo remain in the doldrums. My sense, though, is that the intense feelings of identification with the new society can offset any loss in the quality of life. After all, forced compliance to a distant ruler will often be a social drain. Fortunately the basic kernels of social existence, such as the social support local communities provide individuals and families, seem resilient to the rise and fall of societies. So, while the division of a band society might affect all the members much the same, the breakup of a state is a blow to the powerful but could make far less difference to everyone else.

In all these matters, geography truly is destiny. For a secession to succeed, the clashing group typically has to populate a specific portion of the country's territory, which they often claim as an ancient homeland. This means the break ends up falling between stretches of terrain heavily populated by ethnicities that once had societies of their own there. Nevertheless, drifts in identity from place to place still have an impact, much as they once did across hunter-gatherer bands. These changes help generate regional cultures found in every nation and influence all manner of things from cuisine to politics.

The likelihood that such geographical variation on its own will tear a country apart appears relatively small, however, even when those differences are played up. For example, slavery, not identity, overwhelmingly drove the American Civil War. At the time, most Southerners thought of themselves first and foremost as American. What we think of now as Southern culture became a point of pride only after the war ended. During the fighting, the Southern intelligentsia did promote the superiority of Southern ways to stir up a regional allegiance, including claims that Southern whites were their own ethnic group—descendants of a refined British bloodline that adhered more closely to America's founding principles than did the Northerners.[38] Yet their appeals to Southern commonalities didn't have much effect. Most Southerners, uncertain of the sensibleness of secession, were inspired less by the Confederacy than by the obligation to defend their families. The absence of an intensely shared identity, separate from that in the North, was the key factor underlying the Confederacy's struggle to stay unified, and ultimately its defeat.[39]

While many secessions have featured issues of identity that include reassertions of former territorial rights, a good example being the tense but ultimately peaceful separation of Norway from Sweden in 1905, there are, of course, exceptions. Political issues can come into play to create nations that bear no connection to any formerly autonomous group.[40] When Venezuela and Panama withdrew from

Colombia (originally Gran Colombia) in 1830 and 1903, respectively, difficulties in travel between their territories sharpened the political disconnect. As a result, they became separate countries without any venerable split in heritage among the people. Alternatively, outside forces can prevail, as in the 1945 partitioning of North and South Korea, which was in no small part an outcome of the rivalry between Russians, Chinese, and Americans; or in the mostly artificial lines separating Pakistan and Bangladesh from India, negotiated by the British with the rulers of the principalities set up during Britain's control of the region until 1947. At other times novelties introduced by colonialism were a factor. That was arguably the case in the separation of Eritrea from culturally more traditional Ethiopia: the otherwise similar Eritreans had been shaped by years of Italian culture and rule by the time their country was annexed to Ethiopia after World War II. They won their war for independence in 1991.

No society can persist unless being a part of it is important to its people. Despots keep dysfunctional nations intact for a time, but people with little attachment to a group are less stalwart and industrious.[41] The USSR is one example of a nationality that deteriorated after being imposed on its people not even a century prior. Like Yugoslavia, it balkanized into states to which people felt more committed—in the Soviet instance, Latvia, Estonia, Lithuania, Armenia, and Georgia were the offshoots with particularly deep historical continuities.[42]

Archaeologist Joyce Marcus has found that state societies across antiquity had a finite life span, normally ranging from two to five centuries.[43] This duration suggests states are no more persistent than hunter-gatherer band societies, which the evidence indicates endured for a few centuries as well. We must ask why nations aren't more stable given the controls they put in place, the services they provide, and the improvements in information flow that expose the citizens to each other's perceptions of proper behavior. Despite the benefits of living in a state society, one recurring flaw is that, as an archaeologist proclaimed, states are "ramshackle contraptions, at best half understood by the people who made them."[44] That courts of law, markets, irrigation, and so on had to exist in some form didn't mean people always put them together well. Much of the cobbled structure of a state is grounded in a feeble understanding of the origins of people's devotion to their country and each other, and how these can be managed. Nations can be administered without undue force so long as people feel they share an identity and common purpose. Yet as conquests thundered into human history, achieving satisfactory bonds across a society grew increasingly problematic. Markers of identity came to vary radically, leaving members struggling to achieve visions for their society that were at cross-purposes.[45]

I have shown that accommodating outsiders at a mass scale is unheard of in other animals. The most other vertebrate societies do is to bring on board the occasional lone mate or refugee. Adding foreigners as a class began in our species with the taking of captives and slaves but ramped up with the wholesale acquisition of populations not by mergers gladly made but by force. Over time state societies emerged that were better able to administer and control their varied populations—including what we now think of as ethnicities and races. How these populations get along together is the topic of the next section. State societies, as we have seen, are ephemeral, and to keep going at all, they have to function in the face of their ramshackle structure and disparate populace. Success requires that they retain their cachet as the group that their people value above all other memberships. This means a state has to impart to its citizens a consciousness of being cut from the same cloth, even though—with the incorporation of outsiders—those citizens are often clearly different. In this, acceptance of domination and control has played a critical role.

From Captive to Neighbor . . . to Global Citizen?

CHAPTER 24

The Rise of Ethnicities

S tepping from my apartment building in Brooklyn, I cross paths
with young men and women walking their dogs or running for
the subway into Manhattan. I turn the corner onto Atlantic Avenue to see Sahadi's, a century-old grocery store that smells of fresh-
ground spices and Mediterranean cuisine. All around are Middle
Eastern restaurants and markets where American citizens of Arabic
descent shift effortlessly between the languages of their ancestors and
American English. I settle down at my favorite café, joining an African
American couple, an Arab American family, and a Mexican American
man talking to one of several white people at the small round tables
near me.

For a New Yorker, this is an everyday experience. And yet from the
point of view of a hunter-gatherer from 10,000 years ago, my splendid
spring day would have been beyond comprehension. The incorporation of massive numbers of different peoples has been the single most
radical innovation in the history of human societies. It gave societies
the option to grow by absorbing formerly separate, often adversarial
groups to create ethnicities—that is, groups that once comprised their
own societies but have come to occupy the same society (and that, with
the passage of time, usually see themselves, often wholeheartedly, as
part of that society). While not everyone lives among the sharply different ethnicities around my neighborhood, all societies today, even
those of apparently homogeneous population, ranging from small nations like Liechtenstein and Monaco to expansive ones such as Japan
and China, are admixtures of people, the distinctions among their

citizens merely dimmed with the passing centuries. The other radical changes that took place over the millennia, from innovations in politics to new religious beliefs to scientific achievements, have required minor adjustments by comparison to this acceptance and intermixing of human kinds.

In chiefdoms and the first small states, the mingled populations wouldn't have been what we consider ethnicities and races today. The differences in identity between the conquerors and subjugated would have been minimal to none. More than likely they were merely adjacent villages of the same tribe, once autonomous but now politically fused. Expand enough, though, and a chiefdom or state society would take in not only peoples of distinct societies but groups defined by distinct languages and traditions. Marked differences presented challenges for communication and control, yet had their advantages: the subjugated were unambiguously *other* and could be treated as such. There was less moral imperative to be concerned with the newcomers' welfare. Indeed, by some definitions an empire is a state stretching far enough afield to control peoples of markedly different cultures—a strategy that would turn to colonialism once conquests leapfrogged abroad.[1]

As more peoples were added, encounters with unfamiliar others within one's own society became par for the course. Societies swarmed with people, most of whom were strangers. More bizarrely still, those strangers, no matter how dissimilar, could be recognized and treated as fellow members. How did societies reconcile all this diversity and turn *them* into *us?*

CONTROL

"When peoples were conquered, incorporated into provinces and, in due course of time, became part of an integrated empire," one scholarly treatment of the emergence of racism in the Roman Empire tells us, "this entailed a process of ethnic disintegration and decomposition."[2] During the gobbling up of foreigners, populations could be transformed into ethnicities that interacted well together. This didn't happen across all societies, however. Consider the following passage from a study of the Incan Empire:

> One of the reasons for the Incas' success was their use of the existing political and social structures of conquered peoples for ruling them. Instead of trying to change the people's lives, they tried to

maintain continuity so the subjects' lives were disrupted as little as possible. . . . They assigned conquered leaders positions of authority in the government, gave them high-status gifts, and honored their religious beliefs and practices. In return, the Incas expected the conquered people to work hard for them to produce food, cloth, pottery, buildings, and other large and small items and to be obedient and loyal subjects.[3]

The peoples subjugated by the Inca essentially retained the identities they had had as an independent society. From this you might gather that the Inca were less domineering toward the defeated than the Romans, but I believe this misreads what was a harsh form of indirect rule. Although the conquered peoples received some food and goods from the Inca, the residents of most Incan provinces had no real social status within the empire. They had almost no contact with the regular Incan population, and were prohibited from modeling themselves after their overlords. With their continued foreignness bringing them social exclusion, those unfortunates were for all time wholly and undeniably *other.*

If the Inca had instead chosen to allow their subjugated peoples into the broader society, as Rome and the Chinese dynasties eventually did with the populations of many conquered provinces, those people may have come to identify not just with the Inca, but as Incans themselves, taking pride in the empire even if their stature in it continued to be less than equal. As it was, the occupants of the outlying provinces were beaten into submission, unrests crushed, and unruly villages transplanted to land other than their own to guarantee their subservience. The only possible good to come to them from this abuse was protection, as the Inca fought back tribes beyond their territories. One supposes the border populations may have preferred to bow down to the enemy they already knew. Nothing matches a common foe in drawing people together, and that's true whether those people are separate societies dependent on each other for security, fully accepted groups of the same society, or—across much of the Incan Empire—residents of districts that are browbeaten by their superiors.

Doubtless the system worked. When the Spanish reached Peru in the sixteenth century, they stumbled on an empire containing perhaps 14 million people. This was remarkable given that the Inca had been a pastoral tribe little more than a century prior (albeit a tribe that had the advantage of being able to build upon the foundations of earlier kingdoms in the region). It's one thing, though, for a country to marginalize small groups within it, as the United States would do later with native tribes; it is quite another to lord over a civilization the bulk of whose population is ill-disposed to its rulers and doesn't identify with them. I question how long the Incan Empire would have stayed intact unless they consolidated their provinces by motivating the people there to get behind the ruling population.

The comparison of the Incan and Roman methods of administering a society highlights the distinction between controlling people through out-and-out domination and incorporating them into the society. In the first instance, obedient local leaders were often kept in place after the society was subjugated. They would be handsomely paid to oversee the extraction of goods and services from their people, who received few benefits in return. Alternatively, the population could be brought into the fold with the expectation that, even if underdogs, they would come to see themselves as a part of the society. In that case the central government typically took on the administration of their needs in recompense for goods and services. The more thoroughly the people were incorporated, their allegiance and identity recast and former government disbanded, the less able they were to try to break away from their conquerors.

State societies have attempted manifold approaches, from cruel domination to generous incorporation, depending on the tractability of a people and the profitability of giving them greater status with fewer constraints. The Romans put up with the cultural and religious differences of freeborn people from regions across the empire.[4] However, they still kept unruly outliers, such as certain African districts, on a tight leash while absorbing acquiescent peoples like the Britons.[5] A state's strategy could change over time, too. Over the centuries the Japanese maintained an ambivalent relationship with the Ainu hunter-gatherers of Hokkaido island, at some points attempting to bully them into renouncing their ways and conform with the general population, at other times keeping them well apart from everyone else.[6]

The China subcontinent fascinates in that conquests there began early and to great success, eventually creating a virtual uniformity among what's now regarded as the Chinese, or Han, race, presently amounting to 90 percent of China's people. The scale of this outcome can be attributed to the policy of early dynasties of accepting anyone converting to their culture, writing, and sometimes their language. This tradition traced to Confucius, who promoted the idea that people could become Han simply by committing to a Han mode of life.[7]

Based on the evidence of ancient texts as well as changes in identity expressed in everything from architecture to lacquer work, archaeologists have teased apart how the Qin (221 to 207 BC) and Han (202 BC to 220 AD) dynasties brought together much of the population that would eventually become modern China.[8] Unlike the Romans, who exported to their subjects a miscellany of improvements in plumbing, lighting, and other basics, the Chinese dynasties offered few quality-of-life benefits to the outlying masses, depending more on their military presence to crush repeated uprisings. Some of the tactics used by the Qin and Han were common to many territorial expansions around the world. Both dynasties concentrated on unifying the districts nearest their imperial centers, believed to have been located around the putative birthplace of the original Han in the north. Trusted subjects were encouraged to colonize those districts to ensure the ascendency of the Han culture. Among the provincials, the wealthiest would be first to realize the desirability of teaching their children Han customs. Over the centuries this education filtered down through the social strata to yield a widespread Han identification by the time of the Ming dynasty in the fourteenth century. The focus of the dynasties on the accessible territories could explain why they repeatedly lost control of their most remote compartments, including what would become Korea and Vietnam.

Within the borders were pockets of indigenous societies the dynasties failed to bring into the mainstream. Often the groups inhabited mountainous regions unsuited for cultivation, offering little in return for the effort of subduing them. Some ethnic areas, notably those of the Tibetans and Uygurs to the west and the Wa near the Burmese border, would in the end come under dynastic control, but even then the authorities held these people at a distance as substandard, mirroring the attitude of the early Japanese, who thought of the Ainu as dogs.[9] An unwritten policy was to leave the languages and practices of such heathens intact. In the sixteenth century, the Ming dynasty even walled the rebellious Miao into their mountain strongholds, repressing these and other peoples as any colonial power would.[10] By retaining their identities, the social outliers fulfilled the role that provinces at the hinterlands perhaps had for the Inca, and that slaves had for societies such as the Cherokee. The titan of Greek poetry Constantine Cavafy was right to ask, "Now what's going to happen to us without barbarians? / They were, those people, a kind of solution."[11] Simply by existing, the barbarians cast light on what was proper and right.

ASSIMILATION

What remains to be illuminated is how interactions among such conjoined societies became realigned from enslavement or subjugation
into mutually beneficial relationships that no longer required force. In
such realignments, societies emerged that, despite any social chasms,
inspired the loyalties exhibited by members of small societies in the
past. Somehow people had to accept those who had been outsiders,
with incompatible identities, even though, as we have seen, societies
spurn being combined as intensely as a body rejects a skin graft that
doesn't match the immunological identity of the host. I see the success of these mergers as an outcome of a human attribute, coopted
for a new purpose: while people cannot simply drop one identity for
another, their markers, and how they are deployed, have always had
to be malleable for humans to adjust to the perennial social changes
within their societies.

In describing human psychology in earlier chapters, I assumed that
research findings about different ethnicities and races within a society
could be extrapolated to attitudes about members of foreign societies.
But this reasoning implies the converse: that the psychological toolkit
our ancestors evolved to interact with foreign societies has been coopted and reconfigured to allow the ethnicities and races of a society
to coexist. Reflect on how this may have occurred. Humans express
pride in their society, a cognizance of its specialness and the difference between their identity and those of foreigners. When conquering tribes began to take those same outsiders into their societies, the
same mental circuits already employed to distinguish, and respond to,
foreign groups would have been pressed into action to make sense of
the relationships between the ethnicities within a society. If this description is correct, membership in a society and in an ethnicity or
race can in many ways be equated. However, there's one important
difference. Ethnic groups have come to invest part of their identity
and social obligations into the greater society of which they become a
part. These groups act, to some extent, like societies *within* a society.

In this final stage of our narrative, humans have taken a path for
which there are few parallels in nature. This societies-within-society
complexity finds its closest approximation in the ocean-wide clans of
the sperm whale, each containing hundreds of unit societies of a few
adult females and calves. Yet the similarity is superficial. Units don't
differ in their behavior and power relationships as ethnicities do.
What membership in the same clan provides is the opportunity for

units to team up to catch prey effectively by using a shared approach to hunting, and that is all. By contrast, ethnic distinctions permeate every aspect of human life.

If ethnicities are to find each other palatable within the greater society, the way forward is for them to align their identities well enough to override the significance of whatever differences remain between them. Since differences between ethnicities, like differences between societies, are written on the body—in the form of markers affecting actions and appearance—then grafting outsiders onto a society requires those people to adjust to the local "writing." That is, to assimilate.[12]

Assimilation becomes more challenging when vast cultural distances need to be bridged and the people have the numbers and resources to fight back, an obstacle that can put the brakes on the expansion of empires. It tends to occur in a particular manner, as I will describe, because the way outsiders are brought into a society guarantees that each society solidifies in great part around one ethnicity. Almost always, this dominant people founded the society, originally occupied its heartland, and came to crown the Great Chain of Being, with other groups suffering negative prejudices by comparison to them.[13]

Much of the power is in the hands of this core group of people, and especially the leaders and nobility, which are mostly of the dominant group and tend to support its interests. Hence assimilation is asymmetrical. The onus is on other ethnicities to comply with the dominant culture. Indeed, from this point on I will use the word ethnicity as it tends to be used in everyday practice, to describe not the dominant people, but such lower-status groups. The dominant group can compel those ethnicities to fit in, or if an ethnic people find that the changes are to their benefit, they may take them up of their own accord. Often a bit of both goes on, as well as some accommodations to still other ethnicities. Whether the changes are forced or freely made, some tolerance on the part of the dominant people—certainly more than the Inca usually showed—is entailed while the newcomers learn what's expected.[14]

No surprise here. The same adjustments would have applied to an individual member of a hunter-gatherer band who had married into another society; he or she, too, would be expected to act within the social boundaries permitted by the adopted society. Until then, the locals had to make some allowances for the newcomer's foreign ways. The very possibility of assimilating an entire population must have had its origins with such elective movements of solitary individuals seeking to become members of another society, the sort of transfer widely seen in animals and which would have occurred throughout our evolutionary past.

An ethnicity's acceptance absolutely requires that it curb anything that could cause its members to be branded black sheep, such as traditions the dominant culture considers immoral or reprehensible. The banning of the potlatches of the Pacific Northwest Indians in 1884 by the Canadian government, which declared them wasteful and uncivilized, is one of countless examples. Dominant populations have perennially taken it upon themselves to "civilize" people by enforcing standards of acceptable behavior as well as submission to their proper social place by those "savages"—a term settlers in the United States often applied to American Indians, although the sentiment existed before recorded history. Columbus spotted slaves the very day the *Santa María* set ashore in the New World, where Native Americans rationalized the taking of captives as "taming" them.[15]

While this "civilizing" carried over to less remarkable behaviors than potlatches, an erasure of the boundary between themselves and other ethnic groups would never have been the goal of a dominant people. The end result of assimilation is a sort of merger, but not one entailing a loss of separate identity. Rather, the identity of what had once been an independent society is refashioned in the image of the dominant group—to a degree.

I say "to a degree" because if outlandish behavior on the part of a subordinate group made the dominant people uncomfortable, so too could excessive conformity. Indeed, too many similarities between ethnicities can infringe on people's desire to uphold valued differences, and ultimately make prejudice worse. By the same token, too much assimilation can shatter the self-esteem of an ethnic people.[16] So while dominant groups drive much of what happens, the outlook of the ethnicities plays a role, too. Adjusting to the expectations of the people in power increases an ethnicity's stature, or at least legitimacy, just so long as its population stays distinct enough not to intrude on the unique identity of the dominant group. Hence Jews were targeted by the Nazis not because of their failure to fit into German culture, but rather at least in part because they were recognized as distinct yet couldn't be told apart from other Germans much of the time. Menorahs and kosher diets went unseen behind closed doors. This uncertainty was used to stoke fears that the Jews were using their assimilation to hide wealth and influence, and malevolent intentions.[17] Their fluency with the dominant culture put the Jews in harm's way as much as their differences did.

Of course, every ethnicity has some influence on the dominant culture, nudging the boundaries of what a society permits when that society embraces its cuisine, music, and other cultural highlights—just as the same society could import such elements from other societies.[18] In the process of Romanizing its outlying districts, for example, Rome also took the best those people had to offer—perfumes, dyes, spices, and wine.[19] Even with this fabulous assortment of ethnic traits, we may nevertheless distill a set of shared features that describe the society in toto (as in *the* Canadian culture). Yet most societies are no longer defined by one culture, in the sense people usually employ this term. Instead the members home in on their commonalities while making allowances for more diversity than people ever did in the past—the ethnic richness itself is a wellspring for national character in places like the United States and Singapore. To describe this tolerance and, indeed, positive attitudes between groups, psychologists speak of a superordinate identity cultivated through society-wide commonalities. Such an identification diminishes Us-versus-Them differences to allow for an inclusive We mentality.[20] It's a point of view necessary for societies-within-society to function.

To be sure, converging on a perfect match between ethnicities would be impossible even if a loss of self-esteem weren't an issue. Individuals who distance themselves from their ethnic background not only fail to lose their accents and other ethnic ties, but usually they unwittingly hold on to the underlying codes of conduct of their ancestors and instill them in their kids.[21] Differences will persist across generations even when a near fusion of groups occurs over the long march of history—in the case of the Han, two millennia. Just as the Romanization of people varied across the far-flung Roman provinces, the locals across any society put their own spin on their identity, intentionally or not—and other members notice this. While the Han living in the earliest Chinese dynasties might have seen the provincials who adopted Han ways as legitimate fellow Han, they would nevertheless have registered differences. From what evidence exists, they would have felt those Han to be inferior, even if far less inferior than the ethnic groups of China that weren't Han.[22]

Earlier I argued that hunter-gatherers should be said to have had nations despite lacking the governmental infrastructure associated with modern states. But the fact is that nations—in the sense many scholars think about them, as independent groups of people sharing the same cultural identity and history—really existed only in hunter-gatherer days, when societies were far more uniform.[23] Every state society, scrutinized carefully enough, is an ethnic mishmash. Still, I will continue to bow to common parlance and refer to today's societies as nations.

DOMINANCE

The dominant group's control of the society's power and identity can reasonably be called the founder advantage. However, what matters most for dominant group membership is sharing the ethnicity of those founders rather than tracing one's actual ancestors back to the birth of the nation. Hence Caucasians have stayed dominant in the United States even though the forefathers of many European Americans came to the US after 1840. Meanwhile, virtually all African Americans are descended from slaves that arrived before that country's independence, which is to say the average black family has been American longer than the average white family.[24] Reportedly Winston Churchill said that history is written by the victors. Indeed, the finely edited histories that nations promulgate put the dominant people in the best light and affirm their claims to power and status. This suggests another reason that hunter-gatherer bands, being egalitarian, had little interest in histories.

What keeps the dominant group in power? Its people are often called the majority. As that term implies, they typically make up most of the population, especially in the dominant group's original territory. That other ethnicities, or minorities, sometimes compose the bulk of the populace, however, suggests that being on top can be a matter of circumstance more than numbers. That was true for the white Afrikaners who ruled over native African peoples in apartheid South Africa, and in an even more bluntly repressive form in "slave societies" such as those of the early Greeks, where captives outnumbered freemen. It was true as well for societies created by nomadic herders such as the Mongols, who exploited their equine skills to take over much larger agricultural societies.

Generally the dominant group is less able to cling to control in places where it is outnumbered, as its people tend to be in outlying provinces. This was an issue in the Roman capital, where slaves grew more numerous than citizens. Roman slaves came from many sources and couldn't be distinguished in a clear way from the freeborn population, which was likewise diverse. The government chose to ignore this complication after realizing that branding their servants would have made their numerical advantage transparent and fostered rebellion.[25]

Minority groups with low populations but relatively high economic success can be seen as threats, too. The Chinese in countries such as Malaysia have faced this difficulty, as have the Jews at various times in history.

While the people of outlying districts, who are likely to be most distinct from the dominant people, have had a high success rate of breaking free, it's been rare for them to seize control over the dominant group and the entire country. "We confront here almost the political equivalent of a Newtonian law," two political scientists submit: "Bodies in power tend to stay in power unless acted upon by outside forces."[26] This indicates how effective the dominant people are at protecting their position, being in control of the police and the military; but it also testifies to how difficult it can be for minorities, often geographically and culturally disparate even if large in combined numbers, to join forces. There have been successful overthrows. One occurred in Rhodesia when the white population lost its exalted position and the country was renamed Zimbabwe in 1980, putting an end to years of guerrilla war. This was an exception that proved the rule: the white "majority," former colonists who had separated from the United Kingdom just fifteen years before, were vastly outnumbered by tribal groups, the disparity in numbers making their hold on power fragile.

The control of the dominant group extends from everyday matters of identity to the symbols that a society holds most dear. Such majority control of societal markers makes the position of minority groups tenuous. People of Asian descent born in the United States may celebrate their country's flag, yet experimental studies show that they more readily associate that flag with white people than with fellow Asian Americans.[27] Even when a member of a minority group may be part of a family of proud citizens going back generations, at some level there is the sense of being "a perpetual foreigner in one's own land," one study says.[28] Meanwhile "the white Protestant American is rarely conscious of the fact that he inhabits a group at all. *He* inhabits America. The *others* live in groups," sociologist Milton Gordon writes.[29] This explains what may be the most significant defining feature of dominant people, and why they are rarely spoken of as an ethnicity. People of the majority enjoy the luxury of expressing themselves as unique individuals more than do minority people, who end up diverting more time and effort into identifying with their own ethnic group.[30]

There are serious consequences to a majority associating "their" country mostly with their own people and less so with other ethnicities.

Laboratory studies suggest that the dominant people often express uncertainty about the loyalty of minorities.[31] Because of how people can
grow to accept the biased views that others repeatedly express about
them, this mistrust can spark the very behavior the majority fears, by
marginalizing minorities. The people of especially denigrated groups
may come to behave as they are accused, even to think of crime as a
reasonable choice when few other options are open to them.[32] This
alienation presents another reason, aside from their disconnection
from the symbols and wealth of the nation, for minorities to express
less attachment to their society.[33]

There's a sense in which today's minorities serve a role little different from that of the indigenous tribes within the Chinese dynasties,
such as the mountain-dwelling Miao: they act as points of comparison
for the dominant people, who remain the "pure" representatives of
the society. Hence the awkwardness Asian Americans feel on being
asked where they are from, when, in contrast to people of European
ancestry, they aren't expected to say a place like Peoria (a common
response to such an answer being, "No, where are you really from?").[34]
The most maligned minorities are kept at the greatest social distance,
as is true for African Americans. Although still stigmatized, the biracial children of Asian and Caucasian parents, for instance, find it easier to fit in than those with "black blood."[35]

The link between the society and its dominant people runs deep.
Ask a person to imagine a citizen of his or her country. If that person
is from the United States, the image that almost always pops immediately into his or her mind, whatever the sex or race of the individual,
is a white male.[36] The British political thinker T. H. Marshall wrote of
citizenship as "a claim to be accepted as full members of the society."[37]
Given what we know about the advantages that the majority people
share, and about human psychological responses to ethnicities and
races, the trick here is the word "full."

STATUS

The status relationships among ethnicities and races are more complex
than the mere acceptance of the majority's dominance; the standing
of the minority groups within the Great Chain of Being turns out to
be fluid when looked at over the decades and centuries.[38] Changes in
status are uncommon among the social networks and few groups present in the societies of other animal species. Infrequently, a matriline
in a troop of baboons wins a fight with a higher-status matriline and its

females end up with access to a better sleeping spot and more food.[39] Human ethnic and racial groups rarely shift rank through outright aggression in this way. Instead their position creeps up or down based on shifts in the society's perceptions of them.[40] Nor are all people of an ethnicity of-a-kind in their station. A family that has moved up— often in part by taking on more of the characteristics of their adopted country—may not associate with the poorest or most recently arrived newcomers among their ethnicity.[41]

One reason that change comes slowly has to do with the way established social standings are accepted, and not just by those in power. People often see the position of their ethnicity and race as natural, unalterable, and warranted, in much the same way they perceive their social status as individual persons to be what they deserve. A default view is that the world is fundamentally just; thus people's troubles, and those of their group, are justified.[42] As a team of leading psychologists put it, "Instead of resentment toward the privileged and sympathy for the underdog, on average, people endorse the apparent meritocracy and infer that (for groups) high status invariably represents competence."[43] As a result, according to another group of authors, "people who suffer the most from a given state of affairs are paradoxically the least likely to question, challenge, reject, or change it."[44]

The strength of this conviction can't be denied. It's true today for members of the lowest caste in India, the untouchables, and doubtless was so for any slave resigned to his or her fate throughout history.[45] This acquiescence to social position must have been pivotal to the success of societies going back to the first chiefdoms and states. What might have been expressed in hunter-gatherers as caution, aversion, or disgust at foreigners was redirected toward classes of people within the society in an effect so pervasive that even the downtrodden maintained a poor opinion of themselves. The outcome, as our overview of psychology revealed, is that ethnicities coexist despite the social stigmas they bear.

In fact, similar power distinctions to those among ethnicities have always existed between societies, too, as in the jostling of the United States with emerging economies like India and China that vie for status on the world stage. When it comes to enduring a low status as an ethnicity or as a society, Pygmies provide a case in point. These Africans occupied, and many still occupy, a position somewhere between independent hunter-gatherer society and minority group in an agricultural community. During the season when they move into villages to serve as field hands, they accomplish the drudge work that the farmers avoid, this longstanding arrangement making them the original migrant workers. Their relationship to the farmers affirms that ethnic interactions within a society retain some of the flavor of alliances between societies. Even though Pygmies treat one another as equals, as bands of hunter-gatherers are wont to do, and even though the farmers have a high regard for the Pygmies as musicians and shamans, their social status is clear. Farmers sometimes say the Pygmies "belong to" them, and the Pygmies know and heed their place. "In the forest, Aka [Pygmies] sing, dance, play and are very active and conversant. In the village their demeanor changes dramatically: they walk slowly, say little, seldom smile and try to avoid eye contact with others," says Barry Hewlett, an anthropologist, describing a subservient bearing which primatologists will instantly recognize among rank-conscious baboons, and which the rest of us have come to associate with broken minorities.[46]

The Pygmies take on the role of servants willingly in exchange for goods they can acquire no other way. Minorities within a society would seem to be similarly motivated to accept their position. Indeed, by preserving their independence and the freedom to go back to their forest territories when they wish, the Pygmies have dodged the unremitting inequalities confronted by ethnicities locked into a society controlled by a majority people. Pygmy bands have been known to abandon ties with one farming village and establish a relationship to a village elsewhere.[47] In effect those Pygmies exchange one dominant group for another that perhaps offers a better relationship.

That an entire group of former outsiders—an ethnic group—can be accepted in a society, and move up in status within it, may trace back to human responses that evolved to survive enslavement. No slaves could improve a bad situation by switching communities as Pygmies do, but their ability to adapt to captivity helped see them through, and at times allowed them to improve their lives. Among the Comanche, slaves could be accepted into the tribe, to a degree, when they were judged to have achieved proper *personhood* by taking on the key markers of their society. That meant mastering Comanche customs and language. This took place most successfully for outsiders captured as children and raised by Comanche families. Children have always been ideal war booty, firstly because they are easy to control, and secondly because, with their plastic identities, they assimilate (achieve personhood) more completely than adults.[48] In fact, according to one study, youngsters go through a critical period before the age of 15 when they are most open and adept at absorbing a culture.[49] Adoption by a family was one thing, however; being fully embraced by the society was another—the latter is a hurdle still faced by cross-culturally adopted children. Generally, though, if a slave had the opportunity to assimilate, his children or grandchildren were received as favorably as many second- or third-generation citizens of countries today.[50] The Comanche offered slaves a shortcut, but the bar was high. To be a true Comanche, and marry a Comanche, a slave had to perform an act of heroism on the battlefield.

A slave's prospects for moving up, and possibly becoming a member, depended on the rules of the society. Some upgraded their status if only to become powerful slaves, as was true in the Ottoman Empire; or earned their freedom as individuals, as did the Comanche slaves, or at times as a class. The Greeks were democratic but rarely freed slaves; the Romans on the other hand took many slaves but readily liberated them. Indeed, Roman slaves could achieve citizenship more easily than foreigners who had resided in Rome for generations.[51] Yet former slaves could be stymied in their effort to raise their status. Evidence of a shackled background, such as the widespread association today of dark skin with past slavery, could be a hindrance. As a result, a formal change in standing didn't always translate into social improvement. Roadblocks to advancement went up after the American Civil War when few people hired emancipated slaves, many of whom ended up in conditions more deplorable than those they had suffered when enslaved.[52] The hurdles that societies put in the way of social acceptance and standing for ethnic groups are less imposing, but exist nonetheless.

INTEGRATION

One cornerstone of the process by which ethnicities become interwoven into a society as valued members is integration, a word I use to describe how an ethnic people who had originally lived spatially apart now come to intermingle with the dominant group. Even simple chiefdoms, made up of a few villages, didn't always permit this easy mixing, but instead kept people separated out onto their village properties. Letting members of an outlying province leave their own land at all, other than as slaves stripped of their identity, would be a concession in that the dominant people would view it as a risk. This tight control on travel within the society is unusual. In other animals, freedom of movement by all the members across the territory is normal. In a few species the members split the territory among themselves, with individual prairie dogs and female chimpanzees and bottlenose dolphins often favoring certain areas. The use of space by band societies was often a bit more compartmentalized than this; the bands stuck to the "home" corner of the overall territory their residents knew best. Yet in all these examples, individuals could choose to go elsewhere in the territory.

Human territoriality became far more complex as societies expanded to take in foreign groups and turned them into ethnicities. The Inca forced most of their subjugated peoples to stay apart, both socially and geographically. Usually, only the few ethnic individuals assigned a high status to collect tributes from their people for the Inca were permitted on the famous highways. Even beneficent societies weren't always open to giving ethnic groups free passage to the capital. Most or all minorities, and especially the lowborn, stayed put in the region of their birth. To earn the right to move about, the provincials would have to consistently demonstrate loyalty and submissiveness until the dominant group was convinced there was no longer a need for force. The imposed separation would continue until the group had assimilated sufficiently for its people to be regarded as trustworthy. Making the changes required that the dominant group condone the ethnic people learning about, and taking on, the country's culture, even if from a safe distance. Chinese dynasties kept provincial subjects from the imperial center until their Han behavior passed inspection. By that time those people might have already achieved the "felt closeness" of being Han. The process primed them to be interwoven into the wider population by cultivating an allegiance that put a higher value on the concerns of their society than on their ethnic interests. For the tolerant and self-confident Romans, integration often came before assimilation, with many distinct groups ensconced in the capital.

Ultimately, however, allowing an ethnicity to spread from its place of origin served to firm up the majority's grip on their power. A dispersed people were likely to end up less identified with their group and with a weaker voice than a concentrated one. At the same time, the dominant people can more easily hold sway over an ethnicity's original territory if that ethnicity has scattered beyond its former homeland. Discord in societies is lowest when ethnicities are either well dispersed or well mixed with the general population.[53] In the latter case, with integration encouraged, the path is opened for positive interactions and fast assimilation. Citizens of the majority group learn about, and ideally adjust to, the aliens in their midst, who become a familiar and nonthreatening presence. The minorities, adhering to the "when in

Rome" dictum, in turn fine-tune what is expected of them, and gain social standing for doing so. Properly accommodating to others is impossible without this firsthand exposure. The absorption of a minority into a society slows, however, if naïve newcomers, still following their old customs, continue to arrive from the home country.[54]

The difference is huge between staying put in a province among those of one's background and roaming among the dominant group as a free person. It's hard to say how far back in time such integration existed. While the Roman Empire is viewed as an early experiment in multiculturalism, the Greek civilizations preceding it were open only to those of Greek origin while other ethnicities were prevented from traveling beyond city ports.[55] For earlier states, such as Babylonia, there simply is no evidence to say whether freemen from outlying provinces lived with impunity among the dominant people.

Of course, integration has never meant random mixing. The ancient city of Teotihuacan, established around 100 AD near what is now Mexico City, had a distinct barrio occupied by the Zapotec from farther south.[56] In Rome, there's evidence of the Jews and eastern immigrants concentrating in certain outer boroughs, and the ethnic temples of that city must have been colorful gathering places.[57] Almost certainly ethnic neighborhoods existed in other early cities, though little proof exists, most likely because the quarters occupied by different groups shift during the life of cities, confounding the archaeological evidence.[58] Indeed, studies have shown that modern ethnic populations expand and contract both locally and regionally over time, contributing to the dynamic range of "flavors" in how national identity can be expressed across the landscape in the same country.[59]

Vestiges of spatial segregation remain even when people move away from their original homelands, sometimes by force. American Indian reservations are a kind of holdover of territorial separation— as once vilified groups kept apart from the general population, tribes were shunted off any valuable land and, beyond this, often had restrictions placed on their movements.[60] In practical terms, some inner-city neighborhoods function the same way.

Not that all spatial segregation is bad, especially between groups of similar prosperity and status, as is the case when a neighborhood of working-class Italians abuts one of working-class Hispanics: ethnic neighborhoods can simply reflect people's desire to seek out like others. This self-sorting is an unintended consequence of personal choices that traces back to the more understated sorting-out of hunter-gatherers among the bands of a society.[61] While the reduced contact caused by this sorting contributes to ignorance about those in other neighborhoods, data from modern communities show there's little harm so long as the interactions between enclaves are positive.[62] A backlash can occur if a neighborhood is too self-contained, though, with the residents cocooning themselves exclusively with like others. In that case, people's insularity can contribute to impressions that they are still *foreign*, stirring up resentment among other society members.[63] Even then, however, ethnic communities can serve as staging points that let the new arrivals sidestep culture shock and yet assimilate, often over the course of generations.[64]

The absolute comingling of different ethnicities and races takes integration a step further, yet presents hazards of its own. Studies suggest that taking down ethnic separations, both social and geographic, has to be handled delicately. The dominant people must be open to having unorthodox families one door over. Their confidence in their own identity and sense of security must be great enough to override a there-goes-the-neighborhood attitude toward minority families crossing urban borders.[65] For minority families, the departure from an ethnic neighborhood into the community at large can represent a social boost, yet also poses risks. The social distance between groups still needs to be respected. Anyone who pushes too hard to fit in with the majority is in danger of becoming a "marginal man," as sociologist Robert Park perceptively wrote, forsaken as "one who lives in two worlds, in both of which he is more or less a stranger."[66]

No matter the spatial distribution and degree of interaction of ethnicities and races, they must work together to enable state societies to function across human lifetimes. In the pages ahead, we shall contemplate these interactions in more detail, with close attention to immigration and group relations within the nations of today.

CHAPTER 25

Divided We Stand

Give me your tired, your poor, your huddled masses yearning to breathe free: thus goes the sonnet by Emma Lazarus mounted on a plaque inside the pedestal of the Statue of Liberty. Yet in view of their vast differences, what's always most defined the masses who come to unfamiliar shores seeking refuge is less their escape from fatigue, poverty, and past oppression than that the immigrants are admitted as total aliens. They are not *us*.

The commingling of ethnicities in my Brooklyn neighborhood came about no differently from the diversity found across many communities around the world today. Most black Americans aside, they didn't come to be there because they, or their parents or family members of earlier generations, had been enslaved or subjugated. Immigration, in the sense of the influx of a substantial population from foreign soil, has become the primary means by which foreign peoples enter a society and become members. This latter, more benevolent practice, and how it has contributed to people functioning together as a society, is the theme of this chapter.

Immigration differs from subjugation in that the parties choose to receive one another. The destination society sometimes even encourages the entry of immigrants, in the spirit of the Lazarus poem, with the anticipation that they will stay of their own volition. Such voluntary acceptance over time of whole populations was all but nonexistent in early human history. Of course, absorbing immigrants is rarely indiscriminate: each person usually undergoes a vetting process. And at times the jarring exposure of the host society population to the new

327

arrivals may set off a backlash—a reaction reflective of concern about immigration in many countries today. All the same, what I find extraordinary is that the receiving society often permits the newcomers to immerse themselves in its general population from the start. Even when many of the incoming people move into ethnic neighborhoods, they have the opportunity to rapidly assimilate.

How did nations come to be open to such a generally cordial influx of so many new members? Hunter-gatherer bands could take in an occasional refugee who had escaped some tragedy with his or her birth society. At its root, immigration is this kind of acceptance of dispossessed peoples, amplified manyfold. I suspect that historically this acceptance occurred in steps. In its earliest form the dispossessed would have originated not from foreign societies, but from within the state itself. After all, when a people were subjugated, there would have followed a period when their status as foreign or native was murky. That's a reason why early states often regulated the movement of their outlying peoples. Until the outliers had assimilated enough to be relied on, their exodus from their birth lands into the rest of the society's territory, if permitted at all, would be treated as a germinal sort of immigration. Maybe the movements seen today between foreign societies got their start from there.

While the process of adding members to a society by immigration isn't manifestly hostile, immigration doesn't amount to a merger of equals. Fundamentally little has changed. The issues of dominance and status described in the previous chapter apply to immigrants, too. The ingroup/outgroup psychology essential to subjugating others remains. This lays bare that the bulk of the huddled masses throughout history have indeed been the beaten people of beaten races that the Lazarus poem suggests. Often the newcomers arrive poor in material wealth and social standing alike, their condition unlikely to change in the foreseeable future. Thus the Chinese who went to California and Indians to Africa in the nineteenth century were cheap labor, paid but detested. The best that newcomers can hope for is to arrive at a time when bigotry is widely discouraged.

The obstacles immigrants face are many. Their passion for their new society and its treasured symbols, their knowledge of how to act, and their personal connections with other citizens—all take time to build, leading to a deep skepticism from other society members about their steadfastness.[1] Insult to injury, like divorcees tainted by past marriages gone wrong, new immigrants can be further stigmatized by the failure to stick with their birth society. In some ways the plight of immigrants is reminiscent of the difficulty other kinds of animals face in transferring between societies. There too, new arrivals, in this instance usually arriving one at a time, go through the wringer until, with time and luck, they are accepted.

New arrivals, then, are strangers in a strange land, and no matter how low their social stature or how they try to fit in, the native population can question whether their strangeness will be damaging. Just as people can recoil from too great an influx of traded goods as a threat to their culture, they can regard immigration as eroding their hold on their society. This is true, ironically, even in the United States, a land of immigrants. Thomas Jefferson fretted that the flood of immigrants in his day would "warp or bias [America's] direction, and render it a heterogeneous, incoherent, distracted mass."[2]

The frequent association people make between those looking for asylum and other foreign arrivals with sickness and disgusting things now takes on layered meanings. An immigrant may carry not just biological but cultural diseases that can corrupt our identity—amoral behaviors prominently among them.[3] Perhaps such nativist fears exist because there are few means to constrain the behavior of immigrants. By comparison, the subjugated people of the past could in effect be quarantined on their own property until assimilation removed any "black sheep" social impurities. Despite these anxieties, as we have seen, societies seem amazingly resilient to the arrival of people and their practices, just as they are to anything brought in through trade with the immigrants' country of origin—witness the worldwide popularity of Chinese, Italian, and French cuisines, which haven't reduced the United States or any other society to the incoherent mass Jefferson predicted. This robustness is evinced by languages, which take words from other tongues without blurring the line between them even after prolonged contact. Aside from pidgins cobbled together to promote communication between groups (formalized into creoles if they turn into the speakers' lingua franca), languages never descend from more than one parent tongue.[4]

ROLE

The heterogeneity of societies that so concerned Jefferson can contribute to their success. That would have been true of states formed by conquest, but it has been amplified greatly by immigration. The strength of heterogeneity can be understood in terms of the optimal distinctiveness model described earlier—the tendency of individuals, ethnic groups, and societies alike to gravitate to a middle ground of being alike and different. By this model, ethnicities search out, and enhance, their commonalities, even while striving against cultural loss, in the end intermixing with the most ease when everyone perceives resemblances, yet not equivalence. The outcome of this dynamic aspect to people's superordinate identity is a balancing act in which they demonstrate their fidelity to their ethnicity and to the society as a whole. As a result the majority's negative reaction toward those who were once seen as foreign is alleviated while the society is kept front and center in its citizens' lives.[5]

For groups perceived as contributing something positive, the social advantages can be impressive.[6] Adding ingredients to a society's culture is important. In addition, a people can fortify their ethnic pride by slipping into the system of role differences. This option has been open to ethnicities for a long time: for example, the marble workers from Bithynia, an ancient province in what is now Turkey, renowned across the Roman Empire, or the Rucana of the Incan Empire, assigned to carry the emperor's royal litter.[7] Such distinctions may offer a psychological edge when a small number of talented individuals influence the wider perception of their ethnic group. But friction is reduced if many individuals of an ethnicity move into occupations that few others do. By contrast, minority people can face reprisals when they choose a career favored by the dominant group. For this reason, someone taking a desirable job is evaluated more negatively, as a rival, should he or she happen to be a member of a minority.[8] This sensitivity to people's choice of profession declines in years of little social competition, or when a society needs an injection to its labor force. Economic booms and wars are times when outsiders may be actively brought into a society and some ethnicities can flourish. Even the Comanche hunter-gatherers gave war captives a social footing as warriors when their fighting force needed replenishment. At the other extreme, minorities can be rejected when competition turns severe. The normally tolerant Romans, for example, turned xenophobic in years of famine, when some ethnic groups were banished from Rome.[9]

Immigrants without a notable skill still had a way forward. The baseline expectation would be that ethnicities—like slaves and the subjugated in earlier times—would assume responsibilities for which the locals had no taste. These could be menial jobs that took no special abilities. Or a group might take to employment that required training but was low in status. An example was the black Americans who became barbers for white clients in the nineteenth century. Offering an expertise with a deep foundation, however, could give anyone a boost. In the case of barbers, Italians who happened to have a family history in that trade became popular and had displaced most black barbers by 1910.[10]

In Chapter 18, describing alliances, we saw the antecedents of this kind of specialization in patterns of commerce between societies. Those with something unique to contribute—boomerangs or blankets among hunter-gatherers; wines, cheeses, and perfumes by the French today—made for favored trade partners.[11] Bringing talented outsiders into the society as members circumvented the need for fickle trade negotiations. In a small society, as mentioned earlier, marrying a foreigner could do the trick. African communities that seasonally employed Pygmies relied on them to get food from the forest as well. A few farmers wedded Pygmies whose proficiency at procuring meat and honey was a boon for the villages, reducing their dependence on Pygmies from outside.[12] Alternatively, skilled labor could be taken by force. The captives most valued by the tribes of the Pacific Northwest were artisans, and slaves brought metalworking, carpentry, painting, and other practical and artistic skills to Islamic regions.[13]

But with the right inducements, no force is required to bring in individuals with a special gift. A small proportion of immigrants have always been highly qualified persons whose arrival is a sign of brain drain from their birth countries.[14] When the academy Plato had founded in Athens was shut down in 529 AD, its scholars went to the Sasanian Empire in what's currently Iran, hauling the scrolls along.[15] Julius Caesar granted citizenship to doctors and teachers, vocations in short supply in parts of the world today as well.[16] But historically the role most frequently filled by qualified outsiders might have been that of merchant. The laws established by King Hammurabi of the first Babylonian dynasty in 1780 BC included the right of foreign tradesmen to set up shop, and in the course of time many of them became naturalized.[17]

Even for sought-after people, however, anything close to equality comes hard. "We see the ugly heads of racism and nativism coexisting with our rhetoric of welcome and tolerance," write two authorities on immigration.[18] That's not to say immigrant and adopted country can't mutually prosper. In fact, what makes any ill treatment tolerable is the anticipation of an improved quality of life despite the abuses people might suffer. Skilled immigrants, often coming from societies too poor to support them, could take unfilled elite jobs, much as the unskilled could turn to humble work without stepping on any toes. In either case new arrivals might rely on fellow ethnic residents to help them get a start.[19]

Of course, societies also cultivate options for temporarily using foreigners to solve short-term labor needs without bringing them on board as citizens, in effect importing them until their task is complete, as witnessed by the seasonal travel of farm hands between Mexico and the United States. Even when migrants are permitted to stay in the receiving country, they may not be granted legal rights, or may earn them only after the slow crawl of time, their alien standing making them all the more easily disposable.[20]

In all these matters it seems that stereotypical beliefs about talents or deficiencies have been significant to the coexistence of ethnicities, much as similar beliefs have been key to the coexistence of social classes going back to the earliest settled hunter-gatherers.

RACE

People assimilating into a society face a quandary—a mismatch between how they think of themselves and how everyone else perceives them. Immigrants, more often than not, discover that the identity they have prized their whole life is meaningless in the new country. Thus the immigrant to Europe or America who grew up a proud member of one of the Tsonga tribes of Mozambique is recast as Mozambican at best, and more often as merely black—a designation that has no social relevance across most of Africa.[21]

By this process of amalgamation, broadly defined races spin out from the groups that originally mattered to people. This simplification of identities reflects that the immigrants' original loyalties tend to be too complex to be understood and appreciated in their adopted country. Immigrants therefore find themselves obliged to undergo the ethnic disintegration and decomposition mentioned earlier.[22] And so it was that the Shoshone, Mohawk, Hopi, Crow, and other tribes "became Indians, all of us more or less identical in practical terms," the Comanche Paul Chaat Smith laments, "even though until that moment, and for thousands of years before, we were as different from one another as Greeks are from Swedes." The overhaul was no easy task. "The truth is that we didn't know a damn thing about being Indian. This information was missing from our Original Instructions. We had to figure it out as we went along."[23]

Such sweeping categories of racial identity call to mind how Europeans put a name to Bushmen as a class while failing to appreciate the diversity of those hunter-gatherers, who didn't see themselves as Bushmen but as members of many distinct societies. In a similar fashion, 20 centuries ago, in what would become modern-day China, the dominant Han spoke of everyone in the south as Yue, describing their tattoos and unbound hair in an oversimplification of what must have been immense, now largely forgotten, diversity.[24]

A crude lumping into groups like "black" echoes the attitude of the conquerors of bygone days toward foreign societies. While as far as we know chimpanzees treat all outsiders with equal dechimpizing venom, humans direct their dehumanization selectively, but the people receiving the brunt of this ill will don't always see themselves connected with the groups they find themselves pigeonholed into. The early Chinese, Greeks, and Romans gathered what intelligence they could against their most threatening foes, but otherwise were uninterested in differentiating outsiders. Ignorance can indeed be bliss. Military might assured their victory, so why make the effort? "Countries whose very names we did not know precisely in former times we now rule," wrote Roman historian Cassius Dio in the second century AD.[25] Human beings don't need to accurately compare themselves to outsiders, or even have the foggiest notion about them, to be confident about their own position in the world and the correctness of their way of life. Even now many Americans and Europeans envision Africa, home to by far the greatest diversity of people on Earth, as the Dark Continent—one monolithic social block.

As for what transpires within a society, the widening of people's identities isn't entirely a loss for minorities—they can in part drive it. Their transformation resembles what happened during the formation of coalescent societies in which refugees from different tribes joined up to survive. By dropping their primary identity for a broader definition of who they are, minorities gain the base of support essential for their social and political survival within the general population of a nation. This is true whether they decide to work alongside the dominant group or rebel against it. The Haitian uprising at the end of the eighteenth century might never have succeeded, for example, if the slaves had stuck firmly with their allegiances to their original African tribes (some of which must have once been enemies). Nor of course are the broadened groups exactly homogenized: if their community has a strong enough presence, aspects of people's identities can remain connected to an ethnicity that is a subset of a race, such as the Japanese or Korean subcultures of Asian Americans.

Also, no matter how irrelevant today's ethnicities and races may have been to a people's ancestors, each group has, in its new setting, created its own way of living. The identities of former immigrants undergo revolutionary alterations, which tie them to other immigrants and at the same time set them apart from the populations of their ancestors. Visiting Israel a few years ago I was surprised how uncommon bagels are there: many Jews learn about bagels, Italians about spaghetti and meatballs, and Chinese about chop suey only after emigrating to North America or Europe, as the case may be.[26] Immigrants and their descendants are a people reinvented, and turning back is difficult or impossible. "I used to worry I would visit China, and people would look at me and not let me go home because I was so Chinese," the writer Amy Tan once told me. She met the opposite response. "The way you walk, the way you look, the way you act—nothing about you is Chinese," she was told. While the Chinese word for nationalism derives from their word for the Chinese race, Tan's ancestral background, to truly be *Chinese* isn't so simple.

This calls up the oft-expressed American ideal that a blending produces a superior sort of person.[27] "Here individuals of all nations are melted into a new race of men," wrote the French-born New York farmer J. Hector St. John de Crevècoeur in an essay from 1782.[28] But for societies composed of different human groups, the purée intimated by the motto *e pluribus unum* is never achieved. Certainly the dictum that "all men are created equal" applied to egalitarian, ethnically uniform hunter-gatherer bands more than to any society since. Even when ethnic relations are positive, no nation is the idealist's melting pot any more than its members are ever completely coequal. This reflects in part how people give up a degree of freedom, and to some extent their equality, to gain the security and social and economic payoffs of belonging to a nation, with some ethnicities relinquishing more than others.

Many sociologists speak as if ethnicities could be thoroughly digested into the whole. Insofar as this occurs at all it takes considerable time, as realized most remarkably with China's ancient and almost monolithic Han majority. The urge to be the same but different guarantees the coalescing doesn't carry on to completion across a society, that diffuse categories like black and white are as much melting as the melting pot entails.

The majority people can likewise widen their identity to absorb other groups, as the Han once did, though only after the newcomers lose some of the trademarks of their past lives. Italian Americans "made the cut" as white a century ago after growing a bit less Italian and more American. In the northern United States this transformation appears to have reflected the psychological need for whites to present a strong ingroup. I say this since the status of Italians changed at the same time black communities were growing at such a rate that the difference between the whites and Italians felt less relevant than it had before.[29] The fact is, faced with increasing minority populations, the dominant group may be obliged to broaden its membership to retain a grip on power, in this instance by exchanging one outgroup (Italian) for another (black). Nowadays, only about one in four Americans traces his or her ancestry to the British Protestants who once formed the country's majority and ethnic core. Yet a gradual acceptance of other people as white, Italians among them, has ensured that Caucasians continue to make up a majority (about two-thirds) of Americans.[30]

At the same time that a society broadens its boundaries to accept outsiders, the ethnicities and races within it remain divided. Perceived distinctions can extend to the very *humanness* of the groups.[31] There's no reason to expect this to change, even with, for example, the increasing acceptability of interracial marriages. Hypersensitive eyes will pick out different ethnicities and races, especially if a group is blighted by ignoble status, for centuries to come—even if inaccurately. Hence a destitute American of unclear race is prone to be registered as black, just as is a non-African person with an Afro.[32]

CITIZENSHIP

Gauging who belongs in a society, usually an easy task in hunter-gatherers, has become a challenge given the scale of immigration over the last centuries. The difficulty is exacerbated in the United States as an outcome of its population being almost entirely composed of immigrants from many sources. To this extent Thomas Jefferson was right about the potential warping influence of immigrants: a society must keep strong a filigree of common culture and collective belonging to sustain itself. Jefferson had such a core identity in mind when formulating American ideals about rights, religious commitment, and work ethic.[33] Since there was little history to be shared among its citizens at first, awareness of being one nation by necessity relied less on people's ethnic origin and communal stories and more on symbols that were put in place almost de novo, making flags and fanfare a prominent part of American life ever since. With sturdy symbols as rallying points, Jefferson's American creed served to generate a spirit of purpose and solidarity.[34]

Still, America's openness to immigration got off to a slow start. The authors and signers of the Declaration of Independence and the Constitution certainly weren't all that diverse: aside from two of Dutch ancestry, a few were from Great Britain—England, Ireland, Scotland, and Wales—and the others were American-born descendants of British citizens. Nor was citizenship dispensed generously at first, and racism was everywhere. As originally manifested by the Founding Fathers, nationality was proffered largely to Europeans, with immigration encouraged from the northern and western parts of that continent. Even this was open-minded for the time.

In its current, widely recognized meaning, citizenship is a form of membership that extends beyond a sense of belonging to include basic rights and legal status and a role in politics.[35] The broad application of citizenship in the US, so defined, built up slowly.[36] Women were enfranchised in 1920, but in practice that right primarily applied only to white women. Native Americans became citizens in 1924, but whether they could vote was left up to the states until as late as 1956. People of Chinese descent, including those born in America, were blocked from citizenship until 1943. Those of Asian Indian ancestry waited until 1946 for the vote, while other Asian Americans earned the right in 1952. The path for African Americans was tough. The Fifteenth Amendment, ratified in 1870, granted black men the vote, but the states' adherence was spotty until the passage of the 1965 Voting Rights Act.

The modern definition of citizenship means that the prerequisites to be a legal citizen for countries worldwide have in practice been reduced to a few conditions, among them minimal requirements, to the extent these can be measured, that immigrants identify with the overall population and follow its basic social mores. All the same, an immigrant's grasp of the adopted land may be quite detailed, if only because of the requirement that he or she pass a civics test to become naturalized. Immigrants are likely to learn more about the principles and symbols of the nation than its long-time citizens, who are liable to have never considered their meanings deeply despite professing—and expecting—a devotion to them. Indeed, most Americans would fail their country's naturalization exam.[37]

Oaths of allegiance, like oaths of marriage, are intended to seal the deal. Still, the complexity of human identities makes fitting in, let alone acceptance, onerous even when the devotion of immigrants to the new country is strong from the start.[38] Knowing facts won't cut it. Membership, at the most intimate level of interactions, isn't an intellectual exercise but a way of *being*. Impossible to prescribe in an immigration law is the deep tissue of national identity: those myriad details like learning how to walk or talk "like an American" (or a French person). Such minutiae aren't things people easily notice, let alone refine from practice the way they learn to ride a bike. A generation or two often goes by before an immigrant's family members adopt such details.[39] That such markers aren't mandated underpins a basic point that integration demands an allowance for differences.[40]

Regardless of the reasonable probability that fellow members pick each other out by gait, accent, or smile, the fact that minority individuals are asked where they are from tells us that the days of near certitude in distinguishing citizen from alien are over. We have ceded the task to state agencies. What this means is that even while our commitment to societies is undiminished, if not ramped up by the rhetoric of governments, possessing a passport has been largely uncoupled from how our brains register who *belongs*: citizenship and our psychological assessments of membership don't always mesh.

That was apparent, and extraordinarily so, way back in the Roman Empire, when in 212 AD virtually all foreign residents were declared citizens by decree. And yet this was done primarily for the practical purpose of taxing them: as a psychologist might predict and historical evidence tells us, the prejudices of the Roman majority remained entrenched. The record is full of disparaging statements about ethnicities by those who felt that Rome was "aglut with the scum of humanity," or so the poet Lucan complained.[41]

Gut responses about who truly belongs can go south quickly, notably when anyone of a marginalized ethnicity commits a crime. So when an American citizen with Afghan parents shot up a Florida nightclub in 2016, killing 49, the horrors set off a different and stronger outrage than if a majority person had pulled the trigger: one directed at a whole group of people perceived by many to share responsibility for the offense. Meanwhile a white committing such an atrocity, like Timothy McVeigh, who killed 168 in the 1995 Oklahoma City bombing, is more often seen as a deviant individual, personally accountable for what happened.[42]

Throughout the history of nations, one loathed group has replaced another in endless succession, their trustworthiness, indeed their merit and citizenship, called to question in a roller-coaster of perceptions. A hunger for scapegoats can lead people to frame ethnicities in sinister light.[43] Tolerance has varied precipitously, and in general tracks the rise and fall of economies. By the late nineteenth century in America, Italians and Irish were thought less desirable than Norwegians, Germans, and English. Considered impossible to assimilate and cultural poison, those immigrants were branded with the word "Irishism," used to describe the perceived depravity of the people from that land.[44]

Discrimination against ethnic citizens often comes to a head during disputes with foreign powers. Those even vaguely associated with the enemy state du jour can face backlashes, if not rampant dehumanization. For the United States, sovereign adversaries have variously included Native Americans, Britain, France, Morocco, Tripoli, Algiers, Mexico, Spain, Japan, Germany, the USSR, Cuba, China, North Korea, and Iran and other Middle Eastern countries. Each time, Americans of those ancestries have suffered, with the contempt for Japanese Americans during World War II being particularly hard to fathom today.

The New Yorkers I've spoken to attest that after the 9/11 attacks were carried out by Muslim extremists, flags were on especially prominent display in the shops of Muslims, or those easily confused with Muslims, who found their standing as countrymen on shaky ground. By this public show of a clear identifier, the shopkeepers made sure no one mistook them for the enemy during a time when the cost of being misidentified was high—the so-called ingroup-overexclusion effect. When groups feel threatened by their own society, displays of patriotism while toning down their ethnic origins is par for the course.

Some countries support only the most tenuous hold on their citizenry. Without a strong core identity for their people to celebrate, they are in danger of fragmenting into the "natural" units from which their members derive their primary identity and primeval ties—the petty subgroups that built up among the bands of our hunter-gatherer predecessors writ large. The reason so much of the globe consists of artificial nations to which the people have little commitment is that their national boundaries were drawn up after the First World War to register not the homogeneity or solidarity of their people, but the economic interests of Britain, France, and the United States.[45] The outcome of those decisions has been that the populations of these regions often remain more attached to their original tribes and ethnic groups than to their country. That's notably true when those groups retain ancient connections to a territory, and may be hostile to other tribes that are now in the same country. When passions are directed keenly toward local peoples in preference to the government, it's a challenge to act together let alone as a functional part of an interconnected world. A country made up of such a regionally fragmented populace can be more of a delicate confederacy in the service of economic gain than a nation.[46]

This sense of a fragile alliance holds to some measure between all ethnic groups, insofar as the dominant people are the preeminent owners of the nation's symbols, power, and wealth. Such inequalities ensure that when a society is under pressure, the feeling of unity is brought to bear with most conviction among the majority, who have more to lose, and more feebly among any ethnicities treated as second-class citizens.[47]

This disparity between majority and minorities in their perception of proprietorship over the society is the Achilles heel of nations. Even in cosmopolitan America, where all men are created equal by decree, respect for citizenship is one thing, for the diversity of those citizens another. Though minorities can have clashing interests and prejudices about each other, support for pro-diversity policies such as affirmative action is greater among them, with diversity reinforcing diversity in ways that bring discord with the monolithic majority.[48] In such matters both the minorities and the majority can be equally self-serving.

NATIONALISTS AND PATRIOTS

Not only did Paleolithic societies lack ethnicities, but their people would not have abided the emergence of radical groups such as the American Tea Party and Occupy Wall Street movements. Had partisan confrontations as fervent as those of the last decade arisen among hunter-gatherers, the society, with its people's conflicts over identity pushed to a crisis point, would have severed in two; yet even with regional differences building up in the United States and other nations, people with varied social and political leanings are too intermixed for their societies to readily fracture. We are stuck with each other.

Some of the convictions people hold dear bear on ethnicities and races. Here's the rub: the predisposition to dominate other groups, whether those groups are other societies or ethnicities within a society, influences people's views concerning whether the primary role of societies should be to protect as opposed to provide for their members.[49]

We are aware that the societies of most animals serve both functions. Protection focuses on hostile outside factors—notably other societies. The provisioning role extends to the care of those in it.

People's outlooks on many pressing social issues betray how these roles are valued differently depending on whether individuals subscribe to patriotism or to nationalism. As most psychologists use the words today, these are habits of thought that represent distinct expressions of how people identify with their society. Sometimes lumped together, patriotism and nationalism become plain, and clash with each other, in troubled times. Depending on the person, perspectives shift to more nationalist or more patriotic viewpoints during periods of stress.[50] Yet each individual usually sticks within a narrow range of attitudes over the course of his or her life; the sentiments emerge in childhood under the dual influence of inheritance and upbringing.[51]

The fundamental difference between nationalism and patriotism is that while individuals with both outlooks are devoted to their society, they relate to it differently.[52] Patriots display pride in their people and a sense of shared identity and particularly of *belonging*; such a feeling comes naturally to those born in a country but can be acquired by immigrants. With most of their passion directed at their own group, patriots prioritize the needs of its members: making sure they have food, housing, an education, and so on. Nationalists have similar emotions but couch their identity in glorification. Their pride connects with prejudice. As obsessed as patriots can be with caring for the members, nationalists are absorbed with preserving a superior way of life by keeping the society safe and sound and putting their own people prominently on the world stage.

Where it gets interesting is that patriots and nationalists have divergent ideas of who constitutes "their own people." Indeed, among the aspects of their identity nationalists admire are those that set the trusted majority apart. It's this position they guard.[53] The extreme nationalist ardently protects each detail of that identity to keep the nation firmly associated with the angels. The priorities of nationalists include staunch demonstrations of loyalty, accepting customary rules of order, obeying leaders whom they see as responsible, and maintaining the established social relationships, most clearly between ethnicities and races.[54] All of these values came to the fore as people settled down and began dominating others. Tradition-driven nationalists believe in their country no matter what. They commit to the status quo, at times at odds with those democratic ideals that allow for transformation: their personalities are less open to new experiences and social change.[55] Compare this *my country right or wrong* stance to the outlook of patriots, who likewise give their country a high standing yet believe it must be earned rather than fought for, allowing that there are possibilities for improvement.

In their attention to differences between groups, nationalists treat both people of other nations and minority citizens as outsiders, taking a narrow view of who is, at the heart, truly part of the society.[56] They're more comfortable with the majoritarian idea of democracy in which the dominant people should have the primary say in governance. Their perspectives on moral and legal issues reflects this. I believe it fair to say that to a nationalist, a person of another ethnicity, citizen or not, is relatively more *foreign*.

Earlier I called ants extreme nationalists because they stick tight to their colony marker—its scent—as a stamp of their identity. Indeed, though in our species a patriot can become as teary-eyed as any nationalist in displays of allegiance to a flag or anthem, nationalists are supersensitive to those symbols.[57] For them brief exposure to a flag or an idolized leader incites an intense reaction—as does the absence of such an emblem when one is expected. Thus the uproar about gymnast Gabby Douglas not placing a hand over her heart while the American national anthem played in the 2012 Olympics, a lapse that to a nationalist made her gold-medal win too much about herself and not about the United States. The reaction was a sign of the sentiment that societies are entities: people don't compete in the games, countries do.

Both the nationalist and the patriot perspectives can be logically consistent, with nationalists being more risk averse and on guard against anything that may contaminate their culture. They prefer to err on the side of separatism, erecting boundaries that might alienate those whose interests could differ from their own, while patriots are more sympathetic to opportunities for trade and cooperation with outsiders.[58]

In short, the nationalist is suspicious of diversity, while patriots often welcome it.[59] Or at least they tolerate it, because even a patriot, no matter how equality-minded, isn't immune to prejudice: the ardor that patriots reserve for fellow society members of their own race or ethnicity still leads to discrimination as they subtly, and unwittingly, treat those like themselves more fairly.[60]

Why did these differences in patriotic and nationalistic attitudes evolve? The fact is that a clash in perspectives within societies, although at times so extreme as to verge on the dysfunctional, may have always been integral to human survival. Our varied expression of social viewpoints probably connects back to "timeless social concerns," as one research team put it.[61] Each outlook is beneficial in certain contexts. This dimension of our social identity may be an adaption to balancing the needs for protecting and provisioning the society. Even though people with opposing perspectives might not see eye to eye, a society with too few or too many individuals at either end of the spectrum could be open to catastrophes. This promotion of behavioral diversity has parallels in unlikely animal species. Social spiders are most successful when their colonies contain both individuals that retreat from danger but fastidiously tend the nest, and bold ones that put more effort into defense against social parasites, which steal the colony's food; the colonies of certain ant species function most efficiently when they contain a similarly effective mix of personality types.[62]

For humans, the hazards of a population overly committed to either the nationalist or patriot extreme are manifest. Nationalists see the patriot's greater openness to weak borders and sharing across ethnicities as promoting social dependence and cheating, fears that reflect the competitive nature of groups present across species. Meanwhile, the prevalence of nationalists across societies, convinced their ways are right and prepared to fight for them, means the dangers that nationalists fear can indeed be realized. Still, by readily espousing oppression and aggression, extreme nationalists bring to mind the historian Henry Adams's description of politics as a systematic organization of hatreds. Their outlook feeds on certain facets of psychology.[63] It's intoxicating to fall in line against an enemy, at times at a whiff of trouble. For those swept up in a nationalist perspective, the swell of group emotions and awareness of common purpose gives life a greater meaning. Not just morale, but mental health improve among civilians when nations face conflict.[64] The fact is, trigger-happy societies have long had an edge, with the impulse for war and the fear of attack critical in driving many social and technical innovations and the expansion of states.[65] What's more, nationalists, adhering to a narrow interpretation about what behaviors are proper, have the advantage of being far more tight-knit and homogeneous than patriots and better able to act together.[66] All this is to say that the patriot's vantage point is and always will be a more onerous path.

Because of the partiality for their group, displayed by patriots and nationalists in different ways, the troubles our societies face go deep. It's bad enough that a wicked act by one minority person—the Florida nightclub shooting, for example—can set off outrage at an entire minority population. But mistreatment can carry over to ethnicities unconnected to the tragedy.[67] That's an outcome of how stereotypes strip away detailed understanding, making it easy to conflate groups to the point of creating such fuzzy and nonsensical categories as "brown people."[68] Even when no conflation exists, prejudices can be linked, with the denigration of one people associated with the devaluation of others.[69] Persons who fear for their safety, jobs, or way of life indiscriminately lump *them* together much as ancient societies did with the "barbarians" beyond their borders. The impulse is so strong that when a sample of Americans was asked what they thought of Wisians, nearly 40 percent regarded them poorly and did not want them as neighbors, even though they could have known nothing about them since the researcher had made the name up.[70]

Societies contain ethnicities and races that stick together despite the members' prejudices about each other. The usual view, voiced by William Sumner more than a century ago, is that friction with outsiders draws a society together. Clearly that's not always true. The external forces that promote civil peace primarily galvanize the dominant people while often straining their ties to a society's other ethnicities when those groups are regarded as part of the problem. This tension among the members can cause a kind of social autoimmune disease, turning a society against itself. For all these tribulations, we may reasonably ask whether societies are necessary at all.

CHAPTER 26

The Inevitability of Societies

C an we discard our societies, combine them into one, or at least make them secondary to a more universal union of humanity?

Here is a snippet of history that reads like a parable. For centuries, the Pacific island of Futuna, a low chunk of volcanic rock, at 46 square kilometers in size, offered space and resources for just two chiefdoms—Sigave and Alo. These societies, claiming opposite ends of the island, were in almost constant conflict, pausing only briefly now and then for island-wide ceremonies featuring a psychoactive drink made from a shrub native to the western Pacific. One wonders if this enabled their people to tolerate each other for the day.[1] I can only imagine their spear-throwing clashes were a primary motivator in their lives, the Arab-Israeli conflict in a microcosm. One might expect that in such a confined space, and over the course of so much time, one chiefdom would have conquered the other. That this never happened might bear on the human craving for an outgroup, if not an outright opponent. Could Alo have continued on without Sigave—a society in a vacuum? Would it, alone in the world, even be what we could call a society?

"Go to the ant, thou sluggard, consider her ways and be wise," advised King Solomon, a keen observer of nature. Indeed, the Argentine ants battling for control of Southern California bring to mind a second hypothetical about what might happen if one society were to extinguish all others. For the ants, the final colony's national emblem (its scent) would no longer represent a marker; rather their odor would become synonymous with being an Argentine ant. By this means, the species would achieve the universal peace that experts believed it had attained before they discovered the boundary wars between the supercolonies.

One lesson is that while we may indeed learn a few things by considering the ways of the ant, such as the value of investing in roads and sanitation, it would be unadvisable to emulate them. Peace would have depended on the ants' nonpareil knack for slaughter, carnage that, in sheer numbers, would exceed the most nightmarish episodes of human history.

Yet a second lesson, to the point here, is that calling a group a society—and recognizing any markers that identify the members as a society—makes sense only when more than one society exists. The implication is that the compulsion to be part of a society must be matched by the imperative to identify an outgroup—Sigave for Alo and vice versa, or at least "others" vaguely rendered, as the barbarians were for the Roman Empire and Chinese dynasties—if only as a standard for comparison and a source of gossip, if not denigration. In this sense Futuna's chiefdoms, as simple and similar as they may be by our standards, exemplify human nature stripped to its heart.

Or is that so? In a study entitled "Us Without Them," psychologist Lowell Gaertner and colleagues found that when people need each other they can build a communal identity without contrasting themselves to outsiders.[2] Such mutually dependent people may feel as if they are functioning as a unit. These feelings can foster conviviality and bonding—something we might expect of a ship's crew struggling together in a storm. But it would be a stretch to call such a group, no matter the interdependence of its members or their level of cooperation, a society. For one thing, the shipmates would already identify with a society, whatever nation or nations they came from.

Take this a step further: Let's say the crew becomes shipwrecked and loses all contact with the outer world. Needless to say their former identities wouldn't vanish overnight. Twenty-five years after the 1789 mutiny aboard the HMS *Bounty*, the crew was found to have fled to Pitcairn Island. Rediscovered by an American ship 18 years later, the mutineers, and the Polynesians and Tahitians who accompanied them, were still recognizably English, Polynesian, and Tahitian. But suppose they had lived and remained undetected. In the passage of time, would they, or their descendants, have reimagined themselves into what we would recognize as one clearly defined and unified society?[3]

It isn't easy to find examples of a single collection of people utterly cut off from the rest of the world for generations. Some of the Vikings reaching Iceland and North America may have survived in isolation. However, they stuck to their trademark Viking mode of life sufficiently to link back up again without difficulty with Vikings in Europe; but at most their separation lasted decades, their origins never passing out of living memory.[4] Prehistoric peoples reached remote islands, too, but in most places either were in contact with tribes on other islands or had room to split into more than one society. Futuna had its two, and Easter Island harbored 17 adversarial giant-rock-head-erecting tribes, while in Australia the hundreds of aboriginal societies that thrived in pre-colonial times all descended from one group that landed on the continent by way of Asia.

One possible historical example of a lone society can be found on Henderson. At 37 square kilometers, this Polynesian island is thought to have offered too little space and too few resources to allow its few dozen residents to divide into two societies like Alo and Sigave. Having no more wood to make boats, the sixteenth-century Hendersonites were cut off from their trading partners on Pitcairn and Mangareva, islands 90 and 690 kilometers distant, respectively. By the time Spanish explorers found the island in 1606, its inhabitants had died out. No one can say how those few souls thought of themselves—whether, for instance, they still considered themselves a tribe and gave themselves a name.[5] My guess is that by holding on to dim recollections of others *out there*, passed down through the generations, those desperate few last islanders retained some shred of identity as a society, with a sense of Us versus Them never expunged from their minds.

The existence of foreigners would have had to wane even from legend and myth over the passage of many generations for the Hendersonites to see themselves as completely alone in the world. Had the Hendersonites lasted that long, might they have still expressed a need to belong, a yearning for oneness and *we*-ness? Or would any commitment they once had to a common identity have petered out? Perhaps in that case what we would find on Henderson would be individuals who continued to be social with friends and family but no society to speak of. That seems to be the view of the anthropologist Anya Peterson Royce: "The hypothetical group on an island with no knowledge of others is not an ethnic group; it does not have an ethnic identity; it does not have strategies based on ethnicity."[6]

One could disagree with Royce, pointing to the human craving to explore commonalities, demonstrated by the ease with which we model ourselves on the quirks of those we admire. Trendsetters drive the popularity of many practices, while the habits of disliked people may be shunned.[7] Copying a leader or venerated individual could result in conventions amounting to a kind of identity, even for the long-suffering Hendersonites. Those islanders' lives must have been stripped to the minimum by the Tasmanian effect, described earlier, whereby people in sparse groups forget aspects of their culture. Even so they no doubt still had a lot in common from growing up with and learning from each other. However, without an *other*, their alikeness would be reduced to such irrelevance that Royce has it right: it would be hard to regard them as a society (and certainly not as an ethnic group).

Their similarities might no longer matter in the profound way that shared characteristics do when they serve as societal markers. But perhaps the people's mere knowledge of each other would be enough to make them a society. Do species with individual recognition societies—which lack markers—perceive themselves as a collective regardless of the existence of outsiders, or do their societies collapse without an "other"? Field biologist Craig Packer tells me such a dissolution is the fate of isolated lion prides: the members disperse into ever smaller groups. No real surprise given that a primary function of societies is to succeed against competitors. This pressure disappears when there are no others. Yet the fate of lions may not be a reliable indicator for humans. Lions, which can hunt alone, get along by themselves better than people, who ache to be around more than spouses to avert loneliness and give their lives significance, if not to stay safe and fed. That's one reason why failing human societies seldom utterly collapse but rather split into smaller societies in which people retain their support networks, even if in a simpler setting. Escaping a broken society doesn't mean giving up on societies entirely.

On that basis we can expect that whether or not their relationships merit the name society, isolated humans will stick together more than lions, even if it's merely the Us-without-Them devotion we see among shipmates. For sure that's true of chimpanzees. A lone chimpanzee community stays interconnected in the same way that chimps do anywhere else, or so research on a group on its own in a gorge at Kyambura, Uganda, tells us.[8]

Yet such isolated groups would be quick to adjust to changing conditions. For the isolated chimpanzees, the Argentine ant supercolony that kills off all enemies, or for the tribal islanders who forget all others, their identity as a society would leap back to significance the moment they encountered outsiders again. The islanders would accentuate whatever meager social features set them apart, firming up a boundary between themselves and the newcomers (that is, if the invaders didn't overrun and kill or assimilate their little collective first).[9]

An experiment in 1954 roughly approximated this situation, demonstrating how rapidly some of the features of societies can come into play when conditions are right. Twenty-two 12-year-old boys at Robbers Cave State Park in Oklahoma were randomly assigned to two groups that were originally kept on their own. Once the groups saw each other, from a distance at first, and then came into contact, they were quick to develop separate identities: favorite songs, propensity for cursing, and so on. The boys gave each group a name, drew its totem animal on their shirts, and waved flags emblazoned with the same symbol when they competed in sports. The violence that the Rattlers and Eagles soon displayed toward each other was alleviated after the researchers had them work together, but their differences remained.[10]

Even if no foreigners were to come along from outside to disturb a lone community—if, for example, a hyperaggressive society conquered the entire globe, so that no outsiders existed—their achievement would be short lived. Once humans in isolation grow to more than a very few in number they seem to privilege some individuals over others, creating conditions under which multiple societies are born. The requisite foreigners would be birthed from within. The union would fragment, as have all nations, and all the societies that came before them.[11]

DREAMS OF A UNIVERSAL SOCIETY

What we know about human cognition therefore doesn't bode well for those who think humanity may someday blanket the earth in one borderless population, leaving no person an outsider. Still, even though societies will never dissolve away, perhaps another scenario could play out that would effectively remove them from the picture. With the number of societies generally declining century after century, we might imagine that all the remaining nations of the world will someday drop their boundaries sufficiently to create a cosmopolitan community that is more significant to people than the societies themselves.

Some claim the internationalization of culture (think McDonald's, Mercedes-Benz, *Star Wars*) and connections (with Facebook bringing people together from Aa, Estonia, to Zu, Afghanistan) are a harbinger of a Berlin Wall–type collapse of state borders. This is false. Societies have never freely merged, and that won't change. Populations worldwide may devour KFC, Starbucks, and Coca-Cola while screening Hollywood blockbusters—and enjoy sushi, flamenco, French couture, Persian carpets, and Italian cars. They may adopt, and at times be swamped by, cosmopolitan trends. But no matter how many exotic influences nations absorb or how many foreign connections they make, such nations do not slip into disarray, and they retain their people's fierce devotion.[12] After all, from time immemorial societies have taken what they wanted from the outside world to claim as their own, and have been all the stronger for doing so. Even the Statue of Liberty, the quintessential symbol of the United States, was designed and first erected on French soil by the same Mr. Eiffel who gave Paris its Tower.

Notwithstanding this rigid boundary maintenance, humankind can erect umbrella organizations comprised of many nations. Yet such a universal group would also fail to completely supersede our bonds to societies, as demonstrated by the most binding association of societies in the anthropological record. In northwest Amazonia reside 20 or so tribes, or language groups, known collectively as the Tukanoans. Each has its own language or dialect, some similar, some mutually unintelligible. The tribes are so cross-connected as to be tied economically, each specializing in goods it exchanges with the others. Between them are what amount to obligatory trading relationships of an unusual sort: marriages within a tribe are improper. "Those who speak the same language with us are our brothers, and we do not marry our sisters," the people say.[13] Thus a bride marries into another tribe, where she learns the local tongue. In case you imagine the arrangement to be an aberration, similar across-the-board spousal interchanges are recorded for New Guinea.

One explanation for this arrangement is that it reduces inbreeding in very small societies. We see this in many nonhumans as well, such as in chimpanzees, where the females avoid mating with kin by similarly transferring between communities. When populations are minuscule, as those of the Tukanoans have been on occasion in the past, the only marriage options actually could be between brothers and sisters, an incestuous act to which our species has an innate abhorrence.[14] This is a far more likely problem for the Tukanoans than for the countries of today. The psychological aversion to marrying siblings seems to overpower any trepidation the Tukanoans might have had about firmly linking their societies. Their compulsory spousal exchanges have created what to my mind are the tightest alliances ever recorded, currently totaling about 30,000 Amazonian souls. Yet for all that, the Tukanoan tribes remain clear and separate, each confined to specific areas.[15]

Putting aside the unusual situation of the Tukanoans, the failure of alliances to supersede people's affiliation to a society holds true universally, the United Nations and the European Union included. These intergovernmental organizations don't earn people's emotional commitment because they lack ingredients that make them *real* for the members. The EU may be the most ambitious attempt at economic integration conceived; yet it will never supplant the nations within it. The members don't see the EU as an entity worthy of their loyalty the way they do their countries, and for several reasons. First off, its borders are not fixed—in fact, are subject to revision as states enter or go. Additionally, its members have a history of conflict dating from the Middle Ages, and a split already exists from east to west between communist and capitalist cultures. To top all that off, the EU offers no grand foundation story, no venerable symbols or traditions, and there's little sense anyone would fight and die for Europe as they might for their nation.[16] That makes the EU a political coalition much like the Iroquois Confederacy was, though with less power. Each member country handles issues relating to its citizens' identity and remains the focus of their self-worth, an outlook that makes membership in the EU secondary and disposable. An analysis of the 2016 Brexit vote shows that those most strongly thinking of themselves as English went against staying with the EU. Those voters saw an economic and peacekeeping tool as a threat to their national identity.[17]

Financial and security issues hold the EU together. The same can be said for Switzerland, a country subject to perennial scrutiny because, as the four languages and complex territoriality of its people attest, its nationhood rests on a detailed social and political alliance between 26 local communities, or cantons. These self-governing settlements act in many respects as miniature nations nestled in a mountain landscape that enhances each one's autonomy. "Each Canton has its own history, constitution, and flag, and some even have an anthem," the political scientist Antoine Chollet reports, such that Swiss "citizenship refers to *one who can vote* and nothing more."[18] The formation of the Swiss confederation required rewriting accounts of the past to maintain a sense of equality between the cantons. This challenging step was necessary for the cantons to survive over the centuries when they were forced to negotiate their interests with far larger and more powerful neighboring countries.

The EU and Switzerland are regional entities kept together by the perceived need to counter hazards from outsiders, which gives both a reasonable chance of success. A global human union would have no such motivation, and would therefore be far more precarious. One possible means of attaining global unity might be to shift people's perception of who is an outsider—a point Ronald Reagan made often: "I occasionally think how quickly our differences worldwide would vanish," he remarked in an address to the United Nations, "if we were facing an alien threat from outside this world."[19] Popular science fiction tales like *The War of the Worlds* depict all of humankind pulling together against a common enemy. Yet our societies will endure even this. Space aliens wouldn't make nations irrelevant any more than Europeans arriving in Australia caused Aborigines to dispense with their tribes. That would be so regardless of how much the aliens shattered the beliefs people held about their own societies, whose beloved differences would now look trivial by comparison. Moreover, when societies of our species turn to one another, whether for commercial advantage or to defend against aliens, that reliance doesn't diminish the weight they place on their differences. The notion of cosmopolitanism, the idea that the people of the world will come to feel a primary connection to the human race, is a pipe dream.

SOCIETIES AND BEING HUMAN

Let me put forward a final question. What might happen if people could forgo their markers or somehow put aside the drive to categorize each other with labels? In such a world the only differences people would perceive would be between individuals—not between groups. One supposes that under such circumstances our nations would disintegrate entirely, but it's hard to predict what would rise in their stead. Perhaps our affiliations would coalesce around local neighborhoods, or around the people we know best, with the global population splintering into millions of micro nations. We might foresee a return to the individual recognition societies of our forebears, where everyone knew everyone else.

Or, by discarding our differences, or our penchant for making judgments about the differences, could we achieve the opposite result, doing away with societies entirely? Would the beehive of networks built up through international travel and Facebook friendships interlink us so indiscriminately that we would actually secure that elusive panhuman unity, encompassing every man, woman, and child?

Our reliance on markers may trace back far into the human past, but what comes naturally isn't always desirable, and fortunately our intelligence gives us some prospect of breaking free from our biology.

and history. When changes concern the matter of how we mark off our identities, though, any alteration would be extremely difficult and take more than education. While casting off ethnic and societal markers may sound good at first blush, the move would undoubtedly mean the loss of much of what humans cherish. Nationalists or patriots, people care about their memberships and few would willingly give them up. Nor could they, because human responses to groups are involuntary. Markers are two-edged swords, causing us to discount those who differ from us, yet at the same time imparting an esprit de corps with complete strangers who fit our expectations. To abandon human markers would strike against timeless psychological yearnings. No doubt if a mass hypnotist caused us to forget our differences, we would scramble to discover new differences to hold dear. The only way to retool this human attribute would be for a surgeon with near-miraculous understanding of the nervous system to ablate portions of the brain. The result of this science-fictional adjustment would be a creature we wouldn't recognize as ourselves. I'm unsure how one could measure whether such people were any happier than we are today, but surely, *they* would no longer be *us*.

For humans in our current condition, the question of whether societies need to exist really boils down to whether people must be in one for their emotional health and viability. I think so. "A man must have a nationality as he must have a nose and two ears," wrote Ernest Gellner, a prominent thinker on nationalism. Gellner—who went on to argue, mistakenly, that the human need to be part of a nation is nothing more than a contrivance of modern times—never fathomed how right his statement was.[20] The mind evolved in an Us-versus-Them universe of our own making. The societies coming out of this psychological firmament have always been points of reference that give people a secure sense of meaning and validation.

To say a person has no country calls to mind dysfunction, trauma, or tragedy. Without such an identity, humans feel marginalized, rootless, adrift: a dangerous condition. A case in point is the homelessness felt by immigrants who have lost connections to their native land only to face the sting of rejection by their adopted country.[21] Social marginalization has been a motivator stronger than religious fanaticism, explaining why many terrorists originally took to extremism only after being excluded from the cultural mainstream. For the socially dispossessed, radical views fill a void.[22] Organized crime groups likewise commandeer some of the properties that give a society its vitality by providing social pariahs with common goals and a sense of pride and belonging. This we saw in a very embryonic form with the socially well-adjusted young boys at Robbers Cave.

The evidence presented in this book points to societies being a human universal. Human ancestors lived in fission-fusion groups that evolved, by simple steps, from individual recognition societies to societies set apart by markers. The ingroup-outgroup boundaries of society membership would have made it through this transition unaltered. That means that humans have always had societies. There was no original, "authentic" human society, no time when people and families lived in an open social network before deciding to set themselves apart into well-defined groups. Being in a society—indeed, multiple, contrasting societies—is more indispensable and ancient than faith or matrimony, having been the way of things from before we were human.

The last Henderson Islanders must have been starved not only physically but socially—apathetic, or so I imagine, about the meaning of their lives. Rest assured, few facets of life match a society in striking passion in the human heart, so long as other societies exist to compare with our own. Provided we don't re-engineer ourselves and instead choose to remain fully human, societies, and the markers that bring us together and set us apart, are here to stay, signifying the boundaries between people in our minds, and setting the borders between them physically, across the earth's surface.

Identities Shift and Societies Shatter

He who should wish that his fatherland might never be greater, smaller, richer, poorer, would be the citizen of the world.

—VOLTAIRE

Making their way across the African savannah, the Australian coastline, and the American plains, our ancestors moved in small affiliated bands of lifelong fellow travelers. Month after month they pitched their camps and searched for food and water. Rarely did they encounter other human souls. I find it hard to imagine seeing as few strangers day to day as they did throughout their entire lives. With the passing of ages, societies have swelled to the point that we now move like ants through an anonymous swarm. Many in the throng are far less like us than the people that hunter-gatherers came across, over hundreds of generations, were like them.

So rarely did our ancestors encounter foreign peoples that outsiders seemed to occupy a realm between reality and myth. Aborigines guessed the first Europeans they met were ghosts.[1] Over time our view of the members of other societies has changed radically; today, foreigners don't seem outlandish or otherworldly, as they once routinely did. As a consequence of global exploration starting in the fifteenth century, and more so today thanks to tourism and social media, contact between people from far-flung parts of the globe is now commonplace. Outright incomprehension of outsiders is no longer the excuse it often was in prehistory. In those times so little was known about foreigners that barbarian hordes could be treated as if they were monsters under the beds of children.[2]

Still, humankind continues to express its relationships to societies in ways that people seldom recognize or avow. Yet our minds, configured to interact with fewer individuals and groups, are overloaded. And so this book has drawn from many fields of study to apprehend the nature of societies and understand how we cope with the overload, with many revelations along the way. Here I distill these down to a few essential conclusions.

The most fundamental of these is that societies are not solely a human invention. Most organisms lack the closed groups we call societies, but in the species that do have them, societies serve in diverse ways to provide for and protect the members. In all such animals, individuals must recognize each other as belonging to the same group. This membership can offer advantages regardless of whether or not those members cooperate or have any other social or biological relationship.

While societies are not unique to humans, they are necessary to the human condition, having existed since ancient prehumans diverged on the evolutionary tree from other apes. A million human societies may well have come and gone. Every one has been a group closed to outsiders, a group that its members were willing to fight for and that at times they died for. Each commanded intense commitments from its members extending from birth to death and through the generations. Until recent millennia, all of those societies were small communities of hunter-gatherers, but that didn't mean their attachments to their societies were any weaker than ours today.

In the societies of our prehuman ancestors, as in those of most other mammals, the members had to recognize each other individually to function as a group. The resulting constraint on memory put an upper limit of roughly 200 on the size of societies in most animals. At some point in our evolution, probably before the origins of *Homo sapiens*, humans broke this glass ceiling by forming anonymous societies. Such societies, found in humans and a few other animals—notably ants and most other social insects—can potentially attain vast sizes because the members no longer must remember each other personally. Instead, they rely on identifying markers to accept both individuals they know and strangers who fit their expectations. Scents serve as markers in the insects, but humans go broader. For us, markers range from accents and gestures to styles of dress, rituals, and flags.

Markers are essential components in all societies of more than a few hundred individuals, whether mighty human or mere insect. At a certain point of growth, however, markers alone are insufficient to hold a society of humans together. Large human populations depend on an interplay between the markers and an acceptance of social control and leadership, along with increasing commitments to specializations, such as jobs and social groups.

In early humans, new societies were born in a two-step process that has its parallels in other vertebrates. The process begins, usually very gradually, with the formation of subgroups within a society, brought on by divergences in identity. Years later, those identities diverge to such an extent that they become incompatible. The factions then separate to permanently form distinct societies.

The ability to accept strangers born in the same society as fellow members does not on its own account for the enormous growth of human societies. What made such vast expansions possible was the acquisition of people from other societies. Outsiders had to make accommodations to the expected identity to be accepted as part of the society. The addition of foreigners in numbers, initially by force through slavery and subjugation and more recently by immigration, generated the mixtures of ethnicities and races found within societies today. The relationships between these groups retain the imprint of differences in power and control that in some cases extend back to before recorded history.

The differentiation of identities among society members continues to be a source of disruption. However, rather than dividing in the manner of hunter-gatherer groups, today's societies more often fragment along geographical fault lines that roughly reflect the claimed ancestral homelands of the ethnic groups that have come to live within them.

The need for societies, by its very ancientness, has shaped all aspects of human experience. Most notably, relationships between societies have profoundly influenced the evolution of the human mind, which would in turn affect the interactions among the ethnic and racial groups that emerged later in our history. While we may not be operating out of pure ignorance of outsiders the way early humans often did, automatic responses reflect our proneness to harbor stereotypes about different groups and about the superiority of our own group. The psychology underlying our identification with societies and ethnicities is registered in our every action. Our reaction to each person we see, how we vote, and whether we approve of our country's decision to go to war have all been shaped by processes deeply embedded in our biology. The buzzing confusion of modernity may simply exacerbate our reactions.

Even as we face this deluge of social interactions as individuals, our nations grow ever more interdependent. Yet we are who we are, and so our societies still put inordinate effort into jockeying for territory, resources, and power, just as societies of animals have always done. We attack. We cajole. We blame. We abuse. We insulate ourselves from foreign powers we don't trust by partnering with those we do. Such alliances, unique to humankind, may save us. Nevertheless, they can bring further uncertainty and destruction, angering or striking fear into those who have been left out.

We can hope the past decades have brought change: aware of their interdependence and leery of the costs of modern conflicts, most nations have ceased trying to conquer each other outright. Our global knowledge of humankind has turned the extraordinary ordinary, creating an everyday reality that would have been unachievable when intergroup contact was rare and limited. Luckily not only can we serenely enter a café full of strangers, but we aren't alarmed if the latte sippers are different from us, whether they are members of ethnic pockets of our society or come from abroad. Given the chance, we may shake those people's hands with barely an increase in heart rate. Yes, among the others that we are compelled to live with there are still people whose identities irritate, repel, offend, or scare us. Yet for all the fraught social navigation and the painful run-ins, seen in an evolutionary and even a historical context the casualness of this mixing is a big deal.

Through it all people's identification with the societies that shelter them from an unpredictable world has remained steadfast. A sense of belonging inoculates us against outside influences. And our commitments have been invigorated by perceptions that our nations and tribes are venerable and timeless. An accurate reading of the past, and of the human social condition, requires us to squarely face the fact that such beliefs about social stability are a comforting illusion. New groups will assuredly gain traction. Tensions over national or ethnic differences won't go away. Human markers have evolved to facilitate not just the forces that bind society members together, but also those that tear them apart. They do the former by enabling strangers to view each other as fellow members; they do the latter because, over the passing of generations and across geographical distances, identities shift and societies shatter. All societies are ephemeral to the core—fleeting, as Chief Seattle foretold. Italy, Malaysia, the United States, Amazonian tribe, or Bushman band society: all act organically, their continuation requiring a dynamic responsiveness never free of conflict and anguish. Even so, the hard reality is that the social warp and woof can support only so many alterations. At some point the social fabric can no longer be mended, and ultimately the territorial integrity of every nation will be challenged, and fail.

As we look to the future, it may be illuminating to reframe the discussion around an idea cherished by people everywhere, and which I have brought up only in passing thus far: their freedom. Americans have declared pride in their freedom ever since winning their independence, and yet at the time of the Revolutionary War the British thought of themselves as free compared to more repressive European societies of the day. Indeed, much of human activity can be understood in terms of our pursuit of the choices a society opens to us. But this freedom for personal expression is never simple. Permissiveness is never without limits. Any society defines itself by what it won't tolerate, and the behavior required from its members; thus, by its nature, membership in a society reduces people's options and entails a loss of freedom. For most species the constraint is simply that individuals are required to interact with other members and consort little if at all with outsiders. Human societies come with further obligations. We must act appropriately, adhering to whatever markers set us apart from outsiders. We are free insofar as our actions fall within the most essential of these rules, and adhere to our commitments to the society and our station and status within it. Generally speaking, the more privations a society has undergone, the more rigid the expectations placed on its people.[3] Extremist regimes aside, by and large citizens everywhere gladly accept such restrictions. They believe in the *rightness* of their society and find comfort within the limitations it imposes on them. In return societies give a great deal, including a feeling of ease, even camaraderie, around likeminded others; the security and social support that comes with membership; and access to resources, opportunities for employment, suitable marriage partners, the arts, and much more.

While people value their freedom, in truth, socially imposed limits on freedom are as indispensable to happiness as freedom itself. If people are overwhelmed or unsettled by the options open to them and other members, and by the acts of those around them, what they feel is chaos, not freedom. What we think of as freedom, then, is invariably quite restricted. Yet only an outsider would suggest that the restrictions imposed by a culture are oppressive. For this reason, societies that promote individualism, like the United States, and those that nurture a collectivist identity, such as Japan or China—where people put greater emphasis on their identification with the collective and the support this gives them—can equally celebrate the opportunities and happiness their society offers.[4] Regardless of a society's permissiveness, unity falters if its citizens have the freedom (or feel they should have the freedom) to act outside the comfort zone of others, whether that means women voting, protestors burning the flag, or nontraditional couples claiming the right to marry.

These are weaknesses of the social fabric many societies struggle with today. But ethnic diversity presents an even greater complication in the pursuit of unity and freedom. The difficulty is balancing one group's pursuit of freedom with another group's comfort. All too often, inequalities in personal freedom emerge between groups. Minorities must fit in with what the society finds acceptable—especially with what the dominant group expects. Yet they are also obliged not to overly emulate the dominant people, leaving the majority to its separate and privileged status. Minorities are thereby put in the position of having to invest in identifying not only with the society where they are citizens, but with their own ethnicity, too. Hispanic Americans, for example, are constantly registered by their fellow Americans, and almost constantly see themselves, as Hispanic. By contrast, the dominant members, in primary control of the symbols and resources of nations, much more rarely need to think about their own ethnicity or race, other than as they need to when pulling together during times of economic trouble. This gives them greater freedom: majority individuals enjoy the luxury of considering themselves as one-of-a-kind, idiosyncratic persons.[5]

The United States, which overran the native peoples of the subcontinent to cobble together a society composed—slaves and their descendants aside—of people who for the most part came voluntarily, is an experiment with few parallels in history. Its population derives from many sources, with no ethnic groups numerically swamping broad geographical regions of the country where they can lay claim to ancient territorial roots. As a result it lacks the fracturing points that riddle many Old World societies brought together by forceful incorporation followed by an interminable history of resentment. A lack of ready-made fault lines may give America a measure of durability despite its political turmoil. Still, what will become of that country, as it approaches the two-and-a-half-century mark, is uncertain. A single question overrides all others: namely, whether the United States can persist as one nation, indivisible, in a productive relationship with the rest of the world as its remaining superpower. With latitude for diversity reducing requirements for citizenship to a few conditions, social understanding and an openness to adapt will be essential, as it is for many countries.

What's the best-case scenario for the future of our societies? What will create social health and longevity? The current trend has been for societies to move away from a support of diversity to focus their national identity more narrowly on their dominant group. Even so, minorities aren't going away. We may slow immigration, but it's no longer tenable to expel ethnic people as the Romans sometimes did when times got bad. Fortunately the United States is among a growing number of nations today that are exceptional not just in their ethnic plurality but in their richness of all kinds. Such nations pride themselves on their wealth of job opportunities, religious choices, sports fandoms, and other interest groups. This cornucopia can amplify a society's strength by giving its citizens many options that add layers to their personal identities and affinities with others. Those able to reach outside their own ethnicity and race, or find commonalities with people like themselves yet with different outlooks, have the chance to bond over other shared enthusiasms; think of the study that shows that a person's race may be overlooked should he or she be wearing the jacket of a beloved team.[6] Such cross-connections can be individually fragile but strong in bulk, keeping a society whole in the face of upheaval.[7] Governance matters, too. Nations comprised of distinct ethnicities function well when their institutions support diversity.[8] As long as interactions stay productive, prejudice declines. This phenomenon is true no matter how insular people may be in their choice of friends or how little face time they spend with other ethnicities.[9]

Diversity presents social challenges at the same time that it brings creative exchange, innovation, and problem solving, fueled by varied talents and perspectives.[10] How long any society can remain strong in the face of the shifting relationships among its people remains a troubling question. For a society to stay intact other than by force, all of its communities must be motivated to rally with equal passion around a core identity—easier said than done given the majority's greater freedoms and power to manipulate the rules of the game in its favor, often through institutions largely under the control of its social upper crust. A nation supporting strong ties among its people and similarly masterful in its dealings with other societies would serve the greater well-being of its citizens, extending its time on Earth and making its legacy a high point in the story of humankind.

It would be the height of foolishness to think such an outcome could be achieved through cheerful goodwill or careful social engineering. However optimistic you feel about our adroitness as problem solvers, human minds—and the societies we fabricate when those minds interact—are malleable only to a degree. Our willingness to enjoy an advantageous social position, even hurt each other in the service of preserving our sense of dominance and superiority, is an abiding human trait.

Our misfortune has been, and will be always, that societies don't eliminate discontent; they simply redirect it toward outsiders—which paradoxically can include the ethnic groups within them. Our improved knowledge of others hasn't always been enough to improve how we treat them. We will need to better understand the urge to view other peoples as less human, even bug-like, if our species is to break from the history of dissonance between groups that traces back beyond antiquity. We also must know more about how people reformulate their identities, and respond to each sea change with the least damage possible. *Homo sapiens* is the only creature on Earth that can do this. If our dispositions toward outsiders vary, some of us inclined to caution, others to trust, nevertheless we share an aptitude for harnessing connections with seemingly incompatible others. Our salvation lies in reinforcing this gift, guided by the increasingly refined discoveries of the sciences covered by this book. The good news is that humans have some capacity to counter our inherited propensities for conflict through deliberate self-correction. Divided we will be, and divided we must stand.

Acknowledgments

One underappreciated aspect of roughing it in remote parts of the world is that one is forced to slow down. From long days spent as a biologist waiting out tropical downpours under a tarp or lumbering along on a camel over bleached bone and sand, I've come to learn that true creative time is the time *between* things, empty intervals a packed calendar rarely affords. In *The Journey* the poet Mary Oliver writes of following one's personal calling. By taking a "road full of fallen branches and stones" one finds that little by little a new world opens up. The poem resonates with me because a lifetime on such roads has given me an abundance of time to contemplate what I have seen and motivated me to write this book.

Because making connections between fields with divergent vocabularies and approaches sometimes required me to simplify arguments to make them accessible for a general audience, I've included a more thorough bibliography than usually found in non-academic writing to give readers with different backgrounds the option to pursue particular threads; even so, I've had to be selective. This book and its technical predecessor (Moffett 2013) would have also been incomplete without going beyond the literature by seeking out the advice of experts who proved generous, from reading drafts to tolerating naïve questions. A gratifying response was "I never thought of that." I freely admit that someone far smarter than me should have written this book; any errors of interpretation are mine.

The people below have kindly helped me on issues ranging from why hamsters get excited about body odor and how radio programs can accelerate a genocide to the devotion of immigrants to their country. Among the names that follow, those in *italics* reviewed parts of the manuscript: Dominic Abrams, Stephen Abrams, Eldridge Adams, Rachelle Adams, *Lynn Addison*, Willem Adelaar, Alexandra Aikhenvald, Richard Alba, Susan Alberts, *John Alcock*, Graham Allan, Francis Allard, Bryant Allen, Warren Allmon, Kenneth Ames, David Anderson, Valerie Andrushko, Gizelle Anzures, Coren

Apicella, Peter Apps, Eduardo Araral Jr., *Elizabeth Archie*, Dan Ariely, Ken Armitage, Jeanne Arnold, *Alyssa Arre*, Frank Asbrock, Filippo Aureli, Robert Axelrod, Leticia Avilés, Serge Bahuchet, Russell Paul Balda, *Mahzarin Banaji*, Thomas Barfield, Alan Barnard, *Deirdre Barrett*, Omer Bartov, Yaneer Bar-Yam, Brock Bastian, Andrew Bateman, *Roy Baumeister*, James Bayman, Isabel Behncke-Izquierdo, *Dan Bennett*, Elika Bergelson, Joel Berger, Luís Bettencourt, Rezarta Bilali, Michał Bilewicz, Andrew Billings, Brian Billman, Thomas Blackburn, Paul Bloom, Daniel Blumstein, Nick Blurton-Jones, Galen Bodenhausen, Barry Bogin, Milica Bookman, Raphaël Boulay, Sam Bowles, Reed Bowman, Robert L. Boyd, Liam Brady, Jack Bradbury, Benjamin Braude, Stan Braude, Anna Braun, Lauren Brent, *Marilynn Brewer*, Charles Brewer-Carias, Charles Brown, Rupert Brown, Allen Buchanan, Christina Buesching, Heather Builth, Gordon Burghardt, and David Butz.

Francesc Calafell, Catherine Cameron, Daniela Campobello, Mauricio Cantor, Elizabeth Cashdan, *Kira Cassidy*, Deby Cassill, Emanuele Castano, Frank Castelli, Luigi Luca Cavalli-Sforza, Richard Chacon, Napoleon Chagnon, Colin Chapman, Russ Charif, Ivan Chase, Andy Chebanne, Jae Choe, Patrick Chiyo, Zanna Clay, Eric Cline, Richmond Clow, Brian Codding, Emma Cohen, Lenard Cohen, *Anthony Collins*, Richard Connor, Richard Cosgrove, Jim Costa, Iain Couzin, Scott Creel, Lee Cronk, Adam Cronin, Christine Dahlin, Anne Dagg, Graeme Davis, Alain Dejean, Irven DeVore, Marianna Di Paolo, Shermin de Silva, Phil deVries, *Frans de Waal*, Oliver Dietrich, Leonard Dinnerstein, Arif Dirlik, Robert Dixon, Norman Doidge, Anna Dornhaus, Ann Downer-Hazell, Michael Dove, Don Doyle, Kevin Drakulich, Carsten De Dreu, Christine Drea, Daniel Druckman, Robert Dudley, *Lee Dugatkin*, *Yarrow Dunham*, Rob Dunn, *Emily Duval*, David Dye, *Timothy Earle*, *Adar Eisenbruch*, Geoff Emberling, Paul Escott, Patience Epps, Robbie Ethridge, Simon Evans, Peter Fashing, Joseph Feldblum, Stuart Firestein, *Vicki Fishlock*, Susan Fiske, Alan Fix, Kent Flannery, Joshua Foer, John Ford, AnnCorinne Freter-Abrams, Doug Fry, and Takeshi Furuichi.

Lowell Gaertner, Helen Gallagher, Lynn Gamble, Jane Gardner, *Raven Garvey*, Peter Garnsy, Azar Gat, Sergey Gavrilets, Daniel Gelo, Shane Gero, Owen Gilbert, Ian Gilby, *Luke Glowacki*, Simon Goldhill, Nancy Golin, Gale Goodwin Gómez, Alison Gopnik, Lisa Gould, Mark Granovetter, Donald Green, Gillian Greville-Harris, *Jon Grinnell*, Matt Grove, Markus Gusset, Mathias Guenther, Micaela Gunther, Gunner Haaland, Judith Habicht-Mauche, Joseph Hackman, *David Haig*, Jonathan Hall, Raymond Hames, Christopher Hamner, Marcus Hamilton, Sue Hamilton, Bad Hand, John Harles, Stevan Harrell, Fred Harrington, John Hartigan, Nicholas Haslam, Ran Hassin, Uri Hasson, Mark Hauber, Kristen Hawkes, John Hawks, *Brian Hayden*, Mike Hearn, Larisa Heiphetz, Bernd Heinrich, Joe Henrich, Peter Henzi, Patricia Herrmann, Barry Hewlett, Libra Hilde, Jonathan Hill, Kim Hill, Lawrence Hirschfeld, Tony Hiss, Robert Hitchcock, Robert Hitlan, Michael Hogg, *Anne Horowitz*, Kay Holekamp, Leonie Huddy, Mark Hudson, Kurt Hugenberg, Stephen Hugh-Jones, Marco

Iacoboni, Yasuo Ihara, Benjamin Isaac, Tiffany Ito, Matthew Frye Jacobson, Vincent Janik, *Ronnie Janoff-Bulman*, Julie Jarvey, Robert Jeanne, Jolanda Jetten, Allen Johnson, Kyle Joly, Adam Jones, *Douglas Jones*, and John Jost.

Alan Kamil, *Ken Kamler*, *Robert Kelly*, Eric Keverne, Katherine Kinzler, Simon Kirby, John Kloppenborg, Nick Knight, Ian Kuijt, Sören Krach, Karen Kramer, Jens Krause, Benedek Kurdi, Rob Kurzban, Mark Laidre, *Robert Layton*, Kang Lee, James Leibold, Julia Lehmann, Jacques-Philippe Leyens, Zoe Liberman, Ivan Light, Wayne Linklater, Elizabeth Losin, Bradley Love, *Margaret Lowman*, Audax Mabulla, Zarin Machanda, *Richard Machalek*, Cara MacInnis, Otto MacLin, Anne Magurran, Michael Malpass, Gary Marcus, *Joyce Marcus*, *Curtis Marean*, Frank Marlowe, *Andrew Marshall*, William Marquardt, José Marques, Anthony Marrian, Ahigail Marsh, Ben Marwick, John Marzluff, Marilyn Masson, Roger Matthews, David Mattingly, John (Jack) Mayer, Sally McBrearty, Brian McCabe, John McCardell, Craig McGarty, William McGrew, Ian McNiven, David Mech, Doug Medin, Anne Mertl-Millhollen, Katy Meyers, Lev Michael, Taciano Milfont, Bojka Milicic, Monica Minnegal, John Mitani, Peter Mitchell, Panos Mitkidis, Jim Moore, Corrie Moreau, Cynthia Moss, Ulrich Mueller, *Paul Nail*, Michio Nakamura, Jacob Negrey, Douglas Nelson, Eduardo Góes Neves, David Noy, and Lynn Nygaard.

Michael O'Brien, Caitlin O'Connell-Rodwell, Molly Odell, Julian Oldmeadow, Susan Olzak, Jane Packard, Craig Packer, Robert Page, Elizabeth Paluck, Stefania Paolini, David Pappano, Colin Pardoe, William Parkinson, Olivier Pascalis, Shanna Pearson-Merkowitz, Christian Peeters, Irene Pepperberg, Sergio Pellis, Peter Peregrine, *Dale Peterson*, Thomas Pettigrew, David Pietraszewski, Nicholas Postgate, Tom Postmes, Jonathan Potts, Adam Powell, Luke Premo, Deborah Prentice, Anna Prentiss, Barry Pritzker, Jill Pruetz, *Jonathan Pruitt*, Sindhu Radhakrishna, Alessia Ranciaro, Francis Ratnieks, Linda Rayor, Dwight Read, Elsa Redmond, Diana Reiss, Ger Reesink, Michael Reisch, Andres Resendez, *Peter Richerson*, Joaquín Rivaya-Martínez, Gareth Roberts, Scott Robinson, David Romano, Alan Rogers, Paul Roscoe, Stacy Rosenbaum, Alexander Rosenberg, Michael Rosenberg, Daniel Rubenstein, Mark Rubin, Richard Russell, Allen Rutberg, Tetsuya Sakamaki, Patrick Saltonstall, Bonny Sands, *Fabio Sani*, Stephen Sanderson, *Laurie Santos*, Fernando Santos-Granero, Robert Sapolsky, Kenneth Sassaman, Jr., Chris Scarre, Colleen Schaffner, *Mark Schaller*, Walter Scheidel, Orville Schell, Carsten Schradin, Jürgen Schweizer, James Scott, Lisa Scott, Tom Seeley, and Robert Seyfarth.

Timothy Shannon, Paul Sherman, Adrian Shrader, Christopher Sibley, James Sidanius, Nichole Simmons, *Peter Slater*, Con Slobodchikoff, David Small, Anthony Smith, *David Livingstone Smith*, Eliot Smith, Michael Smith, Noah Snyder-Mackler, Magdalena Sorger, Lee Spector, Elizabeth Spelke, Paul Spickard, Göran Spong, *Daniel Stahler*, Charles Stanish, Ervin Staub, Lyle Steadman, *Amy Steffian*, Fiona Stewart, Mary Stiner, Ariana Strandburg-Peshkin, Thomas Struhsaker, *Andy Suarez*, Yukimaru Sugiyama, Frank

Sulloway, Martin Surbeck, Peter Sutton, Maya Tamir, Jared Taglialatela, John Terborgh, Günes Tezür, John and Mary Theberge, Kevin Theis, Elizabeth Thomas, Barbara Thorne, Elizabeth Tibbetts, *Alexander Todorov*, Nahoko Tokuyama, Jill Trainer, *Neil Tsutsui*, Peter Turchin, Johannes Ullrich, Sean Ulm, Jay Van Bavel, Jojanneke van der Toorn, Jeroen Vaes, Rene van Dijk, Vivek Venkataraman, Jennifer Verdolin, Kathleen Vohs, Chris von Rueden, Marieke Voorpostel, Athena Vouloumanos, Lyn Wadley, Robert Walker, Peter Wallensteen, Fiona Walsh, David Lee Webster, *Randall Wells*, Tim White, Hal Whitehead, *Harvey Whitehouse*, Polly Wiessner, Gerald Wilkinson, Harold David Williams, Edward O. Wilson, John Paul Wilson, *Michael Wilson*, Mark Winston, George Wittemyer, Brian Wood, *Richard Wrangham*, *Patricia Wright*, Tim Wright, Frank Wu, *Karen Wynn*, Anne Yoder, Norman Yoffee, Andrew Young, Anna Young, Vincent Yzerbyt, and João Zilhão.

For my research on animal societies in the field, Melissa and I were generously hosted by Elizabeth Archie for baboons, Filippo Aureli for spider monkeys, Anthony Collins for chimpanzees, Kay Holekamp for hyenas, Cynthia Moss for savanna elephants, Daniel Stahler for wolves, and Randall Wells for dolphins. Special thanks go to the National Geographic Society for a long history of funding my research; Gerry Ohrstrom for assistance when I needed it early in my writing; Richard Wrangham for supporting my visiting scholar position at Harvard's Department of Human Evolutionary Biology; Allen Rodrigo and team for a sabbatical at the former National Evolutionary Synthesis Center (the now sadly defunct NESCent); Lynn Addison for much wise counsel; and Ted Schultz and the Department of Entomology for my ongoing status as a research associate at the National Museum of Natural History (Smithsonian Institution). I am fortunate to have so many true friends.

I am also grateful to my agent and adviser, Andrew Stuart, for guiding me through the complexities of modern book publishing, and to my editor, Thomas "T.J." Kelleher, as well as Roger Labrie, Bill Warhop, and editor-in-chief Lara Heimert at Basic, for envisioning *The Human Swarm* as a book of major scope.

Finally, I have been thankful every day for Melissa and her extraordinary tolerance of my single-mindedness with this book when we could have been going about our usual business of tracking down astonishing creatures in wondrous places.

Notes

Introduction

1   Breidlid et al. (1996), 14. Different accounts exist of what Seattle said (Gifford 2015).
2   Sen (2006), 4.
3   For perspectives on "tribe" in its looser meanings, I recommend Greene (2013).
4   Of course humans can have these impulses diverted to develop powerful bonds to other groups such as cults (see Chapter 15 and Bar-Tal & Staub 1997).
5   Dunham (2018). Taken further, people divided into groups with the flip of a coin almost immediately value their group-mates more than those assigned to the other group (Robinson & Tajfel 1996).
6   Quoted in Dukore (1966), 142.
7   For evidence of ongoing human evolution, see Cochran & Harpending (2009).

Chapter 1

1   The self-sacrifice in humans often demands cultural indoctrination (Alexander 1985).
2   Anderson (1982).
3   My thanks to Emilio Bruna, the editor of *Biotropica*, for permitting me to adapt my thoughts in this paragraph from Moffett (2000), 570–571.
4   Thanks to David Romano and Günes Tezcür for advice on the citizenship and individual rights of the Kurds across the Middle East.
5   For example, Wilson (1975, 595) defines a society as "a group of individuals belonging to the same species and organized in a cooperative manner," adding the diagnostic criterion of "reciprocal communication of a cooperative nature, extending beyond mere sexual activity." One can imagine societies where mutual benefits accrue that aren't reciprocally communicated, however.
6   Émile Durkheim (1895), who established the discipline more than a century ago, saw cooperation as a key element of societies. He thought of cooperation arising when people share similar sentiments and points of view, and certainly beliefs and moral principles are essential elements of what I will talk about regarding identification in societies of humans. See also Turner & Machalek (2018).
7   For a few of many fascinating discussions: Axelrod (2006), Haidt (2012), Tomasello et al. (2005), Wilson (2012).

8   Dunbar et al. (2014). The premise of the social brain hypothesis has recently been put in doubt by data showing that brain size is better predicted by ecology than sociality (DeCasien et al. 2017).

9   For the accuracy of applying the word friendship to animals, see Seyfarth & Cheney (2012).

10   Dunbar (1996), 108. Dunbar's number is usually described in terms of positive relationships, but arguably one's knowledge of enemies should be factored in as well (De Ruiter et al. 2011).

11   This is true for other species, too. As biologist Schaller (1972, 37) points out, companionship among lions has no influence on the composition of a pride.

12   Dunbar (1993), 692. It is worth repeating the sentence in full: "How is it that, despite these apparent cognitive constraints on group size, modern human societies are nonetheless able to form super-large groups (e.g., nation states)?" Dunbar presented the human ability to categorize society members by their social roles as a solution to this question, but knowing what people do doesn't explain the memberships of societies and the distinct boundaries between them.

13   Turnbull (1972). Some doubt his interpretations (e.g., Knight 1994).

14   At least as of the date of the following study: European Values Study Group and World Values Survey Association (2005).

15   Simmel (1950).

16   A chimpanzee will sometimes be generous to an individual that's likely to return the favor in other ways (Silk et al. 2013).

17   Jaeggi et al. (2010). Tomasello (2011; see also 2014) finds hunter-gatherers to be more cooperative than apes in every domain: "Cooperation is simply a defining feature of human societies in a way that it is not for the societies of other great apes" (p 36).

18   Ratnieks & Wenseleers (2005).

19   e.g., Bekoff & Pierce (2009); de Waal (2006).

20   Social life can offer benefits for one group over another even when the individuals in it don't profit directly and aren't related (this is group selection) or when there are advantages for individual and group both (multilevel selection) (e.g., Gintis 2000; Nowak 2006; Wilson & Wilson 2008; Wilson 2012). I won't dwell on these alternatives because the controversies are well covered elsewhere. Group selection would seem to require the stability of societies. Still, my judgment is that, as far as most species go, societies offer sufficient benefits to individual members without the need to invoke group or kin selection.

21   Allee (1931); Clutton-Brock (2009); Herbert-Read et al. (2016).

22   Females sometimes carry another's infant or gang up on a female they don't like (Nakamichi & Koyama 1997).

23   Daniel Blumstein and Christina Buesching, pers. comm.; Kruuk (1989), 109. A male marmot will also drive off competing males, but whether this is advantageous to anyone other than him is hard to say. The badgers live in distinct closed groups, but it is unclear to me whether marmots do(e.g., Armitage 2014).

24   Henrich et al. (2004); Hogg (1993).

25   Quoted from Columbus's log, in Zinn (2005), 1–2.

26   Erwin & Geraci (2009). For more on canopy biodiversity, see Moffett (1994).

27   Wilson (2012).

28   Caro (1994).

29   In this case different species may join the group (e.g., Sridhar et al. 2009). Flocks that are distinct societies will be addressed in Chapter 6.

30   e.g., Guttal & Couzin (2010); Krause & Ruxton (2002); Gill (2006); Portugal et al. (2014).

31   Anne Magurran, pers. comm.; Magurran & Higham (1988).

32   Hamilton (1971).
33   This kind of behavior had also been described in an insect before (Ghent 1960).
34   Costa (2006), 35.
35   Rene van Dijk, pers. comm.; van Dijk et al. (2013 & 2014).
36   For descriptions of how societies deal with issues of fairness and "free riders," see for
     example Boyd & Richerson (2005).

Chapter 2

1   My thanks to Stephen Abrams, Ivan Chase, and Carsten Schradin for general advice
    on fish. Bshary et al. (2002); Schradin & Lamprecht (2000 & 2002).
2   Barlow (2000), 87.
3   I have space to suggest just a few key books, when any are available, for the species
    that follow. For meerkats, I thank Andrew Bateman, Christine Drea, Göran Spong,
    and Andrew Young; for horses, Joel Berger, Wayne Linklater, Dan Rubenstein, Allen
    Rutberg (see *The Domestic Horse* by Mills & McDonnell 2005); for gray wolves, Dan
    Stahler, David Mech, and Kira Cassidy (see *Wolves: Behavior, Ecology, and Conservation*
    by Mech & Boitani 2003); for painted dogs, Scott Creel, Micaela Gunther, Markus
    Gusset, and Peter Apps (see *The African Wild Dog* by Creel & Creel 2002); for li-
    ons, Jon Grinnell and Craig Packer (see *The Serengeti Lion* by Schaller 1972); for
    hyenas, Christina Drea, Kay Holekamp, and Kevin Theis (see *The Spotted Hyena* by
    Kruuk 1972); for the bottlenose dolphin of the eastern US seaboard, Randall Wells
    (who has published many technical articles about them); for ringtailed lemurs, Lisa
    Gould, Anne Mertl-Millhollen, Anne Yoder, and the sadly deceased Alison Jolly (see
    Ringtailed Lemur Biology, Jolly et al. 2006); for baboons (by which in this book I mean
    the savanna species—the yellow, chacma, and olive baboons), Susan Alberts, An-
    thony Collins, and Peter Henzi (see *Baboon Metaphysics* by Cheney & Seyfarth 2007,
    and *A Primate's Memoir* by Sapolsky 2007); for mountain gorillas, Stacy Rosenbaum;
    for chimpanzees, Michael Wilson and Richard Wrangham (see *The Chimpanzees of
    Gombe* by Goodall 1986 and *The Mind of the Chimpanzee* by Lonsdorf et al. 2010); and
    for bonobos, Isabel Behncke-Izquierdo, Takeski Furuichi, Martin Surbec, Nahoko
    Tokuyama, and Frans de Waal (see *Behavioural Diversity in Chimpanzees and Bonobos*
    by Boesch et al. 2002, and *The Bonobos* by Furuichi and Thompson 2007).
4   I thank Jennifer Verdolin, Linda Rayor, and Con Slobodchikoff for prairie dog ad-
    vice. Rayor (1988); Slobodchikoff et al. (2009); Verdolin et al. (2014).
5   I thank Elizabeth Archie, Patrick Chiyo, Vicki Fishlock, Diana Reiss, and Shermin
    de Silva for help with elephants. About everything one needs to know about the
    savanna species is summarized in Moss et al. (2011).
6   De Silva & Wittemyer (2012); Fishlock & Lee (2013).
7   Benson-Amram et al. (2016).
8   Macdonald et al. (2004); Russell et al. (2003).
9   Silk (1999).
10 Laland & Galef (2009); Wells (2003).
11 Mitani et al. (2010); Williams et al. (2004).
12 e.g., Cheney & Seyfarth (2007), 45.
13 Only the Florida dolphins studied by Randall Wells are covered in this book: the
    bottlenose dolphins elsewhere may differ in their behavior and sometimes belong
    to other species.
14 Linklater et al. (1999).
15 Palagi & Cordoni (2009).
16 e.g., Gesquiere et al. (2011); Sapolsky (2007).
17 Van Meter (2009).

18 That's not to say fantasies of success don't keep people going, but the daydreams of James Thurber's Walter Mitty rarely play out in reality and obsessions with being king are pathological unless one is in line for the throne. People are prone to believe in their potential to achieve goals wildly in excess of actual chances, yet seem no less happy when they don't succeed (Gilbert 2007; Sharot et al. 2011).

19 That bonobos sometimes work together to catch large prey is shown by Surbeck & Hohmann (2008).

20 Hare & Kwetuenda (2010).

21 Brewer (2007), 735.

Chapter 3

1   Unlike Aureli et al. (2008), I see little confusion as to the species for which the term "fission-fusion societies" is justified.

2   The difficulty of attacking large parties applies to predators as well as enemies, though leopards are indifferent to the attempts of chimps to stop them and may be an exception (Boesch & Boesch-Achermann 2000; Chapman et al. 1994).

3   Marais (1939).

4   Strandburg-Peshkin, pers. comm.; Strandburg-Peshkin et al. (2015).

5   Bates et al. (2008); Langbauer et al. (1991); Lee & Moss (1999).

6   East & Hofer (1991); Harrington & Mech (1979); McComb et al. (1994).

7   Fedurek et al. (2013); Wrangham (1977).

8   Wilson et al. (2001 & 2004).

9   The loud bonobo calls have complex functions (Schamberg et al. 2017).

10 Slobodchikoff et al. (2012).

11 e.g., Thomas (1959), 58.

12 Bramble & Lieberman (2004).

13 Evans (2007).

14 Stahler et al. (2002).

Chapter 4

1   Leticia Avilés, pers. comm.; Avilés & Guevara (2017).

2   King & Janik (2013).

3   Boesch et al. (2008).

4   Zayan & Vauclair (1998).

5   Seyfarth & Cheney (2017), 83.

6   Pokorny & de Waal (2009).

7   de Waal & Pokorny (2008).

8   Miller & Denniston (1979).

9   Struhsaker (2010).

10 Schaller (1972), 37, 46.

11 This fact has been overlooked, e.g., Tibbetts & Dale (2007) review individual recognition but curiously ignored its role as a requirement and aid for living in societies.

12 Breed (2014).

13 Lai et al. (2005).

14 Jouventin et al. (1999).

15 de Waal & Tyack (2003) and Riveros et al. (2012) call these "individualized societies."

16 Mentioned by Furuichi (2011). Randall Wells tells me female bottlenose dolphins, who largely stick to a contained part of their community's territory, sometimes can

be almost as isolated from other group members as the female chimpanzees described here.

17 Rodseth et al. (1991).

18 Quoted in Jenkins (2011).

19 Berger & Cunningham (1987).

20 e.g., Beecher et al. (1986).

21 The excretion differs from one mongoose to the next and is used by the animals to distinguish individuals, but the intriguing possibility exists that the scent also has a group specific component (Rasa 1973, Christensen et al. 2016).

22 Estes (2014), 143.

23 Joel Berger, Jon Grinnell, and Kyle Joly, pers. comm.; Lott (2002).

24 Whether societies generally attain this maximum population or achieve a lower size limit owing to factors other than memory will be determined by the rules of society reproduction for the species in question, a subject for Chapter 19.

25 "Troops" is a sensible word, since they seem to be of a kind with—are homologous to—the troops of other monkeys (Bergman 2010). The animals may also recognize some of the members of the troop that most recently split off from their own (such divisions are described in Chapter 19). See also le Roux & Bergman (2012).

26 Machalek (1992).

Chapter 5

1 For more about leafcutters see Moffett (1995 & 2010). For a detailed treatment of ants generally, see Hölldobler & Wilson (1990). I thank University of California Press for permission to adapt some passages from Moffett (2010) for use in this book.

2 de Waal (2014). For example, typical of articles comparing chimpanzees and humans, Layton & O'Hara (2010) spends far more time discussing differences than substantial similarities.

3 Monkeys and human infants younger than 18 months fail this test of self-awareness. For these and other issues, see Zentall (2015).

4 Tebbich & Bshary (2004).

5 de Waal (1982).

6 Beck (1982).

7 McIntyre & Smith (2000), 26.

8 e.g., Sayers & Lovejoy (2008); Thompson (1975).

9 Bădescu et al. (2016).

10 I give termites and honeybees short shrift in favor of my beloved ants, but for those who want to know more about them I suggest Bignell et al. (2011) and Seeley (2010).

11 The pivotal role of scaling applies to the size of organisms as well as to societies. See Bonner (2006) and other works by that author.

12 The description of ant market economies is adapted, with permission, from Moffett (2010). See Cassill (2003); Sorensen et al. (1985); and, for honeybees, Seeley (1995).

13 Wilson (1980).

14 For some other insect societies depending on agriculture and domesticated foods, see Aanen et al. (2002); and Dill et al. (2002).

15 Bot et al. (2001); Currie & Stuart (2001).

16 Moffett (1989a).

17 Branstetter et al. (2017); Schultz et al. (2005); Schultz & Brady (2008).

18 Mueller (2002).

Chapter 6

1   Barron & Klein (2016).

2   A few other ants have supercolonies, including one discovered by Magdalena Sorger and myself in Ethiopia with colonies kilometers wide (Sorger et al. 2017). For more details on Argentine ants and a critical review of the literature on which this chapter is based, see Moffett (2010 & 2012).

3   Violence within a supercolony occurs in one situation. Each spring, for reasons unclear, workers mass-execute the queens, sparing enough to maintain colony growth. This exception proves the rule: social integrity is reflected in how well the ants manage this situation, with a colony operating smoothly as its queens are butchered—even the queens don't protest (Markin 1970).

4   Injaian & Tibbetts (2014).

5   There are however rare cases of ant foundress queens recognizing each other individually, by scent (d'Ettorre & Heinze 2005).

6   The ants can also distinguish workers by the tasks they are doing at the time (Gordon 1999).

7   Dangsheng Liang, pers. comm.; Liang & Silverman (2000).

8   After first using the term anonymous society (Moffett 2012), I found Eibl-Eibesfeldt (1998) had employed this phrase to describe any society with a big population. In my usage, even a small society can be anonymous if it is demarcated through the use of labels that *potentially* allow some members not to know others.

9   Brandt et al. (2009). Fighting between supercolonies continues unabated even after the ants are fed an identical diet in a uniform lab setting for a year (Suarez et al. 2002).

10   Haidt (2012).

11   Czechowski & Godzińska (2015).

12   In some cases the colony odors arise primarily from the queens (Hefetz 2007).

13   Slaves are less aggressive toward outsiders than are free-living ants; one interpretation of this is that the diversity of markers in their colonies may lead to sloppiness in identification (Torres & Tsutsui 2016).

14   Elgar & Allan (2006).

15   Thanks to Stan Braude and Paul Sherman for advice on this species. Braude (2000); Bennett & Faulkes (2000); Judd & Sherman (1996); Sherman et al. (1991).

16   Braude & Lacey (1992), 24.

17   Burgener et al. (2008).

18   I thank Russell Paul Balda, John Marzluff, Christine Dahlin, and Alan Kamil for insights into this species. See Marzluff & Balda (1992); Paz-y-Miño et al. (2004).

19   I thank Mauricio Cantor and Shane Gero for advice on sperm whales. Cantor & Whitehead (2015); Cantor et al. (2015); Christal et al. (1998); Gero et al. (2015, 2016a, 2016b).

20   Unlike the sperm whales, Florida's bottlenose dolphins don't seem to use vocalizations to identify their societies (which reach a population of a couple hundred and seem to rely on individual recognition). Still it's within the realm of possibility that differences in culture—learned behavior such as fishing techniques—can similarly act to separate the communities in some populations. For many years one Australian dolphin community, belonging to a different species than the Sarasota dolphins, followed trawlers to scrounge off of fish. These boat-dependent dolphins lived near another community that hunted down fish the normal way, far from boats. The two groups became one after the trawling stopped (Ansmann et al. 2012; Chilvers & Corkeron 2001).

21   The ants are densest at the leading edge of an invasion, but this may reflect the rich untapped food there rather than indicate a weakening of a supercolony away from its borders. Elsewhere in the world some Argentine ant populations have declined,

albeit prognoses of eventual collapse of supercolonies (Queller & Strassmann 1998) appear at best premature (Lester & Gruber 2016).

Chapter 7

1   The gelada is an exception to the tentativeness or discomfort between societies, moving with general indifference among other "units" (Chapter 17).
2   e.g., Cohen (2012). McElreath et al. (2003), Riolo et al. (2001).
3   Womack (2005). Synonyms for "marker" include words like "label" and "tag."
4   de Waal & Tyack (2003), Fiske & Neuberg (1990); Machalek (1992).
5   For details on different levels of human social connections, see Buys & Larson (1979); Dunbar (1993); Granovetter (1983); Moffett (2013); Roberts (2010).
6   This is the cultural version of the idea of an extended phenotype proposed by Dawkins (1982).
7   Wobst (1977).
8   Alessia Ranciaro, pers. comm.; Tishkoff et al. (2007).
9   Simoons (1994).
10   Wurgaft (2006).
11   Baumard (2010); Ensminger & Henrich (2014).
12   Poggi (2002).
13   Iverson & Goldin-Meadow (1998).
14   Darwin (1872).
15   Marsh et al. (2003). People in prolonged social contact can converge in facial appearance as well, perhaps through repeated similar use of the same facial muscles (Zajonc et al. 1987).
16   Marsh et al. (2007).
17   Sperber (1974).
18   Eagleman (2011).
19   Bates et al. (2007).
20   Allport (1954), 21.
21   Watanabe et al. (1995).
22   Nettle (1999).
23   Pagel (2009), 406.
24   Larson (1996).
25   Tajfel et al. (1970).
26   Dixon (2010), 79.
27   Sometimes the language spoken by a Pygmy group does not correspond to the farmers to which they are currently connected, suggesting the Pygmies sometimes migrate (Bahuchet 2012 & 2014). Equally mysterious are Bushmen who parted with their mother tongue to speak versions of the language of Khoikhoi cattle herders, once called Hottentots (Barnard 2007).
28   Giles et al. (1977); van den Berghe (1981).
29   Fitch (2000); Cohen (2012).
30   Flege (1984); Labov (1989).
31   JK Chambers (2008).
32   Cited by Edwards (2009), 5.
33   Dixon (1976).
34   Barth (1969); McConvell (2001).
35   Heinz (1975), 38. In fact, hunter-gatherer societies are considered "loose" because of this wide latitude in allowable behavior (Lomax & Berkowitz 1972).
36   Guibernau (2013). Of course, any form of membership puts some expectations on behavior, as the conclusion of this book makes clear.

37 Kurzban & Leary (2001); Marques et al. (1988).
38 Vicki Fishlock and Richard Wrangham, pers. comm.
39 Despite their normal tight conformity, even ants that don't take slaves can be manip-
    ulated experimentally to receive foreign individuals into their society—different ant
    species included (Carlin & Hölldobler 1983).
40 My viewpoint on superorganisms—which ties societies to the key feature underlying
    all organisms, the united identity of their components—is adapted from Moffett
    (2012).
41 Berger & Luckmann (1966), 149.
42 For the research using tokens, see Addessi et al. (2007).
43 Darwin (1871), 145.
44 Tsutsui (2004).
45 Gordon (1989). Chimpanzees, and likely many other animals, can do the same
    thing (Herbinger et al. 2009), though, again, in the ant instance, it's by recognizing
    their group rather than identifying an individual as a foreigner the way a chimp
    does.
46 Spicer (1971), 795–796.
47 e.g., see discussions in Henshilwood & d'Errico (2011).
48 Geertz (1973).
49 Womack (2005), 51.
50 Leafcutter ants buck this trend with their unusually large brains (Riveros et al.
    2012).
51 Geary (2005); Liu et al. (2014); e.g., relative to body size, Bushmen have exception-
    ally large braincases (Beals et al. 1984).
52 Pointed out by Gamble (1998), 431.
53 Among the theories of social groups framed by psychologists, the inductive and
    deductive groups of Postmes et al. (2005) may be most similar to my distinction
    between individual recognition and anonymous societies. The distinction between
    common bond and common identity groups is also intriguing (Prentice et al. 1994).
54 Berreby (2005).

Chapter 8

1   As with many terms of anthropology, there is a mind-numbing range of definitions
    for "band" and alternative choices for the word, such as hordes, overnight camps,
    or local groups.
2   There have been many other names applied to these mobile hunter-gatherer socie-
    ties, most of them confusing, but "band societies" has a good pedigree (e.g., Leacock
    & Lee 1982) and emphasizes the primacy of fission-fusion condition over egalitari-
    anism, hunting, gathering, or proficiency with tools or fire. Elsewhere I called them
    "multiband societies" (Moffett 2013), a term I simplify here.
3   Binford (1980), 4.
4   e.g., Headland et al. (1989); Henn et al. (2011).
5   Roe (1974); Weddle (1985).
6   Behar et al. (2008).
7   Ganter (2006).
8   Meggitt (1962), 47.
9   Curr (1886), 83–84.
10 Thanks to Thomas Barfield for advice. While Asia's "horse nomads" had leaders,
    they acted in a more egalitarian, hunter-gatherer-like manner while in their scat-
    tered camps (Barfield 2002).
11 Hill et al. (2011).

12 Wilson (2012).
13 Pruetz (2007).
14 I thank Fiona Stewart and Jill Pruetz for advice on savanna chimpanzees. Hernan-dez-Aguilar et al. (2007); Pruetz et al. (2015).
15 For a description of the importance of fire and food sharing, see Wrangham (2009).
16 See, e.g., Ingold (1999) and the "unbounded social landscape" of Gamble (1998).
17 Wilson (1975), 10.
18 Birdsell (1970).
19 Wiessner (1977, xix) points out that "even San [Bushmen] of a different language group . . . are foreign people and to be regarded with suspicion."
20 Arnold (1996); Birdsell (1968); Marlowe (2010). Thanks to Brian Hayden for advice on the size of the Western Desert band societies.
21 e.g., Tonkinson (2011). Note that because of the hardships, the people of the region gave up their original mode of hunting and gathering long ago.
22 Meggitt (1962), 34.
23 Tonkinson (1987), 206. We consider some of these issues in Chapters 17 and 18.
24 These societies were distinguished by language, the importance placed on spiritual tales, or dreamings, and other matters (Brian Hayden and Brian Codding, pers. comm.). Their alliances were fragile, it would seem, because some took to fighting anyway (Meggitt 1962).
25 Renan (1990).
26 Johnson (1997).
27 Dixon (1076), 231.
28 e.g., Hewlett et al. (1986); Mulvaney (1976); Verdu et al. (2010).
29 Murphy & Murphy (1960).
30 e.g., Heinz (1994); Mulvaney & White (1987).
31 Stanner (1979), 230.
32 Stanner (1965) introduced the term "estate" to describe the area for which each band had primary rights. This sense of a local home did vary. The Hadza transferred more fluidly from band to band and the bands moved more readily across expanses of the Hadza territory; but even for them, individuals still stayed within that part of the overall region they knew best (Blurton-Jones 2016).
33 As described for the !Kõ Bushmen by Heinz (1972). See also Chapter 17.
34 For ants, e.g., Tschinkel (2006). We see the same separation today with the en-trenched troops of people during battles (Hamilton 2003).
35 e.g., Smedley & Smedley (2005).
36 Malaspinas et al. (2016).
37 Bowles & Gintis (2011), 99; see also Bowles (2006).
38 Guenther (1976).
39 Lee & DeVore (1976). The word "San" continues to have derogatory connotations in the Kalahari. I prefer Bushmen, which had few negative associations when ini-tially coined by Dutch explorers, although another alternative is the less familiar Bantu word "Basarwa."
40 Schapera (1930), 77.
41 Coren Apicella, pers. comm.; Hill et al. (2014).
42 Silberbauer (1965), 62.
43 Schladt (1998) estimates that a century ago, the number of Khoisan (Bushman and the related Khoikhoi cattle herder) languages was 200.
44 Such traits convey an emblemic style (Wiessner 1983). Wiessner (1984) found less of a connection to particular Bushmen peoples for the styles of their beaded head-bands, but the beads don't have an ancient pedigree; rather they were acquired through trading with Europeans.
45 Wiessner (1983), 267.

46 Sampson (1988).
47 Gelo (2012).
48 Broome (2010), 17.
49 Spencer & Gillen (1899), 205.
50 Cipriani (1966).
51 Fürniss (2014).
52 Clastres & Auster (1998), 36.

Chapter 9

1   Tonkinson (2002), 48; Hayden (1979).
2   So while outsiders experienced hunter-gatherers asking for possessions as begging, hunter-gatherers saw such gifts as generous invitations to partake in relationships of sharing that assured everyone was taken care of (Earle & Ericson 2014; Peterson 1993).
3   Endicott (1988).
4   See arguments in Wiessner (2002).
5   Sahlins (1968) first described hunter-gatherers as "affluent," an idea that has been disputed (Kaplan 2000), the difference in opinion arising in part because work and leisure are impossible to distinguish for people who socialize as they carry out many of their tasks.
6   Morgan & Bettinger (2012).
7   Elkin (1977).
8   Bleek (1928), 37.
9   Chapman (1863), 79.
10 Keil (2012).
11 They also favor learning from dominant individuals (Kendal et al. 2015).
12 Wiessner (2002).
13 Blurton-Jones (2016); Hayden (1995).
14 Baumeister (1986).
15 Pelto (1968): Witkin & Berry (1975).
16 In the Ache, men from different bands got together to fight, though even then not as a team, and the combatants often ended up fighting others in their own band (Hill & Hurtado 1996).
17 Ellemers (2012).
18 Finkel et al. (2010).
19 Lee (2013), 124.
20 Lee & Daly (1999), 4.
21 e.g., Marshall (1976). One practical matter was that a band had too few children of a similar age who could play together in a competitive way (Draper 1976).
22 Boehm (1999).
23 de Waal (1982). In a similar fashion, lower-ranking olive baboons have jointly driven a repressive alpha female entirely out of their troop (Anthony Collins, pers. comm.).
24 Ratnieks et al. (2006).
25 Dominance of one sex over the other exists in many species, with females coming out on top in spotted hyenas, ringtail lemurs, and the bonobo, while males take control in chimpanzees and baboons.
26 Tuzin (2001), 127.
27 Schmitt et al. (2008).
28 Thomas-Symonds (2010).
29 Bousquet et al. (2011).

30 Hölldobler & Wilson (2009), Seeley (2010); Visscher (2007). The dominant painted dogs have somewhat more influence than the other pack members (Walker et al. 2017).
31 Rheingold (2002); Shirky (2008).

Chapter 10

1 Ian McNiven and Heather Builth, pers. comm.; Broome (2010); Builth (2014); Head (1989); McNiven et al. (2015).
2 Cipriani (1966).
3 Brink (2008).
4 In the elephant instance the gatherings are made up of many core groups, or societies, though in hunter-gatherers multiple band societies could similarly get together to forge alliances and trade (Hayden 2014).
5 Guenther (1996).
6 The inevitable could be put off. The Cheyenne put together police squads to oversee the fairness of their joint buffalo hunts, a force that disbanded when the hunts ended (MacLeod 1937).
7 Rushdie (2002), 233.
8 Denham et al. (2007).
9 Mitchell (1839), 290–291.
10 Clastres (1972). This article uses an alternative name for the Ache, the Guayaki.
11 Lee (1979), 361.
12 Hawkes (2000).
13 Morgan & Bettinger (2012).
14 Roscoe (2006).
15 My descriptions of coastal Northwest Indians owes considerably to correspondence with Kenneth Ames and Brian Ferguson. Ames (1995); Ames & Maschner (1999); Sassaman (2004).
16 Some tribes also managed the environment, for example by stocking salmon for a time in artificial pools or raising butter clams in rock terraces still visible at low tides (Williams 2006).
17 Patrick Saltonstall and Amy Steffian, pers. comm.; Steffian & Saltonstall (2001). American naturalist Edward Nelson, who lived in the 1870s among Yupik-speaking peoples of southwestern Alaska, wrote that the labrets, made of stone and thus painful to keep in place, "were removed and carried in a small bag until we approached a village at night, when they were taken out and replaced, that the wearer might present a proper appearance before the people" (Nelson 1899, 50)—equivalent to a person bringing out a national flag during an international event.
18 Townsend (1983).
19 Johnson (1982).
20 Silberbauer (1965).
21 Van Vugt & Ahuja (2011).
22 Bourjade et al. (2009).
23 Peterson et al. (2002).
24 Fishlock et al. (2016).
25 Watts et al. (2000).
26 Baumeister et al. (1989).
27 e.g., Hold (1980).
28 Dawson (1881); Fison & Howitt (1880), 277.
29 Hann (1991), xv.

30   William Marquardt, pers. comm.; Gamble (2012); Librado (1981).

31   Hayden (2014).

32   Van Vugt et al. (2008).

33   Hogg (2001); Van Knippenberg (2011).

34   For example, Passarge (1907) wrote that Bushmen told him they once had hered-
     itary chiefs who went unnoticed by colonists because they displayed little of the
     pomp and circumstance of European leaders. Another anthropologist wrote, "Both
     Naron and Auen [Bushmen societies] had chiefs when the old men were young.
     They seem to have directed the movement of their people from place to place, to
     have ordered the burning of the veld, and in particular to have led in war. Fights
     were frequent both between the opposing Bushman tribes, Naron and Auen, and
     against other natives who were gradually encroaching from all sides" (Bleek 1928,
     36–37).

35   Andersson (1856), 281.

36   Rather than chiefs, a better word for the leaders of the =Au//ei might be "Big Men"
     (Chapter 22), though at least in some cases their position was inherited (Mathias
     Guenther, pers. comm.; Guenther 1997 & 2014).

37   Ames (1991).

38   Testart (1982).

39   Durkheim (1893) distinguished the "mechanical solidarity" of people who perform
     similar work in technologically simple societies from the "organic solidarity" of soci-
     eties with division of labor.

40   This is a by-product of what's described as self-domestication: apes such as humans
     and bonobos have evolved to be tolerant of, and ineffectual without, others of their
     kind (Hare et al. 2012). Self-domestication is associated with a reduction of impul-
     sive violence (Wrangham 2019). Baumeister et al. (2016) argue that by specializa-
     tion, people have made themselves increasingly irreplaceable. Yet because many
     individuals are likely to perform all but the most specialized tasks across a large
     population, only a tiny minority of people is truly irreplaceable today.

41   Originally proposed by Brewer (1991).

42   Even today societies differ in the most comfortable, or optimal, level of distinctive-
     ness their people strive for. The focus on *differences* is sharpest in Western cultures
     where individualism and capitalism reign, though even among them a marketer can
     tell you people fall in categories and are less distinct than they think (JR Chambers
     2008).

43   Hayden (2011).

44   Fried (1967), 118. Potlatches existed pre-contact and may have grown more elab-
     orate after Europeans put a stop to the chronic wars of the Pacific Northwest, sug-
     gesting the feasts became an alternative to fighting as demonstrations of a chief's
     importance.

45   Tyler (2006).

46   An early statement of this point of view was presented by Hayden et al. (1981).

47   To my knowledge this point was first made in Testart (1982).

48   For examples from South America, see Bocquet-Appel & Bar-Yosef (2008); Gold-
     berg et al. (2016).

49   Berndt & Berndt (1988), 108.

50   Cipriani (1966), 36.

51   Mummert et al. (2011).

52   O'Connell (1995).

53   Roosevelt (1999).

54   There is an irony in the contempt industrialized nations display toward hunter-gath-
     erer peoples, given that our gut reaction to outsiders evolved over millennia spent
     as hunter-gatherers. So-called primitive cultures have been associated with animals

and children, as if a reliance on hunting and gathering were evidence of the arrested mental facilities of a bygone era (Jahoda 1999; Saminaden et al. 2010).

Chapter 11

1   Marean (2010).
2   Behar et al. (2008).
3   Mercader et al. (2007).
4   Villa (1983).
5   Curry (2008).
6   Harlan (1967) collected enough wild wheat with stone tools to show that a prehistoric family in Turkey could have gathered a year's supply of this grain, so would have had the option to put down roots.
7   Price & Bar-Yosef (2010); Trinkaus et al. (2014)
8   Jerardino & Marean (2010).
9   d'Errico et al. (2012).
10 Henshilwood et al. (2011).
11 This view has been effectively countered by McBrearty & Brooks (2000).
12 Kuhn & Stiner (2007), 40–41.
13 Wadley (2001).
14 The most convincing evidence of markers differentiating societies comes late—the multiplicity of ivory, antler, wood, tooth, and shell jewelry across Europe between 37,000 and 28,000 years ago (Vanhaeren & d'Errico 2006).
15 Brooks et al. (2018).
16 Rendell & Whitehead (2001); Thornton et al. (2010)
17 Coolen et al. (2005).
18 van de Waal et al. (2013).
19 Bonnie et al. (2007); Whiten (2011). The grooming behavior can be passed on from mother to offspring (Wrangham et al. 2016).
20 McGrew et al. (2001).
21 One female has failed to "correctly" clasp the hands of other chimps since joining her community two decades ago, yet her companions groom her all the same (Michio Nakamura, pers. comm.).
22 Brown & Farabaugh (1997); Nowicki (1983).
23 Paukner et al. (2009).
24 It remains to be firmly proven whether chimpanzees react to the group specific characteristics of the pant-hoots, recognize the callers as individuals from slight differences in their pant-hoots, or both (Marshall et al. 1999; Mitani & Gros-Louis 1998).
25 Crockford et al. (2004). The pinyon jays, with their durable flock societies (Chapter 6), may learn their *rack* or *kaw* vocalizations this way.
26 Boughman & Wilkinson (1998); Wilkinson & Boughman (1998). Meerkats have contact calls that differ from one clan to another but in this species the animals themselves don't seem to grasp the difference (Townsend et al. 2010).
27 Herbinger et al. (2009).
28 Taglialatela et al. (2009). One essential experiment, comparing the chimp's reactions to the pant-hoots of group members versus foreigners, has not yet been done.
29 Fitch (2000). A colony password is also hypothesized for a bird species (Feekes 1982).
30 Zanna Clay, pers. comm.; Hohmann & Fruth (1995). In fact the spider monkey learns a whinny that is similarly specific to its community (Santorelli et al. 2013).
31 The protolanguage would consist of "nothing more than an inventory of calls expressing unanalyzed meanings" (Kirby 2000, 14).

32   Steele & Gamble (1999).

33   Aiello & Dunbar (1993).

34   Grove (2010).

35   I prefer this expression to "release from proximity" (see Chapter 4 and Gamble 1998) given, for example, that the low density of savanna chimps demonstrates that distance per se is not the issue.

36   Or at least ordinarily: cheaters like the colony-infiltrating spider are rare in the Argentine ant, suggesting identification with a supercolony is a tough nut to crack—maybe their members are so precisely alike that even the slightest deviation from the "norm" sets off an alarm.

37   Fiske (2010); Boyd & Richerson (2005).

38   Johnson et al. (2011).

39   Near-hairlessness is open to other explanations, such as making swimming easier, reducing parasites, or keeping the body cool (Rantala 2007).

40   Lewis (2006), 89.

41   Turner (2012), 488. See also Thierry (2005).

42   Gelo (2012).

43   Kan (1989), 60.

44   The tattoos stop the kidnapping of the women (White 2011); for other examples of marking the skin, see Jablonski (2006).

45   Pabst et al. (2009) argues the tattoos could have had medicinal value; even so they could have been associated with the man's tribe.

46   Alan Rogers, pers. comm.; Rogers et al. (2004).

47   Berman (1999).

48   Jolly (2005).

49   Chance & Larsen (1976).

50   Boyd & Richerson (2005).

51   Foley & Lahr (2011).

52   Tennie et al. (2009). Chimpanzees also create rudimentary symbolic cultures, in the sense that a behavior can have different meanings in different communities; for example, noisily tearing leaves with the teeth is an invitation for sex in one community and for play in another (Boesch 2012).

53   Tindale & Sheffey (2002). In the last decade, to give an example, our reliance on GPS has made us less adept at a talent honed by hunter-gatherers, spatial navigation (Huth 2013).

54   Henrich (2004b); Shennan (2001). There is some controversy over how to interpret the simplicity of Tasmanian culture, including their inability to make fire (Taylor 2008).

55   Finlayson (2009); Mellars & French (2011).

56   Hiscock (2007).

57   Aimé et al. (2013).

58   Powell et al. (2009). Some deny connections between social complexity and population density and interaction rates, and certainly other factors may come into play (Vaesen et al. 2016).

59   Wobst (1977).

60   Moffett (2013), 251.

Chapter 12

1   Wiessner (2014).

2   Hasson et al. (2012). Such coupling also occurs between monkey brains (Mantini et al. 2012).

3   Harari (2015).
4   For an excellent review of many general issues, see Banaji & Gelman (2013).
5   Eibl-Eibesfeldt (1998), 38.
6   Callahan & Ledgerwood (2013).
7   Testi (2005); see also Testi (2010).
8   Bar-Tal & Staub (1997); Butz (2009); Geisler (2005). In fact, the mere sight of a na-
     tional flag can cause people to feel more nationalistic (Hassin et al. 2007), though
     this response differs between societies (Becker et al. 2017).
9   Helwig & Prencipe (1999); Weinstein (1957); Barrett (2007).
10   Billig (1995), Ferguson & Hassin (2007), Kemmelmeier & Winter (2008).
11   Barnes (2001).
12   Thanks to Sören Krach and Helen Gallagher for confirming this expectation of
     human responses to playing a game on a computer. As robots become more lifelike,
     we treat them increasingly as human (Chaminade et al. 2012; Takahashi et al. 2014;
     Wang & Quadflieg 2015).
13   Parr (2011).
14   Henrich et al. (2010b).
15   e.g., Ratner et al. (2013).
16   Schaal et al. (2000).
17   Cashdan (1998); Liberman et al. (2016).
18   Or at least the race of his or her parents (Kelly et al. 2005).
19   Kinzler et al. (2007); Nazzi et al. (2000); Rakić et al. (2011).
20   Kelly et al. (2009); Pascalis & Kelly (2009). This effect is reversible in older children
     with radical changes in environment, but is difficult and requires considerable time
     (Anzures et al. 2012; Sangrigoli et al. 2005). We can capitalize on an appropriately
     aged baby's skills to beef up how well it recognizes those of other races or ethnici-
     ties—exposure to as few as three faces of the group can do the trick (Sangrigoli &
     de Schonen 2004).
21   Oddly enough given the effort that chicks invest in imprinting, it's unknown
     whether the mother hen can recognize her own brood (Bolhuis 1991).
22   Of course, the chick only learns its mother rather than a social group, and the
     ants don't use their imprinting as a starting point to learn the members of their
     colony individually. Still, though the complexity of how people distinguish groups
     is greater, the genetic underpinnings of how we do it might not be much different
     (e.g., Sturgis & Gordon 2012).
23   Pascalis et al. (2005); Scott & Monesson (2009); Sugita (2008).
24   Rowell (1975).
25   Atran (1990).
26   Hill & Hurtado (1996).
27   Keil (1989).
28   Gil-White (2001).
29   This reaches extremes in situations of identity fusion (Chapter 15; Swann et al.
     (2012).
30   Martin & Parker (1995).
31   Different cultures and ethnicities have different ways of categorizing the offspring
     of mixed marriages (e.g., Henrich & Henrich 2007).
32   Hammer et al. (2000).
33   Madon et al. (2001) looks at how stereotypes have changed over the last few decades.
34   MacLin & Malpass (2001).
35   Appelbaum (2015).
36   Levin & Banaji (2006).
37   MacLin & MacLin (2011).
38   Ito & Urland (2003), Todorov (2017).

39   Asch (1946), 48.
40   Castano et al. (2002)
41   Jewish Telegraphic Agency (1943).
42   Greene (2013).
43   Wiessner (1983), 269.
44   Silberbauer (1981), 2.
45   German sociologist Georg Simmel contributed to the confusion by defining "stranger" in a nontraditional way, as a group member who doesn't fit in—that is, someone who acts strangely (e.g., McLemore 1970). Most dictionaries give priority to using "xenophobia" in describing negative reactions to a *foreigner*, whether or not that individual has been encountered previously, and that's how I prefer to use the word.
46   Azevedo et al. (2013). The more distinct the races, the less empathy (Struch & Schwartz 1989). Responses to pain are also stronger for animals that we see as more like us (Plous 2003).
47   Campbell & de Waal (2011).

Chapter 13

1   Macrae & Bodenhausen (2000), 94.
2   Lippmann (1922), 89.
3   Devine (1989).
4   Bonilla-Silva (2014) provides one fascinating discussion.
5   Banaji & Greenwald (2013), 149. The test has its critics (e.g., Oswald et al. 2015).
6   Baron & Dunham (2015).
7   Hirschfeld (2012), 25.
8   Aboud (2003); Dunham et al. (2013).
9   Harris (2009).
10   Bigler & Liben (2006); Dunham et al. (2008)
11   Hirschfeld (1998).
12   Karen Wynn, pers. comm.; Katz & Kofkin (1997).
13   Edwards (2009).
14   Kinzler et al. (2007), 12580
15   Amodio (2011), 104; see also, e.g., Phelps et al. (2000).
16   Thus far this work has been done on people of differing political points of view, rather than national identities per se (Nosek et al. 2009).
17   Beety (2012); Rutledge (2000).
18   Cosmides et al. (2003); Kurzban et al. (2001); Pietraszewski et al. (2014). These authors propose that this cognitive machinery evolved to detect alliances within societies, but among recent hunter-gatherers and in early humans such alliances were likely to be fluid and not at all connected to any identifying traits (markers, including racial differences).
19   Wegner (1994).
20   Monteith & Voils (2001).
21   MacLin & MacLin (2011).
22   Haslam & Loughnan (2014), 418.
23   Greenwald et al. (2015). To give just one of many circumstances where this occurs, minorities often receive medical care inferior to whites under the same doctor (Chapman et al. 2013).
24   Treating ethnic friends as special is called subtyping (Wright et al. 1997).
25   Mild disgust can be a means of differentiation rather than discrimination (Brewer 1999; Douglas 1966; Kelly 2011).

26   See Chapter 17. Bandura (1999); Jackson & Gaertner (2010); Vaes et al. (2012); Viki et al. (2013).
27   Steele et al. (2002).
28   Fiske & Taylor (2013), Phelan & Rudman (2010).
29   Gilderhus (2010)
30   Kelley (2012); e.g., many Norwegian traditions are similar inventions (Eriksen 1993).
31   Leibold (2006).
32   Beccaria (1764).
33   Haslam et al. (2011b).
34   Renan (1990), 11. See also Hosking & Schöpflin (1997); Orgad (2015).
35   For more on memory, see Bartlett & Burt (1933); Harris et al. (2008); Zerubavel (2003).
36   Gilderhus (2010); Lévi-Strauss (1972).
37   Berndt & Berndt (1988).
38   Billig (1995); Toft (2003).
39   Maguire et al. (2003); Yates (1966).
40   Lewis (1976).
41   Joyce (1922), 317. We will see in section IX that, given a history of conquest and immigration, what counts as "the same" people has become a complicated question.
42   Bar-Tal (2000).
43   Hence the mandate of ISIS to lay claim to a state (Wood 2015). See Section IX for how this plays out for races and ethnicities.
44   McDougall (1920).
45   Bigelow (1969).

Chapter 14

1   Wilson (1978), 70. See also Read (2011). As Claude Lévi-Strauss (1952, 21) wrote, "Humankind ceases at the border of the tribe."
2   Gombrich (2005), 278.
3   Giner-Sorolla (2012), 60.
4   Freud (1930).
5   Smith (2011).
6   Aristotle describes slavery as justified for captives taken in battle (Walford & Gillies 1853, 12).
7   Orwell (1946), 112. Anthropomorphism and dehumanization are related (Waytz et al. 2010).
8   David Livingstone Smith, pers. comm.; Haidt & Algoe (2004); Lovejoy (1936); Smith (2011).
9   Costello & Hodson (2012) studied how six-to-ten-year-old white Canadians perceived black children. Those who saw a greater divide between people and animals tended to be more prejudiced.
10   This reinforces my point that it was the society, not the camp or band, upon which these people erected their primary identities. In general when the names of outgroups are simplified descriptions, say of their "humanness," relations with outsiders tend to be more acrimonious than if the names have a more nuanced meaning (Mullen et al. 2007).
11   Haslam & Loughnan (2014).
12   Ekman (1992). Some claim different categories of basic emotions, e.g., Jack et al. (2014) propose that surprise can't be distinguished from fear, or disgust from anger. Yet as psychologist Paul Bloom points out to me, disgust and anger are both

aversive, negative emotions, but they are elicited by different stimuli, evoke different reactions and brain responses, and have different evolutionary histories and
developmental trajectories.

13   Haidt (2012).
14   Bosacki & Moore (2004).
15   Chimpanzees read each other's expressions (Buttelmann et al. 2009; Parr 2001;
     Parr & Waller 2006).
16   Haslam (2006) proposes that we dehumanize everyone we meet to varying degrees.
     He sees the belief that outsiders lack basic human traits as *animalistic dehumaniza-
     tion*, or dehumanization to the animal level. This creates species-like boundaries
     between groups not seen in the *mechanistic dehumanization* of people, such as when
     doctors or lawyers are viewed as calculating and dehumanized to the level of inani-
     mate objects or, more accurately, machines. Martínez et al. (2012) show that mech-
     anistic dehumanization can emerge at a society level.
17   Wohl et al. (2012).
18   Haidt (2003); Opotow (1990).
19   Jack et al. (2009); Marsh et al. (2003).
20   Both races more often misidentify an object as a weapon when it is in a black per-
     son's hands than when a white person is holding it. Ackerman et al. (2006); Correll
     et al. (2007); Eberhardt et al. (2004); Payne (2001).
21   Hugenberg & Bodenhausen (2003).
22   That was true at least if the liar was quiet, when onlookers apparently missed subtle
     cues of affect that differ across cultures; when speaking, awkward pauses in uttering
     sentences may give liars away (Bond et al. 1990). See also Al-Simadi (2000).
23   Ekman (1972).
24   Kaw (1993).
25   The "stereotype content" account of dehumanization described here (Fiske et al.
     2007) was developed separately from the infrahumanization model, with its focus
     on secondary emotions.
26   Vaes & Paladino (2010).
27   Clastres (1972).
28   Koonz (2003).
29   Goff et al. (2008); Smith & Panaitiu (2015).
30   Haslam et al. (2011a).
31   Haidt et al. (1997).
32   Amodio (2008); Kelly (2011); Harris & Fiske (2006).
33   Fear of contamination, expressed as disgust and a belief that group members share
     "some fundamental bodily essence in common" (Fiske 2004), may be an ancient
     kind of "behavioral immune system" (Schaller & Park 2011, 30; O'Brien 2003). Peo-
     ple show more fear of immigrants after being shown pictures of diseased persons
     (Faulkner et al. 2004).
34   Freeland (1979). This idea doesn't work with all parasites, since diseases that spread
     by feces rather than direct contact might transfer readily between territories; also,
     once a disease breaches a territory, the concentrated population tied to that space
     can promote its spread.
35   McNeill (1976). Syphilis may have been brought back to Europe from America, but
     to a much less widely destructive effect than smallpox had in the Americas.
36   Heinz (1975), 21.
37   Tajfel & Turner (1979).
38   Bain et al. (2009).
39   Koval et al. (2012).
40   Reese et al. (2010); Taylor et al. (1977).

41 As far as I can tell there is little research on this question, although one study shows that infants maintain more eye contact with same-race individuals (Wheeler et al. 2011) and there is a classic study demonstrating that white people maintain less eye contact with black job applicants (Word et al. 1974).
42 Mahajan et al. (2011); but see also Mahajan et al. (2014).
43 Another option is that the link we make between outsiders and the sense of disgust arose in humans (D Kelly 2013).
44 Henrich (2004a); Henrich & Boyd (1998); Lamont & Molnar (2002); Wobst (1977).
45 Gil White (2001).
46 Reviewed by Kleingeld (2012).
47 Leyens et al. (2003), 712.
48 Castano & Giner-Sorolla (2006).
49 Wohl et al. (2011).

Chapter 15

1 Orwell (1971), 362.
2 Goldstein (1979).
3 Bloom & Veres (1999); Campbell (1958).
4 Assessments of the warmth and competence of social groups were described in the last chapter. Callahan & Ledgerwood (2016).
5 Those alternative responses depend on the relative strength of the groups and how much they compete (Alexander et al. 2005).
6 McNeill (1995); Seger et al. (2009); Tarr et al. (2016); Valdesolo et al. (2010).
7 Barrett (2007); Baumeister & Leary (1995); Guibernau (2013).
8 Atran et al. (1997); Gil-White (2001).
9 Brewer & Caporael (2006); Caporael & Baron (1997).
10 The idea that a society is more than the sum of its members, now widely considered to be correct, was originally called the "nationalistic fallacy" by Allport (1927).
11 Sani et al. (2007).
12 Castano & Dechesne (2005).
13 Best (1924), 397.
14 Wilson (2002).
15 de Dreu et al. (2011); Ma et al. (2014).
16 People with a strong group identity express group emotions most strongly (Smith et al. 2007).
17 Adamatzky (2005).
18 Hayden (1987). Gatherings of bands from different societies for trading and alliances would have proceeded with greater caution (Chapter 18).
19 Marco Iacoboni pers. comm. and Iacoboni (2008). When we consciously mimic others, however, perceived status may trump race (Elizabeth Losin, pers. comm. and Losin et al. 2012).
20 Rizzolatti & Craighero (2004).
21 Field et al. (1982). Dance could have gotten its start through such imitation (Laland et al. 2016).
22 Parr & Hopkins (2000).
23 Animal rallies grade into more immediate responses to a stimulus, such as the screams that race through a a chimpanzee community, mobilizing the animals to scare off an enemy or predator. Preston & de Waal (2002); Spoor & Kelly (2004).
24 Wildschut et al. (2003).
25 For explanations other than "rallying" for the ants' behavior, see Moffett (2010).
26 Watson-Jones et al. (2014).

27 Chimpanzees and bonobos ape each other when something useful comes out of it, for example, using a stick to catch a termite meal, but seldom imitate an act completely unconnected with a practical purpose. Some behaviors come close, though—as when the bonobos at the San Diego Zoo took to the custom of clapping hands at intervals when grooming each other (de Waal 2001).

28 Thanks to Harvey Whitehouse for advice on identity fusion. Whitehouse et al. (2014a); Whitehouse & McCauley (2005).

29 Documented for the Libyan civilians turned revolutionaries who rose up against Gaddafi (Whitehouse et al. 2014b).

30 The sting feels "like walking over flaming charcoal with a 3-inch nail embedded in your heel" (Schmidt 2016, 225).

31 Bosmia et al. (2015).

32 Fritz & Mathewson (1957); Reicher (2001); Willer et al. (2009).

33 Hood (2002), 186.

34 Barron (1981).

35 Hogg (2007).

36 Caspar et al. (2016); Milgram (1974).

37 Mackie et al. (2008).

38 Kameda & Hastie (2015).

39 Fiske et al. (2007).

40 Staub (1989).

41 People who wish to believe something, prejudices included, ignore contrary evidence so long as they can cling to *anything* that corroborates their point of view (Gilovich 1991).

42 Especially troublesome were radio programs that described the violence as everyday behavior (Elizabeth Paluck, pers. comm.; Paluck 2009).

43 Janis (1982).

44 This is called Asch conformity after the psychologist Solomon Asch (e.g., Bond 2005).

45 Redmond (1994), 3.

46 Hofstede & McCrae (2004).

47 Wray et al. (2011).

48 Masters & Sullivan (1989); Warnecke et al. (1992).

49 Silberbauer (1996).

Chapter 16

1 Marlowe (2000).

2 Non-kin are found in cores across all populations, but join cores most often where there has been poaching (Wittemyer et al. 2009).

3 Sometimes one or even more of these outsiders breed under the noses of the senior, "alpha" pair (Dan Stahler, pers. comm.; Lehman et al. 1992; Vonholdt et al. 2008).

4 The matter is controversial for Gunnison's prairie dogs. Hoogland et al. (2012) found adult females of a coterie tend to be maternal close kin in Colorado, but Verdolin et al. (2014) encountered few relatives among adults in Arizona, a possible regional difference.

5 Horses can leave a band if something goes very wrong—perhaps an overbearing stallion drives the females away (Cameron et al. 2009).

6 Bohn et al. (2009); McCracken & Bradbury (1981); Gerald Wilkinson (pers. comm.).

7 Many of a male chimpanzee's allies are likely to be childhood friends having another mother, though this needs study (Ian Gilby, pers. comm.; Langergraber et al. 2007 & 2009).

8   Masson & Kooki (2011).
9   Sai (2005).
10   Heth et al. (1998). Note that hamsters are social, but have no societies.
11   That's true at least when the baby is a male (Parr & de Waal 1999).
12   Alvergne et al. (2009); Bressan & Grassi (2004).
13   Cheney & Seyfarth (2007).
14   Chapais (2008); Cosmides &Tooby (2013); Silk (2002).
15   Thanks to Elizabeth Archie for insights on baboon matrilines. Because what counts
     as a matrilineal "group" ("network" is a better word) depends on the perspective of
     each female, the one and only category every baboon shares is the troop itself—its
     society.
16   The possibility that the males recognize their physical similarity to their offspring
     has also been suggested (Buchan et al. 2003).
17   This is especially true for women, but for men, too, when they unite against outsid-
     ers (Ackerman et al. 2007).
18   Weston (1991); Voorpostel (2013).
19   Apicella et al. (2012); Hill et al. (2011).
20   Schelling (1978). People often develop an affinity with genetically similar others,
     kin or not, perhaps because subtle resemblances in attitude grease the path to
     friendship (Bailey 1988; Christakis & Fowler 2014).
21   Silberbauer (1965), 69.
22   Lieberman et al. (2007). Hence children raised together on a kibbutz don't marry
     each other even though no prohibitions exist against it (Shepher 1971).
23   Hill et al. (2011).
24   Hirschfeld (1989).
25   Tincoff & Jusczyk (1999). These words might have originated to match the first
     babbling sounds uttered by most infants (Matthey de l'Etang et al. 2011).
26   As it turns out David Haig was paraphrasing himself (Haig 2000). See also Haig
     (2011).
27   Everett et al. (2005); Frank et al. (2008).
28   Frank et al. (2008). Chagnon (1981) describes people recognizing kin categories
     for which they lack words.
29   Woodburn (1982).
30   Gould (1969).
31   Cameron (2016). While just a small proportion of Comanche were captives at any
     one time, for example, a need for fresh warriors led to many of this tribe having
     foreign blood (Murphy 1991).
32   So says Ferguson (2011, 262), referring to the work of Chaix et al. (2004).
33   Barnard (2011). Not that being described as kin is always positive. Some Africans
     used kin metaphors not to suggest closeness but to convey dominance over slaves
     (Kopytoff 1982).
34   Tanaka (1980), 116.
35   As suggested by studies like that on monkeys by Chapais et al. (1997).
36   Often those who point to the entitativity of families (e.g., Lickel et al. 2000) have
     allowed each person to decide what, to them, is a family. I find this problematic.
     Perceiving the particular family members you happen to be close to as a tight group
     seems trivial, and no different than imagining your close buddies as a tight group.
37   This sense of obligation, rather than genetic relatedness as such, is how I would ex-
     plain the results of Hackman et al. (2015) that show high levels of sacrifice toward
     mates and kin.
38   West et al. (2002).
39   This can make it hard to disown just part of a family (Jones et al. 2000; Uehara
     1990).

40   Many societies simplify the issue of inheritance by the people keeping track of their descent from a particular ancestor (Cronk & Gerkey 2007).

41   Johnson (2000). Meanwhile, longer life spans have created more intricate networks of kin than had lived together before (Milicic 2013).

42   e.g., Eibl-Eibesfeldt (1998).

43   Barnard (2011).

44   Johnson (1987); Salmon (1998). Unlike van der Dennen (1999), I believe such metaphors tap our beliefs in essences (Chapter 12) rather than kinship as such.

45   Breed (2014). Hannonen & Sundström (2003) describe an instance of nepotism (favoring kin) for ants, but their evidence is weak.

46   Eibl-Eibesfeldt (1998); Johnson (1986). In my view that ape couldn't have confounded its embryonic society with its kin and allies—it would have kept separate track of them from the start.

47   Barnard (2010).

Chapter 17

1   Voltaire (1901), 11. Thanks to Michael Wilson for the notion of a "Pankind."

2   Toshisada Nishida (1968), a remarkable Japanese researcher working in Uganda, was first to detect the communities.

3   Wrangham & Peterson (1996).

4   Mitani et al. (2010); Wilson & Wrangham (2003); Williams et al. (2004).

5   Aureli et al. (2006).

6   Douglas Smith, Kira Cassidy (pers. comm.); Mech & Boitani (2003); Smith et al. (2015).

7   Quoted in McKie (2010).

8   Wrangham et al. (2006).

9   Wendorf (1968).

10   Morgan & Buckley (1852), 42–44.

11   Europeans perverted this spiritual component of trophy taking by paying for scalps (Chacon & Dye 2007).

12   Boehm (2013).

13   e.g., Allen & Jones (2014); Gat (2015); Keeley (1997); LeBlanc & Register (2004); Otterbein (2004); DL Smith (2009).

14   Moffett (2011).

15   Gat (1999); Wrangham & Glowacki (2012).

16   This cycle is often driven by spur-of-the-moment gut responses, although some societies, Bedouin tribes for example, codify it (Cole 1975).

17   Genetic analysis indicates Aborigines have stuck to the general regions they first occupied upon settling Australia despite environmental changes since that time (Tobler et al. 2017), although that obviously doesn't mean the individual societies didn't move around within an area. Still, many accounts suggest that land tenure by band societies was ancient and respected (LeBlanc 2014).

18   Burch (2005), 59.

19   de Sade (1990), 332.

20   Guibernau (2007); van der Dennen (1999).

21   Bender (2006), 171.

22   Sumner (1906), 12.

23   Johnson (1997).

24   Bar-Tal (2000), 123.

25   This bias is reflected even in the behavior of small groups of children (Dunham et al. 2011).

26 For general descriptions of our ineptitude with risk, see Gigerenzer (2010); Slovic (2000).
27 Fabio Sani, pers. comm.; Hogg & Abrams (1988).
28 e.g., "War is conditioned by human symbol systems," Huxley (1959), 59.
29 e.g., Wittemyer et al. (2007).
30 At least in captivity (Tan & Hare 2013).
31 Furuichi (2011).
32 Wrangham (2014 & 2019).
33 Hrdy (2009), 3.
34 Hare et al. (2012); Hohmann & Fruth (2011).
35 The closest thing to friendship between units can be detected after one unit splits in two. These so-called "teams" remain near each other and grunt pleasantly in each other's direction—but even these feeble signs of a shared bond fade away over several months (Bergman 2010).
36 Pusey & Packer (1987).
37 Boesch (1996); Wrangham (1999). Killings are most common in parts of population-dense Kibale, Uganda, where this strategy of keeping to large parties doesn't exist (Watts et al. 2006). A possible explanation for the violence of chimpanzees is that most populations under study today are restricted to patches of forests where resources and space are limited, but a recent analysis discounts this hypothesis (Wilson et al. 2014).
38 Wrangham (2019) describes this reduction of "reactive aggression" and the traits that accompany it.
39 Pimlott et al. (1969); Theberge & Theberge (1998).
40 Mahajan et al. (2011 & 2014).
41 Brewer (2007); Cashdan (2001); Hewstone et al. (2002).

Chapter 18

1 Sperm whales are the exception that proves the rule that societies have a difficult time working together. However in this case the collaborating societies (units) are part of a greater social entity—clans of whales that share the same hunting traditions (Chapter 6).
2 The collaboration between the eel-fishing people extended to lavish intergroup gatherings, and the eels themselves were widely traded. Other names for the Gunditjmara are the Gournditch-mara or the somewhat more inclusive term Manmeet (Howitt 1904; Lourandos 1977).
3 Timothy Shannon, pers. comm.; Shannon (2008).
4 Dennis (1993); Kupchan (2010).
5 e.g., Brooks (2002).
6 Rogers (2003).
7 e.g., Murphy et al. (2011).
8 Gudykunst (2004).
9 Barth (1969); Bowles (2012).
10 Yellen & Harpending (1972).
11 Marwick (2003); Feblot-Augustins & Perlès (1992); Stiner & Kuhn (2006).
12 Dove (2011).
13 Laidre (2012).
14 Moffett (1989b).
15 Breed et al. (2012).
16 Whallon (2006).

17   Bushmen have been said to rarely steal since they recognize each other's few possessions and can identify the thief from their spoor, though I suspect this applies only to thievery *within* a society (or "ethnolinguistic group": Marshall 1961; Tanaka 1980).

18   Cashdan et al. (1983).

19   Dyson-Hudson & Smith (1978).

20   Bruneteau (1996); Flood (1980); Helms (1885).

21   Bushmen cultivated special partners they could depend on for exchanging goods when traveling through their society's territory (Wiessner 1982).

22   Binford (2001); Gamble (1998); Hamilton et al. (2007).

23   Cane (2013).

24   Jones (1996).

25   Pounder (1983).

26   Mulvaney (1976); Roth & Etheridge (1897). Words traveled, too: In advance of Europeans exploring the interior of Australia, Aborigines speaking many languages had heard about domestic animals and had already adopted words such as "yarra-man" for horse and "jumbuk" for sheep (Reynolds 1981).

27   Fair (2001); Lourandos (1997); Walker et al. (2011).

28   Kendon (1988); Silver & Miller (1997).

29   Newell et al. (1990).

30   An interaction sphere could continue step by step over long distances (Caldwell 1964).

31   Perhaps these differences ameliorated competition during years of resource shortages (Milton 1991).

32   Blainey (1976), 207.

33   Haaland (1969).

34   Franklin (1779), 53.

35   Gelo (2012).

36   Orton et al. (2013).

37   Bahuchet (2014), 12.

38   Boyd & Richerson (2005); Richerson & Boyd (1998); Henrich & Boyd (1998).

39   Leechman (1956), 83. See van der Dennen (2014).

40   Vasquez (2012).

41   Turner (1981); Wildschut et al. (2003).

42   Homer-Dixon (1994); LeVine & Campbell (1972).

43   Pinker (2011); Fry (2013).

Chapter 19

1   Durkheim (1982 [1895]), 90.

2   Best known are examples among the primates, e.g., Malik et al. (1985); Prud'Homme (1991); Van Horn et al. (2007).

3   Serengeti authority Craig Packer filled me in on the subject: "Lions definitely restrict their cooperative behavior to individuals they know and recognize. When prides become too large, they don't all seem to know each other that well anymore and thus split apart."

4   Confusingly, the breakups of societies have been referred to as "fissions." Because fission connotes the functionally very different workaday fission events in fission-fusion societies, in which groups routinely separate but then freely come back together, another term is in order, which is why I use division, albeit Sueur et al. (2011) offers another option, irreversible fission.

5   Joseph Feldblum, pers. comm.; Feldblum et al. (2018).

6   Williams et al. (2008); Wrangham & Peterson (1996).

7   See for example the references in Van Horn et al. (2007).
8   See Sueur et al. (2011).
9   Takeshi Furuichi, pers. comm.; Furuichi (1987); Kano (1992).
10 e.g., Henzi et al. (2000); Ron (1996); Van Horn et al. (2007).
11 In the honeybee, the younger workers swarm off with the original queen to start a new nest, leaving the old ones behind to await the birth of her successor, who takes over the original hive; there's no dispute about who goes where. Sometimes the bees ally themselves to more than one new queen and the hive breaks into several parts. I appreciate the advice of Raphaël Boulay, Adam Cronin, Christian Peeters, and Mark Winston. Cronin et al. (2013); Winston & Otis (1978).
12 Jacob Negrey, pers. comm.; Mitani & Amsler (2003).
13 Stan Braude, pers. comm.; O'Riain et al. (1996).
14 Sugiyama (1999). Bonobo males are also known to leave their community, but are thought to join neighboring groups, a behavior believed impossible for aggressive chimps (Furuichi 2011).
15 Brewer & Caporael (2006).
16 Dunbar (2011) implies this form of societal founding in humans by claiming that early societies tended to have 150 members that represented the living descendants of one couple five generations in the past, but there is no evidence that this ant-like (or termite-like) founding was common.
17 Peasley (2010).
18 Confusingly "budding" is also used (e.g. for Argentine ants, see Chapter 5), when a society's own members move into unoccupied areas, rather than to form a distinct and separate society.
19 McCreery (2000); Sharpe (2005).
20 Generally called "coalescent societies," see Chapter 22 (Kowalewski 2006; Price 1996).
21 For humans, e.g., Cohen (1978).
22 Fletcher (1995); Johnson (1982); Lee (1979); Woodburn (1982). This is also true for tribal villages and settled hunter-gatherers (Abruzzi 1980; Carneiro 1987).
23 Marlowe (2005).
24 A hint of such rebelliousness is seen these days in corporate splits, when employees forced into new relationships by their superiors continue to value their former identities and endeavor not to lose them (Terry et al. 2001).
25 e.g., Hayden (1987).

Chapter 20

1   The Aborigines retained this view at least until the 1950s (Meggitt 1962, 33).
2   Barth (1969).
3   Alcorta & Sosis (2005), 328.
4   Diamond (2005). Rather than starving, they may have simply moved elsewhere, some experts believe (Kintisch 2016; McAnany & Yoffee 2010).
5   Karen Kramer, pers. comm.; Kramer & Greaves (2016). For another example, consider Chapter 6 on the Pathans in Barth (1969).
6   There turns out to be no particular location for "standard American English," which is better interpreted as an absence of extreme speech patterns than a specific accent (Gordon 2001).
7   Proposed for language changes by Deutscher (2010).
8   Menand (2006), 76.
9   Thaler & Sunstein (2009). Leaders, in other words, tend to be "prototypical" (Hais et al. 1997).

10   Cipriani (1966), 76.
11   Bird & Bird (2000).
12   Pagel (2000); Pagel & Mace (2004).
13   e.g., Newell (1990).
14   Langergraber et al. (2014).
15   Boyd & Richerson (2005). Borderland residents also had to put their identities on conspicuous display to accentuate their dissimilarities with outsiders (Bettinger et al. 2015; Conkey 1982, 116; Giles et al. 1977, Chapter 1). While contact with outsiders can cause people to flaunt their identities, and sometimes form alliances with other societies for their protection, it isn't true that "tribes make states and states make tribes," proposed by Whitehead (1992), who thought that distinct tribes came into being only after colonialism forced locals to build their group identities for their own self-protection.
16   This boundary regulation brings to mind the situation described for chimpanzees. Those apes show this sort of adjustment to neighbors with their pant-hoots, with the calls appearing most dissimilar for communities that are immediate neighbors and consequently that have to take the most care to avoid any befuddlement about who's who. Even then, there's no evidence the pant-hoots show regional variation within the territory depending on which neighbor the apes confront—that's not likely because male chimps in particular tend to move around their territory rather than stay within part of it most of the time, as human bands do (Crockford et al. 2004).
17   cf. Read (2011).
18   Poole (1999), 16.
19   Packer (2008).
20   This is suggested by the fact that trust is higher in smaller groups, which may tolerate greater deviance among people who are known well (Jolanda Jetten, pers. comm.; La Macchia et al. 2016), whereas lesser-known members of a society may be trusted less (Hornsey et al. 2007).
21   Psychologists call this pluralistic ignorance (Miller & McFarland 1987). An example was the assumption by American whites in the 1960s that other whites supported segregation, which paradoxically led to bigoted practices few thought right (O'Gorman 1975).
22   Forsyth (2009).
23   The Comanche subgroups could be described as being well on their way toward acting as separate societies (Daniel Gelo, pers. comm.; Gelo 2012, 87).
24   With no prior experience even a Chihuahua will pick out a 100 kilogram mastiff as a member of its own kind from a photograph alone (Autier-Dérian et al. 2013).
25   Dollard (1937), 366.
26   Note, however, that if our ancestors distinguished their societies by vocalizations prior to the existence of a vocabulary, a scenario put forward in Chapter 11, there may never have been a time when all humans spoke one language.
27   e.g., Birdsell (1973).
28   Dixon (1972).
29   Cooley (1902), 270.
30   Presuming that the Ypety Ache people started to eat human flesh before that split.
31   Birdsell (1957).
32   Kim Hill, pers. comm.; Hill & Hurtado (1996).
33   Quoted in Lind (2006), 53.
34   Sani (2009).
35   Shown experimentally, e.g., by Bernstein et al. (2007).
36   e.g., Hornsey & Hogg (2000).
37   Erikson (1985).

38   Pagel (2009); Marks & Staski (1988).
39   Abruzzi (1982); Boyd & Richerson (2005).
40   Darwin (1859), 490.

Chapter 21

1   Atkinson et al. (2008); Dixon (1997).
2   Billig (1995); Butz (2009).
3   The idea originated with Tajfel & Turner (1979); e.g., see Van Vugt & Hart (2004).
4   Connerton (2010); van der Dennen (1987).
5   Goodall (2010), 128–129.
6   Russell (1993), 111.
7   Goodall (2010), 210.
8   See also Roscoe (2007).
9   Prud'Homme (1991).
10   Gross (2000).
11   Gonsalkorale & Williams (2007); Spoor & Williams (2007).
12   As is true of oppressed races in modern nations (e.g., Crocker et al. 1994; Jetten et al. 2001).
13   Boyd & Richerson (2005); studied for small groups by Hart & van Vugt (2006).
14   For primates, see Dittus (1988); Widdig et al. (2006). For hunter-gatherers, see Walker (2014), Walker & Hill (2014).
15   e.g., Chagnon (1979).
16   That's true even when the group in question is far more trivial than a society: kids joining a new clique of playmates, for example, overwhelmingly find new friends within that clique even when the members of the clique were chosen at random by the researchers (Sherif et al. 1961). As Muzafer Sherif (1966, 75), pioneer in studying such trivial competing groups, put it, "Freedom to choose friends on the basis of personal preferences turns out to be freedom to choose among persons selected according to the rules of membership of the organization"—with the membership of interest in this book being that of the society itself.
17   Taylor (2005).
18   Binford (2001); RL Kelly (2013a); Lee & Devore (1968).
19   This would lead to a band society breaking down most often beyond that size: Birdsell (1968) gave the typical number for division as a thousand.
20   Wobst (1974); Denham (2013).
21   How these dolphin communities form in the first place is a mystery (Randall Wells, pers. comm.; Sellas et al. 2005).
22   Eventually the amoebas run out of steam for dividing at all, and then languish in the same culture dish indefinitely (Bell 1988; Danielli & Muggleton 1959).
23   Birdsell (1958).
24   Hartley (1953), 1.
25   For estimates of numbers of languages in the past, see Pagel (2000).

Chapter 22

1   Kennett & Winterhalder (2006); Zeder et al. (2006).
2   There is a slight increase in population size for some species, but hardly what might be expected from the resource opportunities in cities (Colin Chapman, Jim Moore, and Sindhu Radhakrishna, pers. comm.; Kumar et al. 2013; Seth & Seth 1983).
3   Bandy & Fox (2010).

4   Wilshusen & Potter (2010). That for example most Yanomami would experience divisions during their lifetimes can be deduced from the graphs in Hunley et al. (2008) and Ward (1972).
5   Olsen (1987).
6   Flannery & Marcus (2012).
7   The Enga tribes are formally called phratries. The clans of a tribe intermarry and tend to be on good terms unless a clan outgrows its gardens. At that point fights can turn nasty: the Enga have never been magnanimous about diplomacy (Meggitt 1977; Wiessner & Tumu 1998).
8   Scott (2009).
9   I thank Luke Glowacki for advice on the Nyangatom. The generations to which the males belong are determined in a curious manner (Glowacki & von Rueden 2015).
10   Chagnon (2013).
11   "Tribe" is a word with a confusing history. I employ it here because the word has been used by some others to describe a group of villages sharing a language and culture, and because no other word is in general use for such a society (Stephen Sanderson, pers. comm.; Sanderson 1999). Hutterites behave much like a village society in North America, where they belong to three sects that still see themselves "of a kind," though each finds the other sects misguided (Simon Evans, pers. comm.).
12   Smouse et al. (1981); Hames (1983).
13   Many tribes carry out swidden, or slash-and-burn agriculture. A village clears a plot of land to garden, then travels to clear another forest patch when harvests decline.
14   Harner (1972). The Jívaro also tried to cajole other tribes to participate in their assault on the Spaniards, but these others contributed little (Redmond 1994; Stirling 1938).
15   With their many spats, villagers seldom stuck together long enough to invent different customs. The pair of villages resulting from a split would be pretty much indistinguishable in mode of life, as was true when people drifted away from a hunter-gatherer band—the split was more like people changing neighborhoods than a shift in their social roots (albeit slight differences in word usage can arise between villages after they split: Aikhenvald, pers. comm.; Aikhenvald et al. 2008).
16   Kopenawa & Albert (2013).
17   e.g., Southwick et al. (1974).
18   e.g., Jaffe & Isbell (2010).
19   Exceptions include fusions of army ant colonies after one looses a queen (Kronauer et al. 2010) and acacia ant colonies after fights (Rudolph & McEntee 2016). In termites, colony mergers have been demonstrated between colonies of some "primitive" (basal) species in which workers transform into reproductives following the death of the original queen and king (e.g., Howard et al. 2013). Claims of mergers for more "advanced" termites (Termitidae) are difficult to evaluate and as far as is known occur rarely if ever between mature colonies in nature (Barbara Thorne, pers. comm.).
20   Moss et al. (2011).
21   Ethridge & Hudson (2008).
22   Gunnar Haaland, pers. comm.; Haaland (1969).
23   Brewer (1999 and 2000).
24   The same is true for other such alliances, including those between some chiefdoms in North America and the league of states formed in China during the sixth century (Schwartz 1985).
25   Robert Carneiro originally took this point of view but later backed off it to allow for the fusion of some groups into chiefdoms (Carneiro 1998). I believe such "mergers" should be interpreted in terms of formerly sovereign groups (e.g., independent

villages), often of the *same* society, uniting under a political umbrella to accomplish a task; however, to fully incorporate those villages into a single entity would have required power plays by a chief that amounted to a kind of subjugation.

26   Bowles (2012).
27   Bintliff (1999).
28   Barth (1969).
29   There have been arguments about how much this would have been so; for example, some claim captives taken by the Iroquois were fully assimilated over time, while others view this as impossible (Donald 1997). I suspect full acceptance would be a more accurate description than full assimilation, in that differences would still be evident.
30   Chagnon (1977), 155.
31   Jones (2007) claims that this infant theft is an antecedent to slavery in humans, which I doubt—for sure the little monkey isn't forced into child labor.
32   Boesch et al. (2008).
33   Anderson (1999).
34   Biocca (1996), xxiv.
35   Brooks (2002).
36   Escapees were typically caught before they got far—often by tribe next door (Donald 1997).
37   Patterson (1982).
38   Cameron (2008).
39   Clark & Edmonds (1979).
40   Mitchell (1984), 46.
41   Some commoners sold themselves into slavery during hard times, especially when the slaves of elites led better lives than the poorest free people (Garnsey 1996).
42   Perdue (1979), 17.
43   Marean (2016).
44   e.g., Ferguson (1984).
45   See Chapter 15 and Abelson et al. (1998).
46   Adam Jones, pers. comm.; Jones (2012).
47   Confino (2014).
48   Haber et al. (2017).
49   Stoneking (2003).
50   Grabo & van Vugt (2016); Turchin et al. (2013).
51   Carneiro (1998 & 2000). See Chapter 22, note 25.
52   Oldmeadow & Fiske (2007). As will come up later, this perceived legitimacy of statuses applies to the relationships between ethnic groups and races as well.
53   e.g., Anderson (1994).

Chapter 23

1   Liverani (2006). Mesopotamian sites of the Ubaid period (5500–4000 BC) show a more rudimentary state organization. For a general discussion of the emergence of early states, see Scarre (2013) and Scott (2017).
2   Spencer (2010).
3   e.g., Alcock et al. (2001); Parker (2003).
4   Tainter (1988).
5   Bettencourt & West (2010); Ortman et al. (2014).
6   Richerson & Boyd (1998); Turchin (2015).
7   Wright (2004), 50–51.

8   Birdsell (1968) called this "density of communication."
9   Freedman (1975).
10   Hingley (2005). Some peripheral areas may have been little inclined to Romanization.
11   Spirituality provided a "moral master plan" that made leaders unnecessary (Hiatt 2015, 62).
12   Atran & Henrich (2010); Henrich et al. (2010a).
13   DeFries (2014).
14   Tilly (1975), 42.
15   I thank Eric Cline for suggesting Minoa as an example of a relatively pacifist ancient society.
16   RL Kelly (2013b), 156.
17   Mann (1986).
18   Such large chiefdoms could approach the level of organization of a state; in fact some experts claim certain ones were states (e.g., Hommon 2013).
19   Carneiro (1970 & 2012) expertly dispatches other theories about the rise of civilizations. I admit to having simplified and adjusted his views as I see fit; for example I agree that issues of social status could also have played into state formation (Chacon et al. 2015; Fukuyama 2011).
20   Brookfield & Brown (1963).
21   Lowen (1919), 175.
22   de la Vega (1966, written 1609), 108.
23   Faulseit (2016).
24   Diamond (2005).
25   Currie et al. (2010); Tainter (1988).
26   e.g., Joyce et al. (2001).
27   Marcus (1989).
28   Chase-Dunn et al. (2010); Gavrilets et al. (2014); Walter et al. (2006).
29   Johnson & Earle (2000).
30   Beaune (1991); Gat & Yakobson (2013); Hale (2004); Reynolds (1997); Weber (1976).
31   Kennedy (1987).
32   Frankopan (2015).
33   Yoffee (1995).
34   More support for her hypothesis comes to light every year (Roosevelt 2013).
35   Some of these wars have undertaken to seize control of the whole society rather than fragment it. Holsti (1991); Wallensteen (2012); Wimmer & Min (2006).
36   Kaufman (2001).
37   Bookman (1994).
38   Southerners did often come from different parts of Great Britain than northerners, giving this distinction a crude basis in reality (Fischer 1989; Watson 2008).
39   Allen Buchanan, Paul Escott, and Libra Hilde, pers. comm.; Escott (2010); McCurry (2010); Weitz (2008).
40   Carter & Goemans (2011).
41   This lack of commitment can hold true for many kinds of groups (Karau & Williams 1993).
42   Kaiser (1994); Sekulic et al. (1994).
43   Joyce Marcus, pers. comm.; Feinman & Marcus (1998). Chiefdoms and early states were more ephemeral, by some measures with the former lasting at most 75 to 100 years (Hally 1996).
44   Cowgill (1988), 253–254.
45   Claessen & Skalník (1978).

Chapter 24

1 Alcock et al. (2001).
2 Isaac (2004), 8. For many insights into the transformations of ethnic groups in societies, I can recommend the classic Van den Berghe (1981).
3 Malpass (2009), 27–28. Thanks to Michael Malpass for advice on the Inca.
4 Noy (2000).
5 My arguments are similar to Cowgill (1988), except I prefer the word "domination," where he uses "subjugation," given that both dominated and incorporated groups would have been subjugated initially.
6 Yonezawa (2005).
7 Brindley (2015).
8 Francis Allard, pers. comm., Allard (2000); Brindley (2015).
9 Hudson (1999).
10 Their great wall similarly kept out (and set apart) the Chinese from the "primitive" steppe nomad foreign societies at the periphery (Fiskesjö 1999).
11 Cavafy (1976).
12 Spickard (2005), 2. Assimilation, and the related word "acculturation," have been employed in varied nuanced ways by anthropologists and sociologists, but I will use only the former term here.
13 Smith (1986). This "group dominance perspective" is well supported (Sidanius et al. 1997).
14 The major exceptions to the subjugated undergoing the preponderance of change were among the pastoral nomads, who were vastly outnumbered by the culturally more intricate societies they took over. Genghis Khan and successors drew freely from the civilizations they conquered. They permitted a wide latitude in behavior among their own people and the peoples they controlled while for the most part clinging to their nomadic traditions (nicely described by Chua 2007).
15 e.g., Chapter 8 of Santos-Granero (2009).
16 Hornsey & Hogg (2000); Hewstone & Brown (1986).
17 Aly (2014). Use of yellow badges to make Jews stand out was discussed in Chapter 12.
18 Mummendey & Wenzel (1999).
19 The capital had by far the most impact (Mattingly 2014).
20 In pronouncedly multiethnic societies there can be a struggle over what factors into the superordinate identity (Packer 2008; Schaller & Neuberg 2012).
21 e.g., Vecoli (1978).
22 Joniak-Lüthi (2015).
23 States that are more manifestly polyethnic, like the United States, aren't considered nations on this basis. I have preferred the colloquial usage of "nation" in this book (Connor 1978).
24 Sidanius et al. (1997).
25 Seneca (1970), written first century AD.
26 Klinkner & Smith (1999), 7.
27 Devos & Ma (2008).
28 Huynh et al. (2011), 133.
29 Gordon (1964), 5.
30 Deschamps (1982).
31 Yogeeswaran & Dasgupta (2010).
32 Jost & Banaji (1994); Kamans et al. (2009).
33 Sidanius & Petrocik (2001).

34 Cheryan & Monin (2005); Wu (2002).
35 Ho et al. (2011).
36 Devos & Banaji (2005).
37 Marshall (1950), 8.
38 Deschamps & Brown (1983).
39 Ehardt & Bernstein (1986); Samuels et al. (1987).
40 Lee & Fiske (2006); Portes & Rumbaut (2014).
41 Bodnar (1985).
42 Jost & Banaji (1994); Lerner & Miller (1978).
43 Fiske et al. (2007), 82. See also Major & Schmader (2001); Oldmeadow & Fiske (2007).
44 Jost et al. (2003), 13.
45 Paranjpe (1998).
46 Hewlett (1991), 29.
47 Moïse (2014).
48 An infant captured by the Comanche could be treated as Comanche right away (Rivaya-Martínez, pers. comm.; Rivaya-Martínez 2012).
49 Cheung et al. (2011).
50 Cameron (2008); Raijman et al. (2008).
51 That was true even before 212 AD (Garnsey 1996).
52 Engerman (2007); Fogel & Engerman (1974).
53 Lim et al. (2007).
54 At the same time interactions between groups can also stimulate people to find new ways to distinguish themselves (Hogg 2006; Salamone & Swanson 1979).
55 For the Romans, see Insoll (2007), 11. The Greeks themselves were of several ethnicities (Jonathan Hall, pers. comm.; Hall 1997).
56 Smith (2010).
57 Noy (2000).
58 Greenshields (1980).
59 Portes & Rumbaut (2014).
60 At first the Indians often needed official permission to travel, with the US government going so far as to crack down on church attendance outside the reserve (Richmond Clow, pers. comm.).
61 Schelling (1978).
62 Christ et al. (2014); Pettigrew (2009).
63 Paxton & Mughan (2006).
64 Thompson (1983).
65 Hawley (1944), Berry (2001).
66 Park (1928), 893.

Chapter 25

1 First-generation American immigrants don't necessarily score high in measures of patriotism, but that typically changes for their offspring (Citrin et al. 2001).
2 Quoted in Beard (2009), 11.
3 Arguably this might be more than analogy, in that many norms govern issues that we see today as matters of health, such as what to eat and how to prepare it, and outsiders failing to comply with the local ways could in fact spread sickness (Fabrega 1997; Schaller & Neuberg 2012).
4 Dixon (1997).
5 Gaertner & Dovidio (2000).
6 This brings us back to the connection between specialization and social cohesion that was developed by Durkheim (1893), brought up in Chapter 10, including endnote 39.

7 This again is positive distinctiveness (Chapter 21). Evidence for such roles is scarce because the information was seldom recorded for early states, for example Roman graves indicated the deceased's ethnicity but not job or vice versa (David Noy, pers. comm.).

8 Esses et al. (2001). While competing with the majority is a road to failure, conflict between minorities is also costly (Banton 1983; Olzak 1992). Majority peoples have often gained by promoting rivalries that kept minorities at loggerheads rather than dissatisfied with the people in power. Of course this works between societies, too. The Romans were masters of divide and rule, splitting troublesome Macedonia into four provinces and provoking them to fight.

9 Noy (2000). The status of black Americans has improved when they are needed as soldiers in wartime (Smith & Klinkner 1999).

10 Boyd (2002).

11 Abruzzi (1982).

12 Turnbull (1965); Zvelebil & Lillie (2000).

13 See Cameron (2016) for this and other examples.

14 Appave (2009).

15 Sorabji (2005).

16 Suetonius (1979, written AD 121), 21.

17 McNeill (1986).

18 Dinnerstein & Reimers (2009), 2.

19 Light & Gold (2000).

20 Bauder (2008); Potts (1990).

21 Meanwhile, colonialism, and the founding of nations that ensued it, led many people to lose their original tribal identities in favor of broad ethnic categories in their native regions as well (e.g., the Ewe, Shona, Igbo, and other ethnic groups in Africa: Iliffe 2007).

22 e.g., Gossett (1963, Chapter 1).

23 PC Smith (2009), 4, 5.

24 Brindley (2010).

25 Dio (2008, written second century AD), 281.

26 Sarna (1978).

27 Curti (1946).

28 Crevècoeur (1782), 93.

29 Matthew Frye Jacobson, pers. comm., Alba (1985); Painter (2010).

30 Alba & Nee (2003); Saperstein & Penner (2012).

31 Leyens et al. (2007).

32 Freeman et al. (2011).

33 Smith (1997).

34 Smith (1986).

35 Bloemraad et al. (2008).

36 Ellis (1997).

37 Levinson (1988); Orgad (2011); Poole (1999).

38 Harles (1993).

39 Gans (2007); Huddy & Khatib (2007). To top that off, given the convenience of travel, communications, and trade nowadays, immigrants are seldom severed from their ancestral homeland and its traditions, though those connections are likely to fade in their children (Levitt & Waters 2002).

40 Bloemraad (2000); Kymlicka (1995).

41 Ironically, Lucan was born in what is modern-day Spain. Quoted in Noy (2000, 34), who discusses racism throughout the history of the Roman Empire.

42 Michener (2012); Volpp (2001).

43 van der Dennen (1991).

44 Jacobson (1999).

45 Alesina & La Ferrara (2005), 31–32.
46 May (2001, 235) makes this point for the tribes of Papua New Guinea: "The welfare of village people today depends in part on their ability to capture a share of the goods and services flowing from the state; chiefly or bigman leadership will only be effective to the extent that it ensures access to these benefits, and that implies its articulation with the state."
47 Harlow & Dundes (2004); Sidanius et al. (1997).
48 Bar-Tal & Staub (1997); Wolsko et al. (2006).
49 Marilynn Brewer pointed out to me some overlap between my views with Shah et al. (2004).
50 e.g., Van der Toorn et al. (2014).
51 Barrett (2007); Feshbach (1991); Lewis et al. (2014); Piaget & Weil (1951).
52 Patriotism and nationalism are loosely associated with liberal and conservative views, respectively, but differ especially in their extreme expressions. Ultra liberals, for example, can violently oppose any free speech going against their ideology, while fiscal conservatives can support free trade and positive group relations. Since nationalists can harbor patriotic feelings, my discussion of patriots applies to those high on patriotism but low on nationalism.
53 Bar Tal & Staub (1997).
54 Feinstein (2016); Staub (1997).
55 See Schatz et al. (1999), who refer to nationalism as "blind patriotism."
56 Blank & Schmidt (2003); Devos & Banaji (2005); Leyens et al. (2003).
57 Andrew Billings, pers. comm.; Billings et al. (2015); Rothì et al. (2005).
58 De Figueiredo & Elkins (2003); Viki & Calitri (2008).
59 e.g., Raijman et al. (2008).
60 Greenwald et al. (2015).
61 Smith et al. (2011), 371.
62 Jandt et al. (2014); Modlmeier et al. (2012).
63 Feshbach (1994).
64 Hedges (2002); Junger (2016).
65 Turchin (2015).
66 In contrast patriots try to pull together varied groups by appealing to their common fate (Li & Brewer 2004).
67 e.g., Banks (2016); Echebarria-Echabe & Fernandez-Guede (2006).
68 Competitiveness between groups only worsens the matter (Esses et al. 2001; King et al. 2010).
69 Bergh et al. (2016); Zick et al. (2008).
70 Described by Sidanius et al. (1999).

Chapter 26

1 Hayden & Villeneuve (2012), 130.
2 Gaertner et al. (2006).
3 The descendants of the mutineers, who might be looked upon as forming a kind of coalescent society, are now part of an overseas territory of the United Kingdom.
4 For ants and humans, this separation could potentially continue for generations because of their employment of markers of identity (Chapters 5–7). Opinions about the length of Viking isolation vary (Graeme Davis, pers. comm.; Davis 2009).
5 Weisler (1995). Isolation, and complete ignorance of outsiders, has been claimed of one tribe on mainland New Guinea (Tuzin 2001).
6 Royce (1982), 12.
7 Cialdini & Goldstein (2004).

8   Nichole Simmons, pers. comm.
9   Jones et al. (1984).
10   Differences between the groups had begun to emerge before they were encouraged to compete (Sherif et al. 1961). There are questions about how much of the behavior of the children was manipulated by the researchers (Perry 2018).
11   Carneiro (2004); Turchin & Gavrilets (2009).
12   e.g., China (Knight 2008).
13   Aikhenvald (2008), 47.
14   Soto (2008).
15   Jackson (1983).
16   McCormick (2017); Reese & Lauenstein (2014).
17   Goodwin (2016).
18   Chollet (2011), 746, 751. See also Linder (2010); Rutherford et al. (2014).
19   Leuchtenburg (2015), 634.
20   Gellner (1983), 6. Gellner went on to say, "Having a nation is not an inherent attribute of humanity, but . . . has now come to appear as such" (ibid.). See also Miller (1995).
21   Many new immigrants face considerable stress in adjusting to their situation (Berry & Annis 1974).
22   Lyons-Padilla & Gelfand (2015).

Conclusion

1   Reynolds (1981).
2   Druckman (2001).
3   Gelfand et al. (2011).
4   Blanton & Christie (2003); Jetten et al. (2002); Maghaddam (1998). There is little difference in people's overall sense of happiness (well-being) across different countries (Burns 2018).
5   Deschamps (1982); Lorenzi-Cioldi (2006).
6   Cosmides et al. (2003).
7   Brewer (2009).
8   Easterly (2001).
9   Christ et al. (2014).
10   Alesina & Ferrara (2005); Hong & Page (2004). Accepting those who seem strange will be the greatest challenge for socially dominating people (Asbrock et al. 2012).

References

Aanen DK, et al. 2002. The evolution of fungus-growing termites and their mutualistic fungal symbionts. *Proc Nat Acad Sci* 99:14887–14892.

Abelson RP, et al. 1998. Perceptions of the collective other. *Pers Soc Psychol Rev* 2:243–250.

Aboud FE. 2003. The formation of in-group favoritism and out-group prejudice in young children: Are they distinct attitudes? *Dev Psychol* 39:48–60.

Abruzzi WS. 1980. Flux among the Mbuti Pygmies of the Ituri forest. In EB Ross, ed. *Beyond the Myths of Vulture*. New York: Academic. pp. 3–31.

———. 1982. Ecological theory and ethnic differentiation among human populations. *Curr Anthropol* 23:13–35.

Ackerman JM, et al. 2006. They all look the same to me (unless they're angry): From out-group homogeneity to out-group heterogeneity. *Psychol Sci* 17:836–840.

Ackerman JM, D Kenrick, M Schaller. 2007. Is friendship akin to kinship? *Evol Hum Behav* 28:365–374.

Adamatzky A. 2005. *A Dynamics of Crowd Minds*. Singapore: World Scientific.

Addessi E, L Crescimbene, E Visalberghi. 2007. Do capuchin monkeys use tokens as symbols? *Proc Roy Soc Lond B* 274:2579–2585.

Aiello LC, RIM Dunbar. 1993. Neocortex size, group size, and the evolution of language. *Curr Anthropol* 34:184–193.

Aikhenvald AY. 2008. Language contact along the Sepik River, Papua New Guinea. *Anthropol Linguist* 50:1–66.

Aikhenvald AY, et al. 2008. *The Manambu Language of East Sepik, Papua New Guinea*. Oxford: Oxford University Press.

Aimé C, et al. 2013. Human genetic data reveal contrasting demographic patterns between sedentary and nomadic populations that predate the emergence of farming. *Mol Biol Evol* 30:2629–2644.

Alba R. 1985. *Italian Americans: Into the Twilight of Ethnicity*. Englewood Cliffs, NJ: Prentice Hall.

Alba R, V Nee. 2003. *Remaking the American Mainstream: Assimilation and Contemporary Immigration*. Cambridge, MA: Harvard University Press.

Alcock SE, et al., eds. 2001. *Empires: Perspectives from Archaeology and History*. Cambridge: Cambridge University Press.

Alcorta CS, R Sosis. 2005. Ritual, emotion, and sacred symbols: The evolution of religion as an adaptive complex. *Hum Nature* 16:323–359.

Alesina A, E La Ferrara. 2005. Ethnic diversity and economic performance. *J Econ Lit* 43:762–800.

Alexander MG, MB Brewer, RW Livingston. 2005. Putting stereotype content in context: Image theory and interethnic stereotypes. *Pers Soc Psychol Bull* 31:781–794.

Alexander RD. 1985. A biological interpretation of moral systems. *J Relig Sci* 20:3–20.

Allard F. 2006. Frontiers and boundaries. The Han empire from its southern periphery. In MT Stark, ed. *Archaeology of Asia*. Malden, MA: Blackwell. pp. 233–254.

Allee WC. 1931. *Animal Aggregations*. Chicago: University of Chicago Press.

Allen MW, TL Jones, eds. 2014. *Violence and Warfare among Hunter-Gatherers*. Walnut Creek, CA: Left Coast Press.

Allport FH. 1927. The nationalistic fallacy as a cause of war. *Harpers*. August. pp. 291–301.

Allport GW. 1954. *The Nature of Prejudice*. Reading: Addison-Wesley.

Al-Simadi FA. 2000. Jordanian students' beliefs about nonverbal behaviors associated with deception in Jordan. *Soc Behav Pers* 28.437–442.

Alvergne A, C Faurie, M Raymond. 2009. Father-offspring resemblance predicts paternal investment in humans. *Anim Behav* 78:61–69.

Aly G. 2014. *Why the Germans? Why the Jews?: Envy, Race Hatred, and the Prehistory of the Holocaust*. New York: Macmillan.

Ames KM. 1991. Sedentism: A temporal shift or a transitional change in hunter-gatherer mobility patterns? In S Gregg, ed. *Between Bands and States. Center for Archaeological Investigations Occasional Paper No. 9*. Carbondale: Southern Illinois University Press. pp. 103–133.

———. 1995. Chiefly power and household production on the Northwest Coast. In TD Price, GM Feinman, eds. *Foundations of Social Inequality*. New York: Springer. pp. 155–187.

Ames K, HDG Maschner. 1999. *Peoples of the Northwest Coast: Their Archaeology and Prehistory*. New York: Thames & Hudson.

Amodio DM. 2000. The social neuroscience of intergroup relations. *Eur Review Soc Psychol* 19:1–54.

———. 2011. Self-regulation in intergroup relations: A social neuroscience framework. In A Todorov, ST Fiske, DA Prentice, eds. *Social Neuroscience: Toward Understanding the Underpinnings of the Social Mind*. New York: Oxford University Press. pp. 101–122.

Anderson B. 1982. *Imagined Communities: Reflections on the Origin and Spread of Nationalism*. New York: Verso.

Anderson DG. 1994. *The Savannah River Chiefdoms: Political Change in the Late Prehistoric Southeast*. Tuscaloosa: University of Alabama Press.

Anderson GC. 1999. *The Indian Southwest, 1580–1830: Ethnogenesis and Reinvention*. Norman: University of Oklahoma Press.

Andersson CJ. 1856. *Lake Ngami: Or, Explorations and Discoveries, during Four Year's Wandering in the Wilds of South Western Africa*. London: Hurst & Blackett.

Ansmann IC, et al. 2012. Dolphins restructure social system after reduction of commercial fisheries. *Anim Behav* 575–581.

Anzures G, et al. 2012. Brief daily exposures to Asian females reverses perceptual narrowing for Asian faces in Caucasian infants. *J Exp Child Psychol* 112:485–495.

Apicella CL, et al. 2012. Social networks and cooperation in hunter-gatherers. *Nature* 481:497–501.

Appave G. 2009. *World Migration 2008: Managing Labour Mobility in the Evolving Global Economy*. Sro-Kundig, Switzerland: International Organization for Migration.

Appelbaum Y. 2015. Rachel Dolezal and the history of passing for Black. *The Atlantic*. June 15.

Armitage KB. 2014. *Marmot Biology*. Cambridge: Cambridge University Press.

Arnold JE. 1996. The archaeology of complex hunter-gatherers. *J Archaeol Meth Th* 3:77–126.

Asch SE. 1946. Forming impressions of personality. *J Abnorm Soc Psychol* 41:258–290.

Asbrock F, et al. 2012. Differential effects of intergroup contact for authoritarians and social dominators. *Pers Soc Psychol B* 38:477–490.

Atkinson QD, et al. 2008. Languages evolve in punctuational bursts. *Science* 319:588.

Atran S. 1990. *Cognitive Foundations of Natural History*. Cambridge: Cambridge University Press.

Atran S, et al. 1997. Generic species and basic levels: Essence and appearance in folk biology. *J Ethnobiol* 17:17–43.

Atran S, J Henrich. 2010. The evolution of religion: How cognitive by-products, adaptive learning heuristics, ritual displays, and group competition generate deep commitments to prosocial religions. *Biol Theory* 5:1–13.

Aureli F, et al. 2006. Raiding parties of male spider monkeys: Insights into human warfare? *Am J Phys Anthropol* 131:486–497.

Aureli F, et al. 2008. Fission-fusion dynamics: New research frameworks. *Curr Anthropol* 49:627–654.

Autier-Dérian D, et al. 2013. Visual discrimination of species in dogs. *Anim Cogn* 16:637–651.

Avilés L, J Guevara. 2017. Sociality in spiders. In DR Rubenstein, R Abbot, eds. *Comparative Social Evolution*. Cambridge: Cambridge University Press. pp. 188–223.

Axelrod R. 2006. *The Evolution of Cooperation*. New York: Basic Books.

Azevedo RT, et al. 2013. Their pain is not our pain: Brain and autonomic correlates of empathic resonance with the pain of same and different race individuals. *Hum Brain Mapp* 34:3168–3181.

Bădescu I, et al. 2016. Alloparenting is associated with reduced maternal lactation effort and faster weaning in wild chimpanzees. *Roy Soc Open Sci* 3:160577.

Bahuchet S. 2012. Changing language, remaining Pygmy. *Hum Biol* 84:11–43.

———. 2014. Cultural diversity of African Pygmies. In BS Hewlett, ed. *Hunter-Gatherers of the Congo Basin*. New Brunswick, NJ: Transaction. pp. 1–30.

Bailey KG. 1988. Psychological kinship: Implications for the helping professions. *Psychother Theor Res Pract Train* 25:132–141.

Bain P, et al. 2009. Attributing human uniqueness and human nature to cultural groups: Distinct forms of subtle dehumanization. *Group Proc Intergr Rel* 12:789–805.

Banaji MR, SA Gelman. 2013. *Navigating the Social World: What Infants, Children, and Other Species Can Teach Us*. Oxford: Oxford University Press.

Banaji MR, AG Greenwald. 2013. *Blindspot: Hidden Biases of Good People*. New York: Delacorte Press.

Bandura A. 1999. Moral disengagement in the perpetration of inhumanities. *Pers Soc Psychol Rev* 3:193–209.

Bandy MS, JR Fox, eds. 2010. *Becoming Villagers: Comparing Early Village Societies*. Tucson: University of Arizona Press.

Banks AJ. 2016. Are group cues necessary? How anger makes ethnocentrism among whites a stronger predictor of racial and immigration policy opinions. *Polit Behav* 38:635–657.

Banton M. 1983. *Racial and Ethnic Competition*. Cambridge: Cambridge University Press.

Barfield T. 2002. Turk, Persian and Arab: Changing relationships between tribes and state in Iran and along its frontiers. In N Keddie, ed. *Iran and the Surrounding World*. Seattle: University of Washington Press. pp. 61–88.

Barlow G. 2000. *The Cichlid Fishes*. New York: Basic Books.

Barnard A. 2007. *Anthropology and the Bushman*. New York: Berg.

———. 2010. When individuals do not stop at the skin. In RIM Dunbar, C Gamble, J Gowlett, eds. *Social Brain, Distributed Mind*. Oxford: Oxford University Press. pp. 249–267.

———. 2011. *Social Anthropology and Human Origins*. New York: Cambridge University Press.

Barnes JE. 2001. As demand soars, flag makers help bolster nation's morale. *New York Times*. September 23.

Baron AS, Y Dunham. 2015. Representing "us" and "them": Building blocks of intergroup cognition. *J Cogn Dev* 16:780–801.

Barrett M. 2007. *Children's Knowledge, Beliefs, and Feelings about Nations and National Groups.* New York: Psychology Press.

Barron AB, C Klein. 2016. What insects can tell us about the origins of consciousness. *Proc Nat Acad Sci* 113:4900–4908.

Barron WRJ. 1981. The penalties for treason in medieval life and literature. *J Medieval Hist* 7:187–202.

Bar-Tal D. 2000. *Shared Beliefs in a Society.* Thousand Oaks, CA: Sage Publishing.

Bar-Tal D, E Staub. 1997. Patriotism: Its scope and meaning. In D Bar Tal, E Staub, eds. *Patriotism in the Lives of Individuals and Nations.* Chicago: Nelson Hall. pp. 1–19.

Barth F, ed. 1969. *Ethnic Groups and Boundaries: The Social Organization of Culture Difference.* Boston: Little, Brown. pp. 9–38.

Bartlett FC, C Burt. 1933. Remembering: A study in experimental and social psychology. *Brit J Educ Psychol* 3:187–192.

Bates LA, et al. 2007. Elephants classify human ethnic groups by odor and garment color. *Curr Biology* 17:1938–1942.

Bates LA, et al. 2008. African elephants have expectations about the locations of out-of-sight family members. *Biol Lett* 4:34–36.

Bauder H. 2008. Citizenship as capital: The distinction of migrant labor. *Alternatives* 33:315–333.

Baumard N. 2010. Has punishment played a role in the evolution of cooperation? A critical review. *Mind Soc* 9:171–192.

Baumeister RF. 1986. *Identity: Cultural Change and the Struggle for Self.* New York: Oxford University Press.

Baumeister RF, SE Ainsworth, KD Vohs. 2016. Are groups more or less than the sum of their members? The moderating role of individual identification. *Behav Brain Sci* 39:1–56.

Baumeister RF, et al. 1989. Who's in charge here? *Pers Soc Psychol B* 14:17–22.

Baumeister RF, Leary MR. 1995. The need to belong: Desire for interpersonal attachments as a fundamental human motivation. *Psychol Bull* 117:497–529.

Beals KL, et al. 1984. Brain size, cranial morphology, climate, and time machines. *Curr Anthropol* 25:301–330.

Beard CA. 2009. *The Republic: Conversations on Fundamentals.* New Brunswick, NJ: Transaction Publishers.

Beaune C. 1991. *Birth of an Ideology: Myths and Symbols of a Nation.* Berkeley: University of California Press.

Beccaria C. 2009 (1764). *On Crimes and Punishments and Other Writings.* A Thomas, ed. Toronto: University of Toronto Press.

Beck BB. 1982. Chimpocentrism: Bias in cognitive ethology. *J Hum Evol* 11:3–17.

Becker JC, et al. 2017. What do national flags stand for? An exploration of associations across 11 countries. *J Cross Cult Psychol* 48:335–352.

Beecher MD, et al. 1986. Acoustic adaptations for parent-offspring recognition in swallows. *Exp Biol* 45:179–193.

Beety VE. 2012. What the brain saw: The case of Trayvon Martin and the need for eyewitness identification reform. *Denver Univ Law Rev* 90:331–346.

Behar DM, et al. 2008. The dawn of human matrilineal diversity. *Am J Hum Genet* 82:1130–1140.

Bekoff M, J Pierce. 2009. *Wild Justice: The Moral Lives of Animals.* Chicago: University of Chicago Press.

Bell G. 1988. *Sex and Death in Protozoa.* New York: Cambridge University Press.

Bender T. 2006. *A Nation among Nations: America's Place in World History.* New York: Hill & Wang.

Bennett NC, CG Faulkes. 2000. *African Mole-rats: Ecology and Eusociality.* Cambridge: Cambridge University Press.

Benson-Amram S, et al. 2016. Brain size predicts problem-solving ability in mammalian carnivores. *Proc Nat Acad Sci* 113:2532–2537.

Berger J, C Cunningham. 1987. Influence of familiarity on frequency of inbreeding in wild horses. *Evolution* 41:229–231.

Berger PL, T Luckmann. 1966. *The Social Structure of Reality: A Treatise in the Sociology of Knowledge.* New York: Doubleday.

Bergh R, et al. 2016. Is group membership necessary for understanding generalized prejudice? A re-evaluation of why prejudices are interrelated. *J Pers Soc Psychol* 111:367–395.

Bergman TJ. 2010. Experimental evidence for limited vocal recognition in a wild primate: Implications for the social complexity hypothesis. *Proc Roy Soc Lond B* 277:3045–3053.

Berman JC. 1999. Bad hair days in the Paleolithic: Modern (re)constructions of the cave man. *Am Anthropol* 101:288–304.

Berndt RM, CH Berndt. 1988. *The World of the First Australians.* Canberra: Aboriginal Studies Press.

Bernstein MJ, SG Young, K Hugenberg. 2007. The cross-category effect: Mere social categorization is sufficient to elicit an own-group bias in face recognition. *Psychol Sci* 18:706–712.

Berreby D. 2005. *Us and Them: Understanding Your Tribal Mind.* New York: Little, Brown.

Berry JW. 2001. A psychology of immigration. *J Soc Issues* 57:615–631.

Berry JW, RC Annis. 1974. Acculturation stress: The role of ecology, culture, and differentiation. *J Cross Cult Psychol* 5:382–406.

Best E. 1924. *The Maori*, vol. 1. Wellington, NZ: HH Tombs.

Bettencourt L, G West. 2010. A unified theory of urban living. *Nature* 467:912–913.

Bettinger RL, R Garvey, S Tushingham. 2015. *Hunter-Gatherers: Archaeological and Evolutionary Theory.* 2nd ed. New York: Springer.

Bigelow R. 1969. *The Dawn Warriors: Man's Evolution Toward Peace.* Boston: Little, Brown.

Bigler RS, LS Liben. 2006. A developmental intergroup theory of social stereotypes and prejudice. *Adv Child Dev Behav* 34:39–89.

Bignell DE, Y Roisin, N Lo, eds. 2011. *Biology of Termites.* New York: Springer.

Billig M. 1995. *Banal Nationalism.* London: Sage Publications.

Billings A, K Brown, N Brown-Devlin. 2015. Sports draped in the American flag: Impact of the 2014 winter Olympic telecast on nationalized attitudes. *Mass Commun Soc* 18:377–398.

Binford LR. 1980. Willow smoke and dog's tails: Hunter-gatherer settlement systems and archaeological site formation. *Am Antiquity* 45:4–20.

———. 2001. *Constructing Frames of Reference.* Berkeley: University of California Press.

Bintliff J. 1999. Settlement and territory. In G Barker, ed. *Companion Encyclopedia of Archaeology 1.* London: Routledge. pp. 505–545.

Biocca E. 1996. *Yanoáma: The Story of Helena Valero.* New York: Kodansha America.

Bird DW, RB Bird. 2000. The ethnoarchaeology of juvenile foragers: Shellfishing strategies among Meriam children. *J Anthropol Archaeol* 19:461–476.

Birdsell JB. 1957. Some population problems involving Pleistocene man. *Cold Spring Harbor Symposia on Quantitative Biology* 22:47–69.

———. 1958. On population structure in generalized hunting and collecting populations. *Evolution* 12:189–205.

———. 1968. Some predictions for the Pleistocene based on equilibrium systems among recent foragers. In R Lee, I DeVore, eds. *Man the Hunter.* Chicago: Aldine. pp. 229–249.

———. 1970. Local group composition among the Australian Aborigines: A critique of the evidence from fieldwork conducted since 1930. *Curr Anthropol* 11:115–142.

———. 1973. The basic demographic unit. *Curr Anthropol* 14:337–356.

Blainey G. 1976. *Triumph of the Nomads: A History of Aboriginal Australia.* Woodstock, NY: Overlook Press.

Blank T, P Schmidt. 2003. National identity in a united Germany: Nationalism or patriotism? An empirical test with representative data. *Polit Psychol* 24:289–312.

Blanton H, C Christie. 2003. Deviance regulation: A theory of action and identity. *Rev Gen Psychol* 7:115–149.

Bleek DF. 1928. *The Naron: A Bushman Tribe of the Central Kalahari.* Cambridge: Cambridge University Press.

Bloemraad I. 2000. Citizenship and immigration. *J Int Migrat Integration* 1:9–37

Bloemraad I, A Korteweg, G Yurdakul. 2008. Citizenship and immigration: Multiculturalism, assimilation, and challenges to the nation-state. *Annu Rev Sociol* 34:153–179.

Bloom P, C Veres. 1999. Perceived intentionality of groups. *Cognition* 71:B1–B9.

Blurton-Jones N. 2016. *Demography and Evolutionary Ecology of Hadza Hunter-Gatherers.* Cambridge: Cambridge University Press.

Bocquet-Appel J-P, O Bar-Yosef, eds. 2008. *The Neolithic Demographic Transition and its Consequences.* New York: Springer.

Bodnar JE. 1985. *The Transplanted: A History of Immigrants in Urban America.* Bloomington: Indiana University Press.

Boehm C. 1999. *Hierarchy in the Forest: The Evolution of Egalitarian Behavior.* Cambridge, MA: Harvard University Press.

———. 2013. The biocultural evolution of conflict resolution between groups. In D Fry, ed. *War, Peace, and Human Nature.* Oxford: Oxford University Press. pp. 315–340.

Boesch C. 1996. Social grouping in Tai chimpanzees. In WC McGrew, LF Marchant, T Nishida, eds. *Great Ape Societies.* Cambridge: Cambridge University Press. pp. 101–113.

———. 2012. From material to symbolic cultures: Culture in primates. In J Valsiner, ed. *The Oxford Handbook of Culture and Psychology.* Oxford: Oxford University Press. pp. 677–694.

Boesch C, H Boesch-Achermann. 2000. *The Chimpanzees of the Taï Forest.* New York: Oxford University Press.

Boesch C, et al. 2008. Intergroup conflicts among chimpanzees in Taï National Park: Lethal violence and the female perspective. *Am J Primatol* 70:519–532.

Boesch C, G Hohmann, L Marchant, eds. 2002. *Behavioural Diversity in Chimpanzees and Bonobos.* Oxford: Cambridge University Press.

Bohn KM, CF Moss, GS Wilkinson. 2009. Pup guarding by greater spear-nosed bats. *Behav Ecol Sociobiol* 63:1693–1703.

Bolhuis JJ. 1991. Mechanisms of avian imprinting: A review. *Biol Rev* 66:303–345.

Bond CF, et al. 1990. Lie detection across cultures. *J Nonverbal Behav* 14:189–204.

Bond R. 2005. Group size and conformity. *Intergroup Relations* 8:331–354.

Bonilla-Silva E. 2014. *Racism without Racists: Color-Blind Racism and the Persistence of Racial Inequality in America.* New York: Rowman & Littlefield.

Bonner J. 2006. *Why Size Matters: From Bacteria to Blue Whales.* Princeton, NJ: Princeton University Press.

Bonnie KE, et al. 2007. Spread of arbitrary customs among chimpanzees: A controlled experiment. *Proc Roy Soc B* 274:367–372.

Bookman MZ. 1994. War and peace: The divergent breakups of Yugoslavia and Czechoslovakia. *J Peace Res* 31:175–187.

Bosacki SL, C Moore. 2004. Preschoolers' understanding of simple and complex emotions: Links with gender and language. *Sex Roles* 50:659–675.

Bosmia AN, et al. 2015. Ritualistic envenomation by bullet ants among the Sateré-Mawé Indians in the Brazilian Amazon. *Wild Environ Med* 26:271–273.

Bot ANM, et al. 2001. Waste management in leaf-cutting ants. *Ethol Ecol Evol* 13:225–237.

Boughman JW, GS Wilkinson. 1998. Greater spear-nosed bats discriminate group mates by vocalizations. *Anim Behav* 55:1717–1732.

Bourjade M, et al. 2009. Decision-making in Przewalski horses (*Equus ferus przewalskii*) is driven by the ecological contexts of collective movements. *Ethology* 115:321–330.

Bousquet CA, DJ Sumpter, MB Manser. 2011. Moving calls: A vocal mechanism underlying quorum decisions in cohesive groups. *Proc Biol Sci* 278:1482–1488.

Bowles S. 2006. Group competition, reproductive leveling, and the evolution of human altruism. *Science* 314:1569–1572.

———. 2012. Warriors, levelers, and the role of conflict in human social evolution. *Science* 336:876–879.

Bowles S, H Gintis. 2011. *A Cooperative Species: Human Reciprocity and its Evolution*. Princeton, NJ: Princeton University Press.

Boyd RL. 2002. Ethnic competition for an occupational niche: The case of Black and Italian barbers in northern US cities during the late nineteenth century. *Sociol Focus* 35:247–265.

Boyd R, PJ Richerson. 2005. *The Origin and Evolution of Cultures*. Oxford: Oxford University Press.

Bramble DM, DE Lieberman. 2004. Endurance running and the evolution of *Homo*. *Nature* 432:345–352.

Brandt M, et al. 2009. The scent of supercolonies: The discovery, synthesis and behavioural verification of ant colony recognition cues. *BMC Biology* 7:71–79.

Branstetter MG, et al. 2017. Dry habitats were crucibles of domestication in the evolution of agriculture in ants. *Proc Roy Soc B* 284:20170095.

Braude S. 2000. Dispersal and new colony formation in wild naked mole-rats: Evidence against inbreeding as the system of mating. *Behav Ecol* 11:7–12.

Braude S, E Lacey. 1992. The underground society. *The Sciences* 32:23–28.

Breed MD. 2014. Kin and nestmate recognition: The influence of WD Hamilton on 50 years of research. *Anim Behav* 92:271–279.

Breed MD, C Cook, MO Krasnec. 2012. Cleptobiosis in social insects. *Psyche* 2012:1–7.

Breidlid A, et al., eds. 1996. *American Culture: An Anthology*. 2nd ed. New York: Routledge.

Bressan P, M Grassi. 2004. Parental resemblance in 1-year-olds and the Gaussian curve. *Evol Hum Behav* 25:133–141.

Brewer MB. 1991. The social self: On being the same and different at the same time. *Pers Soc Psychol B* 5:475–482.

———. 1999. The psychology of prejudice: Ingroup love or outgroup hate? *J Soc Issues* 55:429–444.

———. 2000. Superordinate goals versus superordinate identity as bases of intergroup cooperation. In R Brown, D Capozza, eds. *Social Identity Processes*. London: Sage. pp. 117–132.

———. 2007. The importance of being we: Human nature and intergroup relations. *Am Psychol* 62:728–738.

———. 2009. Social identity and citizenship in a pluralistic society. In E Borgida, J Sullivan, E Riedel, eds. *The Political Psychology of Democratic Citizenship*. Oxford: Oxford University Press. pp. 153–175.

Brewer MB, LR Caporael. 2006. An evolutionary perspective on social identity: Revisiting groups. In M Schaller et al., eds. *Evolution and Social Psychology*. New York: Psychology Press. pp. 143–161.

Brindley EF. 2010. Representations and uses of Yue identity along the southern frontier of the Han, ca. 200–111 BCE. *Early China* 33:2010–2011.

———. 2015. *Ancient China and the Yue: Perceptions and Identities on the Southern Frontier, c. 400 BCE–50 CE*. Cambridge: Cambridge University Press.

Brink JW. 2008. *Imagining Heads-Smashed-In: Aboriginal Buffalo Hunting on the Northern Plains*. Edmonton: Athabasca University Press.

Brookfield HC, P Brown. 1963. *Struggle for Land: Agriculture and Group Territories among the Chimbu of the New Guinea Highlands*. Melbourne: Oxford University Press.

Brooks AS, et al. 2018. Long-distance stone transport and pigment use in the earliest Middle Stone Age. *Science* 360: 90–94.

Brooks JF. 2002. *Captives and Cousins: Slavery, Kinship, and Community in the Southwest Borderlands.* Chapel Hill: University of North Carolina Press.

Broome R. 2010. *Aboriginal Australians: A History since 1788.* Sydney: Allen & Unwin.

Brown ED, SM Farabaugh. 1997. What birds with complex social relationships can tell us about vocal learning. In CT Snowdon, M Hausberger, eds. *Social Influences on Vocal Development.* Cambridge: Cambridge University Press. pp. 98–127.

Bruneteau J-P. 1996. *Tukka: Real Australian Food.* Sydney: HarperCollins Australia.

Bshary R, W Wickler, H Fricke. 2002. Fish cognition: A primate's eye view. *Anim Cogn* 5:1–13.

Buchan JC, et al. 2003. True paternal care in a multi-male primate society. *Nature* 425:179–181.

Builth H. 2014. *Ancient Aboriginal Aquaculture Rediscovered.* Saarbrucken: Omniscriptum.

Burch ES Jr. 2005. *Alliance and Conflict: The World System of the Iñupiaq Eskimos.* Lincoln: University of Nebraska Press.

Burgener N, et al. 2008. Do spotted hyena scent marks code for clan membership? In JL Hurst, RJ Beynon, SC Roberts, TD Wyatt, eds. *Chemical Signals in Vertebrates 11.* New York: Springer. pp. 169–177.

Burns RA. 2018. The utility of between-nation subjective wellbeing comparisons amongst nations within the European Social Survey. *J Happiness Stud* 18:1–23.

Buttelmann D, J Call, M Tomasello. 2009. Do great apes use emotional expressions to infer desires? *Devel Sci* 12:688–698.

Butz DA. 2009. National symbols as agents of psychological and social change. *Polit Psychol* 30:779–804.

Buys CJ, KL Larson. 1979. Human sympathy groups. *Psychol Reports* 45:547–553.

Caldwell J. 1964. Interaction spheres in prehistory. In J Caldwell, R Hall, eds. *Hopewellian Studies, Scientific Paper 12.* Springfield: Illinois State Museum. pp. 134–143.

Callahan SP, A Ledgerwood. 2013. The symbolic importance of group property: Implications for intergroup conflict and terrorism. In TK Walters et al., eds. *Radicalization, Terrorism, and Conflict.* Newcastle: Cambridge Scholars. pp. 232–267.

———. 2016. On the psychological function of flags and logos: Group identity symbols increase perceived entitativity. *J Pers Soc Psychol* 110:528–550.

Cameron CM. 2008. Captives in prehistory as agents of social change. In CM Cameron, ed. *Invisible Citizens: Captives and Their Consequences.* Salt Lake City: University of Utah Press. pp. 1–24.

———. 2016. *Captives: How Stolen People Changed the World.* Lincoln: University of Nebraska Press.

Cameron EZ, TH Setsaas, WL Linklater. 2009. Social bonds between unrelated females increase reproductive success in feral horses. *Proc Nat Acad Sci* 106:13850–13853.

Campbell DT. 1958. Common fate, similarity, and other indices of the status of aggregates of persons as social entities. *Syst Res Behav Sci* 3:14–25.

Campbell MW, FBM de Waal. 2011. Ingroup-outgroup bias in contagious yawning by chimpanzees supports link to empathy. *PloS ONE* 6:e18283.

Cane S. 2013. *First Footprints: The Epic Story of the First Australians.* Sydney: Allen & Unwin.

Cantor M, et al. 2015. Multilevel animal societies can emerge from cultural transmission. *Nat Comm* 6:8091.

Cantor M, H Whitehead. 2015. How does social behavior differ among sperm whale clans? *Mar Mammal Sci* 31:1275–1290.

Caporael LR, RM Baron. 1997. Groups as the mind's natural environment. In J Simpson, D Kenrick, eds. *Evolutionary Social Psychology.* Mahwah: Lawrence Erlbaum. pp. 317–343.

Carlin NF, B Hölldobler. 1983. Nestmate and kin recognition in interspecific mixed colonies of ants. *Science* 222:1027–1029.

Carneiro RL. 1970. A theory of the origin of the state. *Science* 169:733–738.

———. 1987. Village-splitting as a function of population size. In L Donald, ed. *Themes in Ethnology and Culture History.* Meerut: Archana. pp. 94–124.

———. 1998. What happened at the flashpoint? Conjectures on chiefdom formation at the very moment of conception. In EM Redmond, ed. *Chiefdoms and Chieftaincy in the Americas.* Gainesville: University Press of Florida. pp. 18–42.

———. 2000. *The Muse of History and the Science of Culture.* New York: Springer.

———. 2004. The political unification of the world: When, and how—some speculations. *Cross-Cult Res* 38:162-77.

———. 2012. The circumscription theory: A clarification, amplification, and reformulation. *Soc Evol Hist* 11:5–30.

Caro T. 1994. *Cheetahs of the Serengeti Plains.* Chicago: University of Chicago Press.

Carter DB, HE Goemans. 2011. The making of the territorial order: New borders and the emergence of interstate conflict. *Int Organ* 65:275–309.

Cashdan E. 1998. Adaptiveness of food learning and food aversions in children. *Soc Sci Inform* 37:613–632.

———. 2001. Ethnocentrism and xenophobia: A cross-cultural study. *Curr Anthropol* 42:760–765.

Cashdan E, et al. 1983. Territoriality among human foragers: Ecological models and an application to four Bushman groups. *Curr Anthropol* 24:47–66.

Caspar EA, et al. 2016. Coercion changes the sense of agency in the human brain. *Curr Biol* 26:585–592.

Cassill D. 2003. Rules of supply and demand regulate recruitment to food in an ant society. *Behav Ecol Sociobiol* 54:441–450.

Castano E, et al. 2002. Who may enter? The impact of in-group identification on in-group-outgroup categorization. *J Exp Soc Psychol* 38:315–322.

Castano E, M Dechesne. 2005. On defeating death: Group reification and social identification as immortality strategies. *Eur Rev Soc Psychol* 16:221–255.

Castano E, R Giner-Sorolla. 2006. Not quite human: Infrahumanization in response to collective responsibility for intergroup killing. *J Pers Soc Psychol* 90:804–818.

Cavafy CP. 1976. *The Complete Poems of Cavafy: Expanded Edition.* New York: Harcourt Brace.

Chacon RJ, DH Dye. 2007. *The Taking and Displaying of Human Body Parts As Trophies by Amerindians.* New York: Springer.

Chacon Y, et al. 2015. From chiefdom to state: The contribution of social structural dynamics. *Soc Evol Hist* 14:27–45.

Chagnon NA. 1977. *Yanomamo: The Fierce People.* New York: Holt, Rinehart & Winston.

———. 1979. Mate competition, favoring close kin, and village fissioning among the Yanomamo Indians. In NA Chagnon, W Irons, eds. *Evolutionary Biology and Human Social Behavior.* North Scituate: Duxbury Press. pp. 86–132.

———. 1981. Terminological kinship, genealogical relatedness, and village fissioning among the Yanomamo Indians. In RD Alexander, DW Tinkle, eds. *Natural Selection and Social Behavior.* New York: Chiron Press. pp. 490–508.

———. 2013. *Yanomamo.* 6th ed. Belmont, CA: Wadsworth.

Chaix R, et al. 2004. The genetic or mythical ancestry of descent groups: Lessons from the Y chromosome. *Am J Hum Genet* 75:1113–1116.

Chambers JK. 2008. *Sociolinguistic Theory: Linguistic Variation and its Social Significance.* 3rd ed. Chichester: Wiley-Blackwell.

Chambers JR. 2008. Explaining false uniqueness: Why we are both better and worse than others. *Soc Pers Psychol Compass* 2:878–894.

Chaminade T, et al. 2012. How do we think machines think? An fMRI study of alleged competition with an artificial intelligence. *Front Hum Neurosci* 6:103.

Chance MRA, RR Larsen, eds. 1976. *The Social Structure of Attention*. New York: John Wiley.

Chapais B. 2008. *Primeval Kinship: How Pair-Bonding Gave Birth to Human Society*. Cambridge, MA: Harvard University Press.

Chapais B, et al. 1997. Relatedness threshold for nepotism in Japanese macaques. *Anim Behav* 53:1089–1101.

Chapman CA, FJ White, RW Wrangham. 1994. Party size in chimpanzees and bonobos. In RW Wrangham, WC McGrew, F de Waal, eds. *Chimpanzee Cultures*. Cambridge, MA: Harvard University Press. pp. 41–58.

Chapman EN, A Kaatz, M Carnes. 2013. Physicians and implicit bias: How doctors may unwittingly perpetuate health care disparities. *J Gen Intern Med* 28:1504–1510.

Chapman J. 1868. *Travels in the Interior of South Africa*, vol. 2. London: Bell & Daldy.

Chase-Dunn C, et al. 2010. Cycles of rise and fall, upsweeps and collapses. In LE Grinin et al., eds. *History and Mathematics: Processes and Models of Global Dynamics*. Volgograd: Uchitel. pp. 64–91.

Cheney DL, RM Seyfarth. 2007. *Baboon Metaphysics: The Evolution of a Social Mind*. Chicago: University of Chicago Press.

Cheryan S, B Monin. 2005. Where are you really from?: Asian Americans and identity denial. *J Pers Soc Psychol* 89:717–730.

Cheung BY, M Chudek, SJ Heine. 2011. Evidence for a sensitive period for acculturation. *Psychol Sci* 22:147–152.

Chilvers BL, PJ Corkeron. 2001. Trawling and bottlenose dolphins' social structure. *P Roy Soc Lond B* 268:1901–1905.

Chollet A. 2011. Switzerland as a "fractured nation." *Nations & Nationalism* 17:738–755.

Christ O, et al. 2014. Contextual effect of positive intergroup contact on outgroup prejudice. *Proc Nat Acad Sci* 111:3996–4000.

Christakis NA, JH Fowler. 2014. Friendship and natural selection. *Proc Nat Acad Sci* 111:10796–10801.

Christal J, H Whitehead, E Lettevall. 1998. Sperm whale social units: Variation and change. *Can J Zool* 76:1431–1440.

Christensen C, et al. 2016. Rival group scent induces changes in dwarf mongoose immediate behavior and subsequent movement. *Behav Ecol* 27:1627–1634.

Chua A. 2007. *Day of Empire: How Hyperpowers Rise to Global Dominance—and Why They Fall*. New York: Doubleday.

Cialdini RB, NJ Goldstein. 2004. Social influence: Compliance and conformity. *Annu Rev Psychol* 55:591–621.

Cipriani L. 1966. *The Andaman Islanders*. London: Weidenfeld and Nicolson.

Citrin J, C Wong, B Duff. 2001. The meaning of American national identity. In RD Ashmore, L Jussim, D Wilder, eds. *Social Identity, Intergroup Conflict, and Conflict Reduction*, vol. 3. New York: Oxford University Press. pp. 71–100.

Claessen HJM, P Skalník, eds. 1978. *The Early State*. The Hague: Mouton.

Clark EE, M Edmonds. 1979. *Sacagawea of the Lewis and Clark Expedition*. Berkeley: University of California Press.

Clastres P. 1972. The Guayaki. In M Bicchieri, ed. *Hunters and Gatherers Today*. New York: Holt, Rinehart & Winston. pp. 138–174.

Clastres P, P Auster. 1998. Cannibals. *The Sciences* 38:32–37.

Clutton-Brock T. 2009. Cooperation between nonkin in animal societies. *Nature* 462:51–57.

Cochran G, HC Harpending. 2009. *The 10,000 Year Explosion: How Civilization Accelerated Human Evolution*. New York: Basic Books.

Cohen E. 2012. The evolution of tag-based cooperation in humans: The case for accent. *Curr Anthropol* 53:588–616.

Cohen R. 1978. State origins: A reappraisal. In HJM Claessen, P Skalnik, eds. *The Early State*. The Hague: Mouton. pp. 31–75.

Cole DP. 1975. *Nomads of the nomads: The Āl Murrah Bedouin of the Empty Quarter.* New York: Aldine.

Confino A. 2014. *A World Without Jews: The Nazi Imagination from Persecution to Genocide.* New Haven, CT: Yale University Press.

Conkey MW. 1982. Boundedness in art and society. In I Hodder, ed. *Symbolic and Structural Archaeology.* Cambridge: Cambridge University Press. pp. 115–128.

Connerton P. 2010. Some functions of collective forgetting. In RIM Dunbar, C Gamble, J Gowlett, eds. *Social Brain, Distributed Mind.* Oxford: Oxford University Press. pp. 283–308.

Connor W. 1978. A nation is a nation, is a state, is an ethnic group is a . . . *Ethnic Racial Stud* 1:377–400.

Coolen I, O Dangles, J Casas. 2005. Social learning in noncolonial insects? *Curr Biol* 15:1931–1935.

Cooley CH. 1902. *Human Nature and the Social Order.* New York: C Scribner's Sons.

Correll J, et al. 2007. Across the thin blue line: Police officers and racial bias in the decision to shoot. *J Pers Soc Psychol* 92:1006–1023.

Cosmides L, J Tooby. 2013. Evolutionary psychology: New perspectives on cognition and motivation. *Annu Rev Psychol* 64:201–229.

Cosmides L, J Tooby, R Kurzban. 2003. Perceptions of race. *Trends Cogn Sci* 7:173–179.

Costa JT. 2006. *The Other Insect Societies.* Cambridge, MA: Harvard University Press.

Costello K, G Hodson. 2012. Explaining dehumanization among children: The interspecies model of prejudice. *Brit J Soc Psychol* 53:175–197.

Cowgill GL. 1988. Onward and upward with collapse. In N Yoffee, GL Cowgill, eds. *The Collapse of Ancient States and Civilizations.* Tucson: University of Arizona Press. pp. 244–276.

Creel S, NM Creel. 2002. *The African Wild Dog.* Princeton, NJ: Princeton University Press.

Crevècoeur JH. 1782. *Letters from an American Farmer.* Philadelphia: Mathew Carey.

Crocker J, et al. 1994. Collective self-esteem and psychological well-being among White, Black, and Asian college students. *Pers Soc Psychol Bull* 20:503–513.

Crockford C, et al. 2004. Wild chimpanzees produce group specific calls: A case for vocal learning? *Ethology* 110:221–243.

Cronin AL, et al. 2013. Recurrent evolution of dependent colony foundation across eusocial insects. *Annu Rev Entomol* 58:37–55.

Cronk L, D Gerkey. 2007. Kinship and descent. In RIM Dunbar, L Barrett, eds. *The Oxford Handbook of Evolutionary Psychology.* Oxford: Oxford University Press. pp. 463–478.

Curr EM. 1886. *The Australian Race: Its Origin, Languages, Customs, Place of Landing in Australia,* vol. 1. Melbourne: J Farnes. pp. 83–84.

Currie CR, AE Stuart. 2001. Weeding and grooming of pathogens in agriculture by ants. *Proc Royal Soc* 268:1033–1039.

Currie TE, et al. 2010. Rise and fall of political complexity in island South-East Asia and the Pacific. *Nature* 467:801–804.

Curry A. 2008. Seeking the roots of ritual. *Science* 319:278–280.

Curti ME. 1946. *The Roots of American Loyalty.* New York: Columbia University Press.

Czechowski W, EJ Godzińska. 2015. Enslaved ants: Not as helpless as they were thought to be. *Insectes Soc* 62:9–22.

Danielli JF, A Muggleton. 1959. Some alternative states of amoeba, with special reference to life-span. *Gerontol* 3:76–90.

Darwin C. 1859. *On the Origin of Species by Means of Natural Selection, or the Preservation of Favoured Races in the Struggle for Life.* London: John Murray.

———. 1871. *The Descent of Man.* London: John Murray.

———. 1872. *The Expression of the Emotions in Man and Animals.* London: John Murray.

Davis G. 2009. *Vikings in America.* Edinburgh: Berlinn Ltd.

Dawkins R. 1982. *The Extended Phenotype.* San Francisco: WH Freeman.

Dawson J. 1881. *Australian Aborigines: The Languages and Customs of Several Tribes of Aborigines in the Western District of Victoria.* Melbourne: George Robertson.

DeCasien AR, et al. 2017. Primate brain size is predicted by diet but not sociality. *Nature Ecol Evol* 1:112.

De Dreu CKW, et al. 2011. Oxytocin promotes human ethnocentrism. *Proc Nat Acad Sci* 108:1262–1266.

De Figueiredo RJ, Z Elkins. 2003. Are patriots bigots? An inquiry into the vices of in-group pride. *Am J Polit Sci* 47:171–188.

DeFries R. 2014. *The Big Ratchet: How Humanity Thrives in the Face of Natural Crisis.* New York: Basic Books.

de la Vega G. 1966. *Royal Commentaries of the Incas and General History of Peru,* Part I. HV Livermore, trans. Austin: University of Texas Press.

Denham TP, J Iriarte, L Vrydaghs, eds. 2007. *Rethinking Agriculture: Archaeological and Ethnoarchaeological Perspectives.* Walnut Creek, CA: Left Coast Press.

Denham WW. 2013. Beyond fictions of closure in Australian Aboriginal kinship. *Math Anthro Cult Theory* 5:1–90.

Dennis M. 1993. *Cultivating a Landscape of Peace: Iroquois-European Encounters in Seventeenth Century America.* New York: Cornell University Press.

d'Errico F, et al. 2012. Early evidence of San material culture represented by organic artifacts from Border Cave, South Africa. *Proc Nat Acad Sci* 109:13214–13219.

de Sade M. 1990. Philosophy in the bedroom. In R Seaver, ed., A Wainhouse, trans. *Justine, Philosophy in the Bedroom, and Other Writings* New York: Grove Press. pp. 177–367.

Deschamps J-C. 1982. Social identity and relations of power between groups. In H Tajfel, ed. *Social Identity and Intergroup Relations.* Cambridge: Cambridge University Press. pp. 85–98.

Deschamps J-C, R Brown. 1983. Superordinate goals and intergroup conflict. *Brit J Soc Psychol* 22:189–195.

De Silva S, G Wittemyer. 2012. A comparison of social organization in Asian elephants and African savannah elephants. *Int J Primatol* 33:1125–1141.

d'Ettorre P, J Heinze. 2005. Individual recognition in ant queens. *Curr Biol* 15:2170–2174.

Deutscher G. 2010. *The Unfolding of Language.* New York: Henry Holt & Co.

Devine PG. 1989. Stereotypes and prejudice: Their automatic and controlled components. *J Pers Soc Psychol* 56:5–18.

Devos T, MR Banaji. 2005. American = white? *J Pers Soc Psychol* 88:447–466.

Devos T, DS Ma. 2008. Is Kate Winslet more American than Lucy Liu? The impact of construal processes on the implicit ascription of a national identity. *Brit J Soc Psychol* 47:191–215.

de Waal F. 1982. *Chimpanzee Politics: Power and Sex Among Apes.* New York: Harper & Row.

———. 2001. *The Ape and the Sushi Master: Cultural Reflections by a Primatologist.* New York: Basic Books.

———. 2006. *Primates and Philosophers: How Morality Evolved.* Princeton, NJ: Princeton University Press.

———. 2014. *The Bonobo and The Atheist: In Search of Humanism Among the Primates.* New York: W.W. Norton.

de Waal FBM, JJ Pokorny. 2008. Faces and behinds: Chimpanzee sex perception. *Adv Sci Lett* 1:99–103.

de Waal FBM, PL Tyack. 2003. Preface. In FBM de Waal, PL Tyack, eds. *Animal Social Complexity: Intelligence, Culture, and Individualized Societies.* Cambridge, MA: Harvard University Press. pp. ix–xiv.

Diamond J. 2005. *Collapse: How Societies Choose to Fail or Succeed.* New York: Penguin.

Dill M, DJ Williams, U Maschwitz. 2002. Herdsmen ants and their mealy-bug partners. *Abh Senckenbert Naturforsch Ges* 557:1–373.

Dinnerstein L, DM Reimers. 2009. *Ethnic Americans: A History of Immigration.* New York: Columbia University Press.

Dio C. 2008. *Dio's Rome,* vol. 3. E Cary, trans. New York: MacMillan.

Dittus WPJ. 1988. Group fission among wild toque macaques as a consequence of female resource competition and environmental stress. *Anim Behav* 36:1626–1645.

Dixon RMW. 1972. *The Dyirbal Language of North Queensland.* Cambridge: Cambridge University Press.

———. 1976. Tribes, languages and other boundaries in northeast Queensland. In N Peterson, ed. *Tribes & Boundaries in Australia.* Atlantic Highlands: Humanities Press. pp. 207–238.

———. 1997. *The Rise and Fall of Languages.* Cambridge: Cambridge University Press.

———. 2010. *The Languages of Australia.* New York: Cambridge University Press.

Dollard J. 1937. *Caste and Class in a Southern Town.* New Haven, CT: Yale University Press.

Donald L. 1997. *Aboriginal Slavery on the Northwest Coast of North America.* Berkeley: University of California Press.

Douglas M. 1966. *Purity and Danger: An Analysis of Concepts of Pollution and Taboo.* London: Routledge.

Dove M. 2011. *The Banana Tree at the Gate: A History of Marginal Peoples and Global Markets in Borneo.* New Haven, CT: Yale University Press.

Draper P. 1976. Social and economic constraints on child life among the !Kung. In RB Lee, I DeVore, eds. *Kalahari Hunter-Gatherers: Studies of the !Kung San and their Neighbors.* Cambridge: Cambridge University Press.

Druckman D. 2001. Nationalism and war: A social-psychological perspective. In DJ Christie et al., eds. *Peace, Conflict, and Violence.* Englewood Cliffs, NJ: Prentice-Hall.

Dukore BF. 1996. *Not Bloody Likely! And Other Quotations from Bernard Shaw.* New York: Columbia University Press.

Dunbar RIM. 1993. Coevolution of neocortical size, group size and language in humans. *Behav Brain Sci* 16:681–735.

———. 1996. *Grooming, Gossip, and the Evolution of Language.* Cambridge, MA: Harvard University Press.

———. 2011. Kinship in biological perspective. In NJ Allen et al., eds. *Early Human Kinship: From Sex to Social Reproduction.* Chichester, W Sussex: Blackwell. pp. 131–150.

Dunbar RIM, C Gamble, J Gowlett. 2014. *Thinking Big: How the Evolution of Social Life Shaped the Human Mind.* London: Thames Hudson.

Dunham Y. 2018. Mere membership. *Trends Cogn Sci,* in press.

Dunham Y, AS Baron, MR Banaji. 2008. The development of implicit intergroup cognition. *Trends Cogn Sci* 12:248–253.

Dunham Y, AS Baron, S Carey. 2011. Consequences of "minimal" group affiliations in children. *Child Dev* 82:793–811.

Dunham Y, EE Chen, MR Banaji. 2013. Two signatures of implicit intergroup attitudes: Developmental invariance and early enculturation. *Psychol Sci* 24:860–868.

Durkheim E. 1982 (1895). *The Rules of Sociological Method and Selected Texts in Sociology and its Methods.* New York: Free Press.

———. 1984 (1893). *The Division of Labor in Society.* New York: Free Press.

Dyson-Hudson R, EA Smith. 1978. Human territoriality: An ecological reassessment. *Am Anthropol* 80:21–41.

Eagleman D. 2011. *Incognito: The Secret Lives of the Brain.* New York: Random House.

Earle TK, JE Ericson. 2014. *Exchange Systems in Prehistory.* New York: Academic Press.

East ML, H Hofer. 1991. Loud calling in a female dominated mammalian society, II: Behavioural contexts and functions of whooping of spotted hyenas. *Anim Behav* 42:651–669.

Easterly W. 2001. Can institutions resolve ethnic conflict? *Econ Dev Cult Change* 49:687–706.

Eberhardt JL, et al. 2004. Seeing black: Race, crime, and visual processing. *J Pers Soc Psychol* 87:876–893.

Echebarria-Echabe A, E Fernandez-Guede. 2006. Effect of terrorism on attitudes and ideological orientation. *Eur J Soc Psychol* 36:259–269.

Edwards J. 2009. *Language and Identity*. Cambridge: Cambridge University Press.

Ehardt CL, IS Bernstein. 1986. Matrilineal overthrows in rhesus monkey groups. *Int J Primatol* 7:157–181.

Eibl-Eibesfeldt I. 1998. Us and the others: The familial roots of ethnonationalism. In I Eibl-Eibesfeldt, FK Salter, eds. *Indoctrinability, Ideology, and Warfare*. New York: Berghahn. pp. 21–54.

Ekman P. 1972. Universals and cultural differences in facial expressions of emotion. In J Cole, ed. *Nebraska Symposium on Motivation*. Lincoln: University of Nebraska Press. pp. 207–282.

———. 1992. An argument for basic emotions. *Cognition Emotion* 6:169–200.

Elgar MA, RA Allan. 2006. Chemical mimicry of the ant *Oecophylla smaragdina* by the myrmecophilous spider *Cosmophasis bitaeniata*: Is it colony-specific? *J Ethol* 24:239–246.

Elkin AP. 1977. *Aboriginal Men of High Degree: Initiation and Sorcery in the World's Oldest Tradition*. St Lucia: University of Queensland Press.

Ellemers N. 2012. The group self. *Science* 336:848–852.

Ellis JJ. 1997. *American Sphinx: The Character of Thomas Jefferson*. New York: Knopf.

Endicott K. 1988. Property, power and conflict among the Batek of Malaysia. In T Ingold, D Riches, J Woodburn, eds. *Hunters and Gatherers 2: Property, Power and Ideology* New York: Berg. pp. 110–127.

Engerman SL. 2007. *Slavery, Emancipation, and Freedom*. Baton Rouge: Louisiana State University Press.

Ensminger J, J Henrich, eds. 2014. *Experimenting with Social Norms: Fairness and Punishment in Cross-cultural Perspective*. New York: Russell Sage Foundation.

Erikson EH. 1985. Pseudospeciation in the nuclear age. *Polit Psychol* 6:213–217.

Eriksen TH. 1993. *Ethnicity and Nationalism: Anthropological Perspectives*. London: Pluto.

Erwin TL, CJ Geraci. 2009. Amazonian rainforests and their richness of Coleoptera. In RG Foottit, PH Adler, eds. *Insect Biodiversity: Science and Society*. Hoboken, NJ: Blackwell. pp. 49–67.

Escott PD. 2010. *The Confederacy: The Slaveholders' Failed Venture*. Santa Barbara, CA: ABC-CLIO.

Esses VM, LM Jackson, TL Armstrong. 2001. The immigration dilemma: The role of perceived group competition, ethnic prejudice, and national identity. *J Soc Issues* 57:389–412.

Estes R. 2014. *The Gnu's World*. Berkeley: University of California Press.

Ethridge R, C Hudson, eds. 2008. *The Transformation of the Southeastern Indians, 1540–1760*. Jackson: University Press of Mississippi.

European Values Study Group and World Values Survey Association 2005. *European and world values surveys integrated data file, 1999–2002*, Release I. 2nd ICPSR version. Ann Arbor, MI: Inter-University Consortium for Political and Social Research.

Evans R. 2007. *A History of Queensland*. Cambridge: Cambridge University Press.

Everett DL, et al. 2005. Cultural constraints on grammar and cognition in Pirahã: Another look at the design features of human language. *Curr Anthropol* 46:621–646.

Fabrega H. 1997. Earliest phases in the evolution of sickness and healing. *Med Anthropol Quart* 11:26–55.

Fair SW. 2001. The Inupiaq Eskimo messenger feast. *J Am Folklore* 113:464–494.

Faulkner J, et al. 2004. Evolved disease-avoidance mechanisms and contemporary xenophobic attitudes. *Group Proc Intergr Rel* 7:333–353.

Faulseit RK, ed. 2016. *Beyond Collapse: Archaeological Perspectives on Resilience, Revitalization, and Transformation in Complex Societies.* Carbondale: Southern Illinois University Press.

Feblot-Augustins J, C Perlès. 1992. Perspectives ethnoarchéologiques sur les échanges à longue distance. In A Gallay et al., eds. *Ethnoarchéologie: Justification, problémes, limites.* Juan-les-Pins: Èditions APDCA. pp. 195–209.

Fedurek P, et al. 2013. Pant hoot chorusing and social bonds in male chimpanzees. *Anim Behavi* 86:189–196.

Feekes F. 1982. Song mimesis within colonies of *Cacicus c. cela.* A colonial password? *Ethology* 58:119–152.

Feinman GM, J Marcus, eds. 1998. *Archaic States.* Santa Fe, NM: SAR Press.

Feinstein Y. 2016. Rallying around the president. *Soc Sci Hist* 40:305–338.

Feldblum JT, et al. 2018. The timing and causes of a unique chimpanzee community fission preceding Gombe's Four Year's War. *J Phys Anthropol* 166:730–744.

Ferguson MJ, RR Hassin. 2007. On the automatic association between America and aggression for news watchers. *Pers Soc Psychol B* 33:1632–1647.

Ferguson RB. 1984. A reexamination of the causes of Northwest Coast warfare. In RB Ferguson, ed. *Warfare, Culture, and Environment.* New York: Academic Press. pp. 267–328.

———. 2011. Born to live: Challenging killer myths. In RW Sussman, CR Cloninger, eds. *Origins of Altruism and Cooperation.* New York: Springer. pp. 249–270.

Feshbach S. 1991. Attachment processes in adult political ideology: Patriotism and Nationalism. In JL Gewirtz, WM Kurtines, eds. *Intersections with Attachment.* Hillsdale, NJ: Erlbaum. pp. 207–226.

———. 1994. Nationalism, patriotism, and aggression: A clarification of functional differences. In LR Huesmann, ed. *Aggressive Behavior.* New York: Plenum Press. pp. 275–291.

Field TM, et al. 1982. Discrimination and imitation of facial expression by neonates. *Science* 218:179–181.

Finkel DN, P Swartwout, R Sosis. 2010. The socio-religious brain. In RIM Dunbar et al., eds. *Social Brain, Distributed Mind.* Oxford: Oxford University Press. pp. 283–308.

Finlayson C. 2009. *The Humans Who Went Extinct: Why Neanderthals Died Out and We Survived.* Oxford: Oxford University Press.

Fischer DH. 1989. *Albion's Seed: Four British Folkways in America.* Oxford: Oxford University Press.

Fishlock V, C Caldwell, PC Lee. 2016. Elephant resource-use traditions. *Anim Cogn* 19:429–433.

Fishlock V, PC Lee. 2013. Forest elephants: Fission–fusion and social arenas. *Anim Behav* 85:357–363.

Fiske AP. 2004. Four modes of constituting relationships. In N Haslam, ed. *Relational Models Theory.* New York: Routledge. pp. 61–146.

Fiske ST. 2010. *Social Beings: Core Motives in Social Psychology.* 2nd ed. New York: John Wiley.

Fiske ST, AJC Cuddy, P Glick. 2007. Universal dimensions of social cognition: Warmth and competence. *Trends Cogn Sci* 11:77–83.

Fiske ST, SL Neuberg. 1990. A continuum of impression formation, from category-based to individuating processes. *Adv Exp Soc Psychol* 23:1–74.

Fiske ST, SE Taylor. 2013. *Social Cognition: From Brains to Culture.* Thousand Oaks, CA: Sage.

Fiskesjö M. 1999. On the "raw" and the "cooked" barbarians of imperial China. *Inner Asia* 1:139–168.

Fison L, AW Howitt. 1880. *Kamilaroi and Kurnai.* Melbourne: George Robertson.

Fitch WT. 2000. The evolution of speech: A comparative review. *Trends Cogn Sci* 4:258–267.

Flannery K, J Marcus. 2012. *The Creation of Inequality.* Cambridge, MA: Harvard University Press.

Flege JE. 1984. The detection of French accent by American listeners. *J Acoust Soc Am* 76:692–707.

Fletcher R. 1995. *The Limits of Settlement Growth.* Cambridge: Cambridge University Press.

Flood J. 1980. *The Moth Hunters: Aboriginal Prehistory of the Australian Alps.* Canberra: AIAS.

Fogel RW, SL Engerman. 1974. *Time on the Cross: The Economics of American Negro Slavery.* vol. 1. New York: Little, Brown & Co.

Foley RA, MM Lahr. 2011. The evolution of the diversity of cultures. *Phil T Roy Soc B* 366:1080–1089.

Forsyth DR. 2009. *Group Dynamics,* 5th ed. Belmont, MA: Wadsworth.

Frank MC, et al. 2008. Number as a cognitive technology: Evidence from Pirahã language and cognition. *Cognition* 108:819–824.

Franklin B. 1779. *Political, Miscellaneous, and Philosophical Pieces.* London: J Johnson.

Frankopan P. 2015. *The Silk Roads: A New History of the World.* London: Bloomsbury.

Freedman JL. 1975. *Crowding and Behavior.* Oxford: WH Freedman.

Freeland WJ. 1979. Primate social groups as biological islands. *Ecology* 60:719–728.

Freeman JB, et al. 2011. Looking the part: Social status cues shape race perception. *PloS ONE* 6:e25107.

Freud S. 1930. *Civilization and its Discontents.* London: Hogarth.

Fried MH. 1967. *The Evolution of Political Society.* New York: Random House.

Fritz CE, JH Mathewson. 1957. *Convergence Behavior in Disasters: A Problem in Social Control.* Washington: National Academy of Sciences.

Fry D, ed. 2013. *War, Peace, and Human Nature.* Oxford: Oxford University Press.

Fukuyama F. 2011. *The Origins of Political Order.* New York: Farrar, Strauss and Giroux.

Fürniss S. 2014. Diversity in Pygmy music: A family portrait. In BS Hewlett, ed. *Hunter-Gatherers of the Congo Basin.* New Brunswick, NJ: Transaction.

Furuichi T. 1987. Sexual swelling, receptivity, and grouping of wild pygmy chimpanzee females at Wamba, Zaire. *Primates* 28:309–318.

———. 2011. Female contributions to the peaceful nature of bonobo society. *Evol Anthropol: Issues, News, and Reviews* 20:131–142.

Furuichi T, J Thompson, eds. 2007. *The Bonobos: Behavior, Ecology, and Conservation.* New York: Springer.

Gaertner L, et al. 2006. Us without them: Evidence for an intragroup origin of positive in-group regard. *J Pers Soc Psychol* 90:426–439.

Gaertner SL, JF Dovidio. 2000. *Reducing Intergroup Bias: The Common Ingroup Identity Model.* Philadelphia: Psychology Press.

Gamble C. 1998. Paleolithic society and the release from proximity: A network approach to intimate relations. *World Archaeol* 29:426–449.

Gamble LH. 2012. A land of power. In TL Jones, JE Perry, eds. *Contemporary Issues in California Archaeology.* Walnut Creek, CA: Left Coast Press. pp. 175–196.

Gans HJ. 2007. Acculturation, assimilation and mobility. *Ethnic and Racial Stud* 30:152–164.

Ganter R. 2006. *Mixed Relations: Asian-Aboriginal Contact in North Australia.* Crawley: University of Western Australia Publishing.

Garnsey P. 1996. *Ideas of Slavery from Aristotle to Augustine.* Cambridge: Cambridge University Press.

Gat A. 1999. The pattern of fighting in simple, small-scale, prestate societies. *J Anthropol Res* 55:563–583.

———. 2015. Proving communal warfare among hunter-gatherers: The quasi-Rousseauan error. *Evol Anthropol: Issues News Reviews* 24:111–126.

Gat A, A Yakobson. 2013. *Nations: The Long History and Deep Roots of Political Ethnicity and Nationalism.* Cambridge: Cambridge University Press.

Gavrilets S, DG Anderson, P Turchin. 2014. Cycling in the complexity of early societies. In LE Grinin, AV Korotayev, eds. *History and Mathematics*. Volgograd: Uchitel. pp. 136–158.

Geary DC 2005. *The Origin of Mind*. Washington, DC: American Psychological Association.

Geertz C, ed. 1973. *The Interpretation of Cultures*. New York: Basic Books.

Geisler ME. 2005. What are national symbols—and what do they do to us? In *National Symbols, Fractured Identities*. Middlebury, CT: Middlebury College Press. pp. xiii–xlii.

Gelfand MJ, et al. 2011. Differences between tight and loose cultures: A 33-nation study. *Science* 332:1100–1104.

Gellner E. 1983. *Nations and Nationalism*. Oxford: Blackwell.

Gelo DJ. 2012. *Indians of the Great Plains*. New York: Taylor & Francis.

Gero S, et al. 2016a. Socially segregated, sympatric sperm whale clans in the Atlantic Ocean. *R Soc Open Sci* 3:160061.

Gero S, J Gordon, H Whitehead. 2015. Individualized social preferences and long-term social fidelity between social units of sperm whales. *Animal Behav* 102:15–23.

Gero S, H Whitehead, L Rendell. 2016b. Individual, unit and vocal clan level identity cues in sperm whale codas. *R Soc Open Sci* 3:150372.

Gesquiere LR, et al. 2011. Life at the top: Rank and stress in wild male baboons. *Science* 333:357–360.

Ghent AW. 1960. A study of the group-feeding behavior of larvae of the jack pine sawfly, *Neodiprion pratti banksianae*. *Behav* 16:110–148.

Gifford E. 2015. *The Many Speeches of Chief Seattle (Seathl)*. Charleston, SC: CreateSpace Independent Publishing Platform.

Gigerenzer G. 2010. *Rationality for Mortals: How People Cope with Uncertainty*. New York: Oxford University Press.

Gilbert D. 2007. *Stumbling on Happiness*. New York: Vintage.

Gilderhus MT. 2010. *History and Historians: A Historiographical Introduction*. New York: Pearson.

Giles H, et al. 1977. Towards a theory of language in ethnic group relations. In H Giles, ed. *Language, Ethnicity and Intergroup Relations*. London: Academic. pp. 307–348.

Gill FB. 2006. *Ornithology*. 3rd ed. New York: WH Freeman.

Gilovich T. 1991. *How We Know What Isn't So: The Fallibility of Human Reason In Everyday Life*. New York: Free Press.

Gil-White FJ. 2001. Are ethnic groups biological "species" to the human brain? *Curr Anthropol* 42:515–536.

Giner-Sorolla R. 2012. *Judging Passions: Moral Emotions in Persons and Groups*. New York: Psychology Press.

Gintis H. 2000. Strong reciprocity and human sociality. *J Theoret Biol* 206:169–179.

Glowacki L, C von Rueden. 2015. Leadership solves collective action problems in small-scale societies. *Phil T Roy Soc B* 370:20150010.

Goff PA, et al. 2008. Not yet human: Implicit knowledge, historical dehumanization, and contemporary consequences. *J Pers Soc Psychol* 94:292–306.

Goldberg A, AM Mychajliw, EA Hadly. 2016. Post-invasion demography of prehistoric humans in South America. *Nature* 532:232–235.

Goldstein AG. 1979. Race-related variation of facial features: Anthropometric data I. *Bull Psychon Soc* 13:187–190.

Gombrich EH. 2005. *A Little History of the World*. C. Mustill, trans. New Haven, CT: Yale University Press.

Gonsalkorale K, KD Williams. 2007. The KKK won't let me play: Ostracism even by a despised outgroup hurts. *Eur J Soc Psychol* 37:1176–1186.

Goodall J. 1986. *The Chimpanzees of Gombe*. Cambridge, MA: Harvard University Press.

———. 2010. *Through A Window: My Thirty Years with the Chimpanzees of Gombe.* Boston: Houghton Mifflin Harcourt.

Goodwin M. 2016. Brexit: Identity trumps economics in revolt against elites. *Financial Times,* June 24.

Gordon DM. 1989. Ants distinguish neighbors from strangers. *Oecologia* 81:198–200.

———. 1999. *Ants at Work: How An Insect Society Is Organized.* New York: Simon & Schuster.

Gordon M. 2001. *Small-Town Values and Big-City Vowels.* Durham, NC: Duke University Press.

Gordon MM. 1964. *Assimilation in American Life.* New York: Oxford University Press.

Gossett TF. 1963. *Race: The History of an Idea in America.* New York: Oxford University Press.

Gould RA. 1969. *Yiwara: Foragers of the Australian Desert.* New York: Scribner.

Grabo A, M van Vugt. 2016. Charismatic leadership and the evolution of cooperation. *Evol Hum Behav* 37:399–406.

Granovetter M. 1983. The strength of weak ties: A network theory revisited. *Soc Theory* 1:201–233.

Greene J. 2013. *Moral Tribes.* New York: Penguin Books.

Greenshields TH. 1980. "Quarters" and ethnicity. In GH Blake, RI Lawless, eds. *The Changing Middle Eastern City.* London: Croom Helm. pp. 120–140.

Greenwald AG, MR Banaji, BA Nosek. 2015. Statistically small effects of the Implicit Association Test can have societally large effects. *J Pers Soc Psychol* 108:553–561.

Gross JT. 2000. *Neighbors: The Destruction of the Jewish Community in Jedwabne, Poland.* Princeton, NJ: Princeton University Press.

Grove M. 2010. The archaeology of group size. In RIM Dunbar, C Gamble, J Gowlett, eds. *Social Brain, Distributed Mind.* Oxford: Oxford University Press. pp. 391–413.

Gudykunst WB. 2004. *Bridging Differences: Effective Intergroup Communication.* Thousand Oaks, CA: Sage.

Guenther MG. 1976. From hunters to squatters. In R Lee, I DeVore, eds. *Kalaharie Hunter-Gatherers: Studies of the !Kung San and Their Neighbors.* Cambridge, MA: Harvard University Press. pp. 120–134.

———. 1996. Diversity and flexibility: The case of the Bushmen of southern Africa. In S Kent, ed. *Cultural Diversity and Twentieth-Century Foragers: An African Perspective.* Cambridge: Cambridge University Press. pp. 65–86.

———. 1997. Lords of the desert land: Politics and resistance of the Ghanzi Basarwa of the nineteenth century. *Botsw Notes Rec* 29:121–141.

———. 2014. War and peace among Kalahari San. *J Aggress Confl Peace Res* 6:229–239.

Guibernau M. 2007. *The Identity of Nations.* Cambridge: Polity Press.

———. 2013. *Belonging: Solidarity and Division in Modern Societies.* Malden, MA: Polity.

Guttal V, ID Couzin. 2010. Social interactions, information use, and the evolution of collective migration. *Proc Nat Acad Sci* 107:16172–16177.

Haaland G. 1969. Economic determinants in ethnic processes. In F Barth, ed. *Ethnic Groups and Boundaries: The Social Organization of Culture Difference.* pp. 58–73. Boston: Little, Brown.

Haber M, et al. 2017. Continuity and admixture in the last five millennia of Levantine history from ancient Canaanite and present-day Lebanese genome sequences. *Am J Hu Genetics* 101:1–9.

Hackman J, A Danvers, DJ Hruschka. 2015. Closeness is enough for friends, but not mates or kin. *Evol Hum Behav* 36:137–145.

Haidt J. 2003. The moral emotions. In RJ Davidson, KR Scherer, HH Goldsmith, eds. *Handbook of Affective Sciences.* Oxford: Oxford University Press. pp. 852–870.

———. 2012. *The Righteous Mind: Why Good People Are Divided by Politics and Religion.* New York: Random House.

Haidt J, S Algoe. 2004. Moral amplification and the emotions that attach us to saints and demons. In J Greenberg, SL Koole, T Pyszcynski, eds. *Handbook of Experimental Existential Psychology*. New York: Guilford Press. pp. 322–335.

Haidt J, P Rozin, C McCauley, S Imada. 1997. Body, psyche, and culture: The relationship between disgust and morality. *Psychol Dev Soc J* 9:107–131.

Haig D. 2000. Genomic imprinting, sex-biased dispersal, and social behavior. *Ann NY Acad Sci* 907:149–163.

———. 2011. Genomic imprinting and the evolutionary psychology of human kinship. *Proc Nat Acad Sci* 108:10878–85.

Hais SC, MA Hogg, JM Duck. 1997. Self-categorization and leadership: Effects of group prototypicality and leader stereotypicality. *Pers Soc Psychol Bull* 23:1087–1099.

Hale HE. 2004. Explaining ethnicity. *Comp Polit Stud* 37:458–485.

Hall JM. 1997. Ethnic identity in Greek antiquity. *Cambr Archaeol J* 8:265–283.

Hally DJ. 1996. Platform-mound construction and the instability of Mississippian chiefdoms. In JF Scarry, ed. *Political Structure and Change in the Prehistoric Southeastern United States*. Gainesville: University Press of Florida. pp. 92–127.

Hames R. 1983. The settlement pattern of a Yanomamo population bloc. In R Hames, W Vickers, eds. *Adaptive Responses of Native Amazonians*. New York: Academic Press. pp. 393–427.

Hamilton J. 2003. *Trench Fighting of World War I*. Minneapolis: ABDO & Daughters.

Hamilton MJ, et al. 2007. The complex structure of hunter-gatherer social networks. *Proc Roy Soc B* 274:2195–2202.

Hamilton WD. 1971. Geometry for the selfish herd. *J Theoret Biol* 31:295–311.

Hammer MF, et al. 2000. Jewish and Middle Eastern non-Jewish populations share a common pool of Y-chromosome biallelic haplotypes. *Proc Nat Acad Sci* 97:6769–6774.

Hann JH. 1991. *Missions to the Calusa*. Gainesville: University Press of Florida.

Hannonen M, L Sundström. 2003. Sociobiology: Worker nepotism among polygynous ants. *Nature* 421:910.

Harari YN. 2015. *Sapiens: A Brief History of Humankind*. New York: HarperCollins.

Hare B, V Wobber, R Wrangham. 2012. The self-domestication hypothesis: Evolution of bonobo psychology is due to selection against aggression. *Anim Behav* 83:573–585.

Hare B, S Kwetuenda. 2010. Bonobos voluntarily share their own food with others. *Cur Biol* 20:230–231.

Harlan JR. 1967. A wild wheat harvest in Turkey. *Archaeol* 20:197–201.

Harles JC. 1993. *Politics in the Lifeboat*. San Francisco: Westview Press.

Harlow R, L Dundes. 2004. "United" we stand: Responses to the September 11 attacks in black and white. *Sociol Persp* 47:439–464.

Harner MJ. 1972. *The Jívaro: People of the Sacred Waterfalls*. Garden City, NJ: Doubleday.

Harrington FH, DL Mech. 1979. Wolf howling and its role in territory maintenance. *Behav* 68:207–249.

Harris CB, HM Paterson, RI Kemp. 2008. Collaborative recall and collective memory: What happens when we remember together? *Memory* 16:213–230.

Harris JR. 2009. *The Nurture Assumption: Why Children Turn Out the Way They Do*. 2nd ed. New York: Simon and Schuster.

Harris LT, ST Fiske. 2006. Dehumanizing the lowest of the low: Neuro-imaging responses to extreme outgroups. *Psychol Sci* 17:847–853.

Hart CM, M van Vugt. 2006. From fault line to group fission: Understanding membership changes in small groups. *Pers Soc Psychol Bull* 32:392–404.

Hartley LP. 1953. *The Go-Between*. New York: New York Review.

Haslam N. 2006. Dehumanization: An integrative review. *Pers Soc Psychol Rev* 10:252–264.

Haslam N, S Loughnan. 2014. Dehumanization and infrahumanization. *Annu Rev Psychol* 65:399–423.

Haslam N, S Loughnan, P Sun. 2011a. Beastly: What makes animal metaphors offensive? *J Lang Soc Psychol* 30:311–325.

Haslam SA, SD Reicher, MJ Platow. 2011b. *The New Psychology of Leadership*. East Sussox: Psychology Press.

Hassin RR, et al. 2007. Subliminal exposure to national flags affects political thought and behavior. *Proc Nat Acad Sci* 104:19757–19761.

Hasson U, et al. 2012. Brain-to-brain coupling: A mechanism for creating and sharing a social world. *Trends Cogn Sci* 16:114–121.

Hawkes K. 2000. Hunting and the evolution of egalitarian societies: Lessons from the Hadza. In MW Diehl, ed. *Hierarchies in Action: Cui Bono?* Carbondale: Southern Illinois University Press. pp. 59–83.

Hawley AH. 1944. Dispersion versus segregation: Apropos of a solution of race problems. *Mich Acad Sci Arts Lett* 30:667–674.

Hayden B. 1979. *Palaeolithic Reflections: Lithic Technology and Ethnographic Excavation among Australian Aborigines*. London: Humanities Press.

———. 1987. Alliances and ritual ecstasy: Human responses to resource stress. *J Sci Stud Relig* 26:81–91.

———. 1995. Pathways to power: Principles for creating socioeconomic inequalities. In T Price, GM Feinman, eds. *Foundations of Social Inequality*. New York: Springer. pp. 15–86.

———. 2011. Big man, big heart? The political role of aggrandizers in egalitarian and transegalitarian societies. In D Forsyth, C Hoyt, eds. *For the Greater Good of All*. New York: Palgrave Macmillan. pp. 101–118.

———. 2014. *The Power of Feasts: From Prehistory to the Present*. New York: Cambridge University Press.

Hayden B, et al. 1981. Research and development in the Stone Age: Technological transitions among hunter-gatherers. *Curr Anthropol* 22:519–548.

Hayden B, S Villeneuve. 2012. Who benefits from complexity? A view from Futuna. In TD Price, G Feinman, eds. *Pathways to Power*. New York: Springer. pp. 95–146.

Head L. 1989. Using palaeoecology to date Aboriginal fishtraps at Lake Condah, Victoria. *Archaeol Oceania* 24:110–115.

Headland TN, et al. 1989. Hunter-gatherers and their neighbors from prehistory to the present. *Curr Anthropol* 30:43–66.

Hedges C. 2002. *War is a Force that Gives Us Meaning*. New York: Anchor Books.

Hefetz A. 2007. The evolution of hydrocarbon pheromone parsimony in ants—interplay of colony odor uniformity and odor idiosyncrasy. *Myrmecol News* 10:59–68.

Heinz H-J. 1972. Territoriality among the Bushmen in general and the !Kō in particular. *Anthropos* 67:405–416.

———. 1975. Elements of !Kō Bushmen religious beliefs. *Anthropos* 70:17–41.

———. 1994. *Social Organization of the !Kō Bushmen*. Cologne: Rüdiger Köppe.

Helms R. 1885. Anthropological notes. *Proc Linn Soc New South Wales* 10:387–408.

Helwig CC, A Prencipe. 1999. Children's judgments of flags and flag-burning. *Child Dev* 70:132–143.

Henn BM, et al. 2011. Hunter-gatherer genomic diversity suggests a southern African origin for modern humans. *Proc Nat Acad Sci* 108:5154–5162.

Henrich J. 2004a. Cultural group selection, coevolutionary processes and large-scale cooperation. *J Econ Behav Organ* 53:3–35.

Henrich J. 2004b. Demography and cultural evolution: How adaptive cultural processes can produce maladaptive losses—the Tasmanian case. *Am Antiquity* 69:197–214.

Henrich J, R Boyd. 1998. The evolution of conformist transmission and the emergence of between-group differences. *Evol Hum Behav* 19:215–241.

Henrich J, et al., eds. 2004. *Foundations of Human Sociality: Economic Experiments and Ethnographic Evidence from Fifteen Small-Scale Societies*. Oxford: Oxford University Press.

Henrich J, et al. 2010a. Markets, religion, community size and the evolution of fairness and punishment. *Science* 327:1480–1484.

Henrich J, SJ Heine, A Norenzayan. 2010b. The weirdest people in the world. *Behav Brain Sci* 33:61–135.

Henrich N, J Henrich. 2007. *Why Humans Cooperate: A Cultural and Evolutionary Explanation.* New York: Oxford University Press.

Henshilwood CS, F d'Errico, eds. 2011. *Homo symbolicus: The Dawn of Language, Imagination and Spirituality.* Amsterdam: John Benjamins. pp. 75–96.

Henshilwood CS, et al. 2011. A 100,000-year-old ochre-processing workshop at Blombos Cave, South Africa. *Science* 334:219–222.

Henzi SP, et al. 2000. Ruths amid the alien corn: Males and the translocation of female chacma baboons. *S African J Sci* 96:61–62.

Herbert-Read JE, et al. 2016. Proto-cooperation: Group hunting sailfish improve hunting success by alternating attacks on grouping prey. *Proc Roy Soc B* 283:20161671.

Herbinger I, et al. 2009. Vocal, gestural and locomotor responses of wild chimpanzees to familiar and unfamiliar intruders: A playback study. *Anim Behav* 78:1389–1396.

Hernandez-Aguilar RA, J Moore, TR Pickering. 2007. Savanna chimpanzees use tools to harvest the underground storage organs of plants. *Proc Nat Acad Sci* 104:19210–19213.

Heth G, J Todrank, RE Johnston. 1998. Kin recognition in golden hamsters: Evidence for phenotype matching. *Anim Behav* 56:409–417.

Hewlett BS. 1991. *Intimate Fathers: The Nature and Context of Aka Pygmy Paterna Infant Care.* Ann Arbor: University of Michigan Press.

Hewlett BS, JMH van de Koppel, LL Cavalli-Sforza. 1986. Exploration and mating range of Aka Pygmies of the Central African Republic. In LL Cavalli-Sforza, ed. *African Pygmies.* New York: Academic Press. pp. 65–79.

Hewstone M, R Brown, eds. 1986. *Contact and Conflict in Intergroup Encounters.* Oxford: Blackwell.

Hewstone M, M Rubin, H Willis. 2002. Intergroup bias. *Annu Rev Psychol* 53:575–604.

Hiatt L. 2015. Aboriginal political life. In R Tonkinson, ed. *Wentworth Lectures.* Canberra: Aboriginal Studies Press. pp. 59–74.

Hill KR, AM Hurtado. 1996. *Ache Life History: The Ecology and Demography of a Foraging People.* Piscataway, NJ: Transaction.

Hill KR, et al. 2011. Co-residence patterns in hunter-gatherer societies show unique human social structure. *Science* 331:1286–1289.

Hill KR, et al. 2014. Hunter-gatherer inter-band interaction rates: Implications for cumulative culture. *PLoS ONE* 9:e102806.

Hingley R. 2005. *Globalizing Roman Culture: Unity, Diversity and Empire.* New York: Psychology Press.

Hirschfeld LA. 1989. Rethinking the acquisition of kinship terms. *Int J Behav Dev* 12:541–568.

———. 1998. *Race in the Making: Cognition, Culture, and the Child's Construction of Human Kinds.* Cambridge, MA: MIT Press.

———. 2012. Seven myths of race and the young child. *Du Bois Rev Soc Sci Res* 9:17–39.

Hiscock P. 2007. *Archaeology of Ancient Australia.* New York: Routledge.

Ho AK, et al. 2011. Evidence for hypodescent and racial hierarchy in the categorization and perception of biracial individuals. *J Pers Soc Psychol* 100:492–506.

Hofstede G, RR McCrae. 2004. Personality and culture revisited: Linking traits and dimensions of culture. *Cross-Cult Res* 38:52–88.

Hogg MA. 1993. Group cohesiveness: A critical review and some new directions. *Eur Rev Soc Psychol* 4:85–111.

———. 2001. A social identity theory of leadership. *Pers Soc Psychol Rev* 5:184–200.

———. 2006. Social identity theory. In PJ Burke, ed. *Contemporary Social Psychological Theories.* Stanford, CA: Stanford University Press. pp. 111–136.

————. 2007. Social identity and the group context of trust. In M Siegrist et al., eds. *Trust in Cooperative Risk Management*. London: Earthscan. pp. 51–72.

Hogg MA, D Abrams. 1988. *Social Identifications: A Social Psychology of Intergroup Relations and Group Processes*. London: Routledge.

Hohmann G, B Fruth. 1995. Structure and use of distance calls in wild bonobos. *Int J Primatol* 15:767–782.

————. 2011. Is blood thicker than water? In MM Robbins, C Boesch, eds. *Among African Apes*. Berkeley: University of California Press. pp. 61–76.

Hold BC. 1980. Attention-structure and behavior in G/wi San children. *Ethol Sociobiol* 1:275–290.

Hölldobler B, EO Wilson. 1990. *The Ants*. Cambridge, MA: Harvard University Press.

————. 2009. *The Superorganism: The Beauty, Elegance, and Strangeness of Insect Societies*. New York: W W Norton.

Holsti KJ. 1991. *Peace and War: Armed Conflicts and International Order, 1648–1989*. Cambridge: Cambridge University Press.

Homer-Dixon TF. 1994. Environmental scarcities and violent conflict: Evidence from cases. *Int Security* 19:5–40.

Hommon RJ. 2013. *The Ancient Hawaiian State: Origins of a Political Society*. Oxford: Oxford University Press.

Hong L, SE Page. 2004. Groups of diverse problem solvers can outperform groups of high-ability problem solvers. *Proc Nat Acad Sci* 101:16385–16389.

Hood B. 2002. *The Self Illusion: How the Social Brain Creates Identity*. New York: New York University Press.

Hoogland JL, et al. 2012. Conflicting research on the demography, ecology, and social behavior of Gunnison's prairie dogs. *J Mammal* 93:1075–1085.

Hornsey MJ, et al. 2007. Group-directed criticisms and recommendations for change: Why newcomers arouse more resistance than old-timers. *Pers Soc Psychol Bull* 33:1036–1048.

Hornsey MJ, M Hogg. 2000. Assimilation and diversity: An integrative model of subgroup relations. *Pers Soc Psychol Rev* 4:143–156.

Hosking GA, G Schöpflin, eds. 1997. *Myths and Nationhood*. New York: Routledge.

Howard KJ, et al. 2013. Frequent colony fusions provide opportunities for helpers to become reproductives in the termite *Zootermopsis nevadensis*. *Behav Ecol Sociobiol* 67:1575–1585.

Howitt A. 1904. *The Native Tribes of South-East Australia*. London: Macmillan and Co.

Hrdy SB. 2009. *Mothers and Others. The Evolutionary Origins of Mutual Understanding*. Cambridge, MA: Harvard University Press.

Huddy L, N Khatib. 2007. American patriotism, national identity, and political involvement. *Am J Polit Sci* 51:63–77.

Hudson M. 1999. *Ruins of Identity: Ethnogenesis in the Japanese Islands*. Honolulu: University of Hawaii Press.

Hugenberg K, GV Bodenhausen. 2003. Facing prejudice: Implicit prejudice and the perception of facial threat. *Psychol Sci* 14:640–643.

Hunley KL, JF Spence, DA Merriwether. 2008. The impact of group fissions on genetic structure in Native South America and implications for human evolution. *Am J Phys Anthropol* 135:195–205.

Huth JE. 2013. *The Lost Art of Finding Our Way*. Cambridge, MA: Harvard University Press.

Huxley A. 1959. *The Human Situation*. New York: Triad Panther.

Huynh Q-L, T Devos, L Smalarz. 2011. Perpetual foreigner in one's own land: Potential implications for identity and psychological adjustment. *J Soc Clin Psychol* 30:133–162.

Iacoboni M. 2008. *Mirroring People: The New Science of How We Connect with Others*. New York: Farrar, Straus and Giroux.

Iliffe J. 2007. *Africans: The History of a Continent*. Cambridge: Cambridge University Press.

Ingold T. 1999. On the social relations of the hunter-gatherer band. In RB Lee, R Daly, eds. *The Cambridge Encyclopedia of Hunters and Gatherers*. Cambridge: Cambridge University Press. pp. 399–410.

Injaian A, EA Tibbetts. 2014. Cognition across castes: Individual recognition in worker *Polistes fuscatus* wasps. *Anim Behav* 87:91–96.

Insoll T. 2007. Configuring identities in archaeology. In T Insoll, ed. *The Archaeology of Identities. A Reader*. London: Routledge. pp. 1–18.

Isaac B. 2004. *The Invention of Racism in Classical Antiquity*. Princeton, NJ: Princeton University Press.

Ito TA, GR Urland. 2003. Race and gender on the brain: Electrocortical measures of attention to the race and gender of multiply categorizable individuals. *J Pers Soc Psychol* 85:616–626.

Iverson JM, S Goldin-Meadow. 1998. Why people gesture when they speak. *Nature* 396:228.

Jablonski NG. 2006. *Skin: A Natural History*. Berkeley: University of California Press.

Jack RE, et al. 2009. Cultural confusions show that facial expressions are not universal. *Curr Biol* 19:1543–1548.

Jack RE, OGB Garrod, PG Schyns. 2014. Dynamic facial expressions of emotion transmit an evolving hierarchy of signals over time. *Curr Biol* 24:187–192.

Jackson JE. 1983. *The Fish People: Linguistic Exogamy and Tukanoan Identity in Northwest Amazonia*. Cambridge: Cambridge University Press.

Jackson LE, L Gaertner. 2010. Mechanisms of moral disengagement and their differential use by right-wing authoritarianism and social dominance orientation in support of war. *Aggressive Behav* 36:238–250.

Jacobson MF. 1999. *Whiteness of a Different Color: European Immigrants and the Alchemy of Race*. Cambridge, MA: Harvard University Press.

Jaeggi AV, JM Stevens, CP Van Schaik. 2010. Tolerant food sharing and reciprocity is precluded by despotism among bonobos but not chimpanzees. *Am J Phys Anthropol* 143:41–51.

Jaffe KE, LA Isbell. 2010. Changes in ranging and agonistic behavior of vervet monkeys after predator-induced group fusion. *Am J Primatol* 72:634–644.

Jahoda G. 1999. *Images of Savages: Ancient Roots of Modern Prejudice in Western Culture*. New York: Routledge.

Jandt JM, et al. 2014. Behavioural syndromes and social insects: Personality at multiple levels. *Biol Rev* 89:48–67.

Janis IL. 1982. *Groupthink*. 2nd ed. Boston: Houghton Mifflin.

Jenkins, M. 2011. A man well acquainted with monkey business. *Washington Post*. Style Section: July 21.

Jerardino A, CW Marean. 2010. Shellfish gathering, marine paleoecology and modern human behavior: Perspectives from cave PP13B, Pinnacle Point, South Africa. *J Hum Evol* 59:412–424.

Jetten J, et al. 2001. Rebels with a cause: Group identification as a response to perceived discrimination from the mainstream. *Pers Soc Psychol Bull* 27:1204–1213.

Jetten J, T Postmes, B McAuliffe. 2002. We're *all* individuals: Group norms of individualism and collectivism, levels of identification and identity threat. *Eur J Soc Psychol* 32:189–207.

Jewish Telegraphic Agency, August 18, 1943. Archived at http://www.jta.org/1943/08/18/ archive/german-refugees-from-hamburg-mistaken-for-jews-executed-in-nazi-death -chambers.

Johnson AW, TK Earle. 2000. *The Evolution of Human Societies: From Foraging Group to Agrarian State*. Stanford, CA: Stanford University Press.

Johnson BR, E van Wilgenburg, ND Tsutsui. 2011. Nestmate recognition in social insects: Overcoming physiological constraints with collective decision making. *Behav Ecol Sociobiol* 65:935–944.

Johnson CL. 2000. Perspectives on American kinship in the later 1990s. *J Marriage Fam* 62:623–639.

Johnson GA. 1982. Organizational structure and scalar stress. In C Renfrew et al., eds. *Theory and Explanation in Archaeology.* New York: Academic. pp. 389–421.

Johnson GR. 1986. Kin selection, socialization, and patriotism. *Polit Life Sci* 4:127–140.

———. 1987. In the name of the fatherland: An analysis of kin term usage in patriotic speech and literature. *Int Polit Sci Rev* 8:165–174.

———. 1997. The evolutionary roots of patriotism. In D. Bar-Tal, E. Staub, eds. *Patriotism in the Lives of Individuals and Nations.* Chicago: Nelson-Hall. pp. 45–90.

Jolly A. 2005. Hair signals. *Evol Anthropol: Issues, News, and Reviews* 14:5.

Jolly A, RW Sussman, N Koyama, eds. 2006. *Ringtailed Lemur Biology.* New York: Springer.

Jones A. 2012. *Crimes Against Humanity: A Beginner's Guide.* Oxford: Oneworld Publishers.

Jones CB. 2007. The Evolution of Exploitation in Humans: "Surrounded by Strangers I Thought Were My Friends." *Ethology* 113:499–510.

Jones D, et al. 2000. Group nepotism and human kinship. *Curr Anthropol* 41:779–809.

Jones EE, et al. 1984. *Social Stigma: Psychology of Marked Relationships.* New York: WH Freeman.

Jones P. 1996. *Boomerang: Behind an Australian Icon.* Kent Town, S Aust.: Wakefield Press.

Joniak-Lüthi A. 2015. *The Han: China's Diverse Majority.* Seattle: University of Washington Press.

Jost JT, MR Banaji. 1994. The role of stereotyping in system-justification and the production of false consciousness. *Brit J Soc Psychol* 33:1–27.

Jost JT, et al. 2003. Social inequality and the reduction of ideological dissonance on behalf of the system. *Eur J Soc Psychol* 33:13–36.

Jouventin P, T Aubin, T Lengagne. 1999. Finding a parent in a king penguin colony: The acoustic system of individual recognition. *Anim Behav* 57:1175–1183.

Joyce AA, LA Bustamante, MN Levine. 2001. Commoner power: A case study from the Classic period collapse on the Oaxaca coast. *J Archaeol Meth Th* 8:343–385.

Joyce J. 1922. *Ulysses.* London: John Rodker.

Judd TM, PW Sherman. 1996. Naked mole-rats recruit colony mates to food sources. *Anim Behav* 52:957–969.

Junger S. 2016. *Tribe: On Homecoming and Belonging.* New York: HarperCollins.

Kaiser RJ. 1994. *The Geography of Nationalism in Russia and the USSR.* Princeton, NJ: Princeton University Press, 1994.

Kamans E, et al. 2009. What I think you see is what you get: Influence of prejudice on assimilation to negative meta-stereotypes among Dutch Moroccan teenagers. *Eur J Soc Psychol* 39:842–851.

Kameda T, R Hastie. 2015. Herd behavior. In R Scott, S Kosslyn, eds. *Emerging Trends in the Social and Behavioral Sciences.* Hoboken, NJ: John Wiley and Sons.

Kan S. 1989. *Symbolic Immortality: The Tlingit Potlatch of the Nineteenth Century.* Washington, DC: Smithsonian Institution Press.

Kano T. 1992. *The Last Ape: Pygmy Chimpanzee Behavior and Ecology.* Palo Alto: Stanford University Press.

Kaplan D. 2000. The darker side of the "original affluent society." *J Anthropol Res* 56:301–324.

Karau SJ, KD Williams. 1993. Social loafing: A meta-analytic review and theoretical integration. *J Pers Soc Psychol* 65:681–706.

Katz PA, JA Kofkin. 1997. Race, gender, and young children. In SS Luthar et al., eds. *Developmental Psychopathology.* New York: Cambridge University Press.

Kaufman SJ. 2001 *Modern Hatreds: The Symbolic Politics of Ethnic War.* Ithaca, NY: Cornell University Press.

Kaw E. 1993. Medicalization of racial features: Asian American women and cosmetic surgery. *Med Anthropol Q* 7:74–89.

Keeley LH. 1997. *War Before Civilization: The Myth of the Peaceful Savage.* New York: Oxford University Press.

Keil FC. 1989. *Concepts, Kinds, and Cognitive Development.* Cambridge, MA: MIT Press.

———. 2012. Running on empty? How folk science gets by with less. *Curr Dir Psychol Sci* 21:329–334.

Kelley LC. 2012. The biography of the Hông Bàng clan as a medieval Vietnamese invented tradition. *J Vietnamese Stud* 7:87–130.

Kelly D. 2011. *Yuck! The Nature and Moral Significance of Disgust.* Cambridge, MA: MIT Press.

———. 2013. Moral disgust and the tribal instincts hypothesis. In K Sterelny et al., eds. *Signaling, Commitment, and Emotion.* Cambridge, MA: MIT Press. pp. 503–524.

Kelly D, et al. 2005. Three-month-olds but not newborns prefer own-race faces. *Dev Sci* 8:F31–36.

Kelly DJ, et al. 2009. Development of the other-race effect during infancy: Evidence toward universality? *J Exp Child Psychol* 104:105–114.

Kelly RL. 2013a. *The Lifeways of Hunter-gatherers: The Foraging Spectrum.* Cambridge: Cambridge University Press.

———. 2013b. From the peaceful to the warlike: Ethnographic and archaeological insights into hunter-gatherer warfare and homicide. In DP Fry, ed. *War, Peace, and Human Nature.* Oxford: Oxford University Press. pp. 151–167.

Kemmelmeier M, DG Winter. 2008. Sowing patriotism, but reaping nationalism? Consequences of exposure to the American flag. *Polit Psychol* 29:859–879.

Kendal R, et al. 2015. Chimpanzees copy dominant and knowledgeable individuals: implications for cultural diversity. *Evol Hum Behav* 36:65–72.

Kendon A. 1988. *Sign Languages of Aboriginal Australia.* Cambridge: Cambridge University Press.

Kennedy P. 1987. *The Rise and Fall of the Great Powers: Economic Change and Military Conflict from 1500 to 2000.* New York: Random House.

Kennett DJ, B Winterhalder. 2006. *Behavioral Ecology and the Transition to Agriculture.* Berkeley: University of California Press.

King EB, JL Knight, MR Hebl. 2010. The influence of economic conditions on aspects of stigmatization. *J Soc Issues* 66:446–460.

King SL, VM Janik. 2013. Bottlenose dolphins can use learned vocal labels to address each other. *Proc Nat Acad Sci* 110:13216–13221.

Kintisch E. 2016. The lost Norse. *Science* 354:696–701.

Kinzler KD, et al. 2007. The native language of social cognition. *Proc Nat Acad Sci* 104:12577–12580.

Kirby S. 2000. Syntax without natural selection. In C Knight et al., eds. *The Evolutionary Emergence of Language.* Cambridge: Cambridge University Press. pp. 303–323.

Kleingeld P. 2012 *Kant and Cosmopolitanism.* Cambridge: Cambridge University Press.

Klinkner PA, RM Smith. 1999. *The Unsteady March: The Rise and Decline of Racial Equality in America.* Chicago: University of Chicago Press.

Knight J. 1994. "The Mountain People" as tribal mirror. *Anthropol Today* 10:1–3.

Knight N. 2008. *Imagining Globalisation in China.* Northampton, MA: Edward Elgar.

Koonz C. 2003. *The Nazi Conscience.* Cambridge, MA: Harvard University Press.

Kopenawa D, B Albert. 2013. *The Falling Sky: Words of a Yanomami Shaman.* Cambridge, MA: Harvard University Press.

Kopytoff I. 1982. Slavery. *Annu Rev Anthropol* 11:207–230.

Koval P, et al. 2012. Our flaws are more human than yours: Ingroup bias in humanizing negative characteristics. *Pers Soc Psychol Bull* 38:283–295.

Kowalewski SA. 2006. Coalescent societies. In TJ Pluckhahn et al., eds. *Light the Path: The Anthropology and History of the Southeastern Indians.* Tuscaloosa: University of Alabama Press. pp. 94–122.

Krakauer J. 1996. *Into the Wild.* New York: Anchor Books.

Kramer KL, RD Greaves. 2016. Diversify or replace: What happens to wild foods when cultigens are introduced into hunter-gatherer diets? In BF Codding, KL Kramer, eds. *Why Forage?: Hunters and Gatherers in the Twenty-First Century.* Santa Fe, NM: SAR/University of New Mexico Press. pp. 15–42.

Krause J, GD Ruxton. 2002. *Living in Groups.* Oxford: Oxford University Press.

Kronauer DJC, C Schöning, P d'Ettorre, JJ Boomsma. 2010. Colony fusion and worker reproduction after queen loss in army ants. *Proc Roy Soc Lond B* 277:755–763.

Kruuk H. 1972. *The Spotted Hyena.* Chicago: University of Chicago Press.

———. 1989. *The Social Badger.* Oxford: Oxford University Press.

Kuhn SL, MC Stiner. 2007. Paleolithic ornaments: Implications for cognition, demography and identity. *Diogenes* 54:40–48.

Kumar R, A Sinha, S Radhakrishna. 2013. Comparative demography of two commensal macaques in India. *Folia Primatol* 84:384–393.

Kupchan CA. 2010. *How Enemies Become Friends: The Sources of Stable Peace.* Princeton, NJ: Princeton University Press.

Kurzban R, MR Leary. 2001. Evolutionary origins of stigmatization: The functions of social exclusion. *Psychol Bull* 127:187–208.

Kurzban R, J Tooby, L Cosmides. 2001. Can race be erased? Coalitional computation and social categorization. *Proc Nat Acad Sci* 98:15387–15392.

Kymlicka W. 1995. *Multicultural Citizenship.* Oxford: Clarendon Press.

Labov W. 1989. The child as linguistic historian. *Lang Var Change* 1:85–97.

Lai WS, et al. 2005. Recognition of familiar individuals in golden hamsters. *J Neurosci* 25:11239–11247.

Laidre ME. 2012. Homes for hermits: Temporal, spatial and structural dynamics as transportable homes are incorporated into a population. *J Zool* 288:99–40.

Laland KN, BG Galef, eds. 2009. *The Question of Animal Culture.* Cambridge, MA: Harvard University Press.

Laland KN, C Wilkins, N Clayton. 2016. The evolution of dance. *Curr Biol* 26:R5–R9.

La Macchia ST, et al. 2016. In small we trust: Lay theories about small and large groups. *Pers Soc Psychol Bull* 42:1321–1334.

Lamont M, V Molnar. 2002. The study of boundaries in the social sciences. *Annu Rev Sociol* 28:167–195.

Langbauer WR, et al. 1991. African elephants respond to distant playbacks of low-frequency conspecific calls. *J Exp Biol* 157:35–46.

Langergraber KE, JC Mitani, L Vigilant. 2007. The limited impact of kinship on cooperation in wild chimpanzees. *Proc Nat Acad Sci* 104:7786–7790.

———. 2009. Kinship and social bonds in female chimpanzees. *Am J Primatol* 71:840–851.

Langergraber KE, et al. 2014. How old are chimpanzee communities? Time to the most recent common ancestor of the Y-chromosome in highly patrilocal societies. *J Hum Evol* 69:1–7.

Larson PM. 1996. Desperately seeking "the Merina" (Central Madagascar): Reading ethnonyms and their semantic fields in African identity histories. *J South Afr Stud* 22:541–560.

Layton R, S O'Hara. 2010. Human social evolution: A comparison of hunter-gatherer and chimpanzee social organization. In RIM Dunbar, C Gamble, J Gowlett, eds. *Social Brain, Distributed Mind.* Oxford: Oxford University Press. pp. 83–114.

Leacock E, R Lee, eds. 1982. *Politics and History in Band Societies.* New York: Cambridge University Press.

LeBlanc SA. 2014. Forager warfare and our evolutionary past. In M Allen, T Jones, eds. *Violence and Warfare Among Hunter-Gatherers.* Walnut Creek, CA: Left Coast Press. pp. 26–46.

LeBlanc SA, KE Register. 2004. *Constant Battles: Why We Fight.* New York: Macmillan.

Lee PC, CJ Moss. 1999. The social context for learning and behavioural development among wild African elephants. In HO Box, ed. *Mammalian Social Learning: Comparative and Ecological Perspectives.* Cambridge: Cambridge University Press. pp. 102–125.

Lee RB. 1979. *The !Kung San: Men, Women, and Work in a Foraging Society.* Cambridge: Cambridge University Press.

———. 2013. *The Dobe Ju/'hoansi.* 4th ed. Belmont, CA: Wadsworth.

Lee RB, R Daly. 1999. Foragers and others. In RB Lee, R Daly, eds. *The Cambridge Encyclopedia of Hunters and Gatherers.* Cambridge: Cambridge University Press. pp. 1–19.

Lee RB, I DeVore, eds. 1968. *Man the Hunter.* Chicago: Aldine.

———, eds. 1976. *Kalaharie Hunter-Gatherers: Studies of the !Kung San and Their Neighbors* Cambridge, MA: Harvard University Press.

Lee TL, ST Fiske. 2006. Not an outgroup, not yet an ingroup: Immigrants in the stereotype content model. *Int J Intercult Rel* 30:751–768.

Leechman D. 1956. *Native Tribes of Canada.* Toronto: WJ Gage.

Lehman N, et al. 1992. A study of the genetic relationships within and among wolf packs using DNA fingerprinting and mitochondrial DNA. *Behav Ecol Sociobiol* 30:83–94.

Leibold J. 2006. Competing narratives of racial unity in Republican China: From the Yellow Emperor to Peking Man. *Mod China* 32:181–220.

Lerner MJ, DT Miller. 1978. Just world research and the attribution process: Looking back and ahead. *Psychol Bull* 85:1030–1051.

le Roux A, TJ Bergman. 2012. Indirect rival assessment in a social primate, *Theropithecus gelada. Anim Behav* 83:249–255.

Lester PJ, MAM Gruber. 2016. Booms, busts and population collapses in invasive ants. *Biological Invasions* 18:3091–3101.

Leuchtenburg WE. 2015. *The American President: From Teddy Roosevelt to Bill Clinton.* Oxford: Oxford University Press.

Levin DT, MR Banaji. 2006. Distortions in the perceived lightness of faces: The role of race categories. *J Exp Psychol* 135:501–512.

LeVine RA, DT Campbell. 1972. *Ethnocentrism: Theories of Conflict, Ethnic Attitudes, and Group Behavior.* New York: John Wiley and Sons.

Levinson S. 1988. *Constitutional Faith.* Princeton, NJ: Princeton University Press.

Lévi-Strauss C. 1952. *Race and History.* Paris: Unesco.

———. 1972. *The Savage Mind.* London: Weidenfeld and Nicolson.

Levitt P, MC Waters, eds. 2002. *The Changing Face of Home: The Transnational Lives of the Second Generation.* New York: Russell Sage Foundation.

Lewis D. 1976. Observations on route finding and spatial orientation among the Aboriginal peoples of the Western Desert Region of Central Australia. *Oceania* 46:249–282.

Lewis GJ, C Kandler, R Riemann. 2014. Distinct heritable influences underpin in-group love and out-group derogation. *Soc Psychol Pers Sci* 5:407–413.

Lewis ME. 2006. *The Flood Myths of Early China.* Albany: State University of New York Press.

Leyens J-P, et al. 2003. Emotional prejudice, essentialism, and nationalism. *Eur J Soc Psychol* 33:703–717.

Leyens J-P, et al. 2007. Infra-humanization: The wall of group differences. *Soc Issues Policy Rev* 1:139–172.

Li Q, MB Brewer. 2004. What does it mean to be an American? Patriotism, nationalism, and American identity after 9/11. *Polit Psychol* 25:727–739.

Liang D, J Silverman. 2000. You are what you eat: Diet modifies cuticular hydrocarbons and nestmate recognition in the Argentine ant. *Naturwissenschaften* 87:412–416.

Liberman Z, et al. 2016. Early emerging system for reasoning about the social nature of food. *Proc Nat Acad Sci* 113:9480–9485.

Librado F. 1981. *The Eye of the Flute: Chumash Traditional History and Ritual.* Santa Barbara, CA: Santa Barbara Museum of Natural History.

Lickel B, et al. 2000. Varieties of groups and the perception of group entitativity. *J Pers Social Psychol* 78:223–246.

Lieberman D, et al. 2007. The architecture of human kin detection. *Nature* 445:727–731.

Light I, SJ Gold. 2000. *Ethnic Economies.* New York: Academic Press.

Lim M, et al. 2007. Global pattern formation and ethnic/cultural violence. *Science* 317:1540–1544.

Lind M. 2006. *The American Way of Strategy*. New York: Oxford University Press.

Linder W. 2010. *Swiss Democracy*. 3rd ed. New York: Palgrave MacMillan.

Linklater WL, et al. 1999. Stallion harassment and the mating system of horses. *Anim Behav* 58:295–306.

Lippmann W. 1922. *Public Opinion*. New York: Harcourt Brace.

Liu C, et al. 2014. Increasing breadth of the frontal lobe but decreasing height of the human brain between two Chinese samples from a Neolithic site and from living humans. *Am J Phys Anthropol* 154:94–103.

Liverani M. 2006. *Uruk: The First City*. Sheffield: Equinox Publishing.

Lomax A, N Berkowitz. 1972. The evolutionary taxonomy of culture. *Science* 177:228–239.

Lonsdorf E, S Ross, T Matsuzawa, eds. 2010. *The Mind of the Chimpanzee*. Chicago: Chicago University Press.

Lorenzi-Cioldi F. 2006. Group status and individual differentiation. In T Postmes, J Jetten, eds, *Individuality and the group: Advances in Social Identity*. London: SAGE. pp. 93–115.

Losin EAR, et al. 2012. Race modulates neural activity during imitation. *Neuroimage* 59:3594–3603.

Lott DF. 2002. *American Bison: A Natural History*. Berkeley: University of California Press.

Lourandos H. 1977. Aboriginal spatial organization and population: South Western Victoria reconsidered. *Archaeol Oceania* 12:202–225.

———. 1997. *Continent of Hunter-Gatherers: New Perspectives in Australian Prehistory*. Cambridge: Cambridge University Press.

Lovejoy AP. 1936. *The Great Chain of Being*. Cambridge, MA: Harvard University Press.

Lowen GE. 1919. *History of the 71st Regiment, N.G., N.Y.* New York: Veterans Association.

Lyons-Padilla S, MJ Gelfand. 2015. Belonging nowhere: Marginalization and radicalization among Muslim immigrants. *Behav Sci Policy* 1:1–12.

Ma X, et al. 2014. Oxytocin increases liking for a country's people and national flag but not for other cultural symbols or consumer products. *Front Behav Neurosci* 8:266.

Macdonald DW, S Creel, M Mills. 2004. Canid society. In DW Macdonald, C Sillero-Zubiri, eds. *Biology and Conservation of Wild Canids*. Oxford: Oxford University Press. pp. 85–106.

Machalek R. 1992. The evolution of macrosociety: Why are large societies rare? *Adv Hum Ecol* 1:33–64.

Mackie DM, ER Smith, DG Ray. 2008. Intergroup emotions and intergroup relations. *Soc Pers Psychol Compass* 2:1866–1880.

MacLeod WC. 1937. Police and punishment among Native Americans of the Plains. *J Crim Law Crim* 28:181–201.

MacLin OH, RS Malpass. 2001. Racial categorization of faces: The ambiguous race face effect. *Psychol Public Pol Law* 7:98–118.

MacLin OH, MK MacLin. 2011. The role of racial markers in race perception and racial categorization. In R Adams et al., eds. *The Science of Social Vision*. New York: Oxford University Press. pp. 321–346.

Macrae CN, GV Bodenhausen. 2000. Social cognition: Thinking categorically about others. *Annu Rev Psychol* 51:93–120.

Madon S, et al. 2001. Ethnic and national stereotypes: The Princeton trilogy revisited and revised. *Pers Soc Psychol B* 27:996–1010.

Maghaddam FM. 1998. *Social Psychology: Exploring the Universals Across Cultures*. New York: WH Freeman.

Maguire EA, et al. 2003. Routes to remembering: The brains behind superior memory. *Nature Neurosci* 6:90–95.

Magurran AE, A Higham. 1988. Information transfer across fish shoals under predator threat. *Ethol* 78:153–158.

Mahajan N, et al. 2011. The evolution of intergroup bias: Perceptions and attitudes in rhesus macaques. *J Pers Soc Psychol* 100:387–405.

———. 2014. Retraction. *J Pers Soc Psychol* 106:182.

Major B, T Schmader. 2001. Legitimacy and the construal of social disadvantage. In JT Jost, B Major, eds. *The Psychology of Legitimacy.* Cambridge: Cambridge University Press. pp. 176–204.

Malaspinas A-S, et al. 2016. A genomic history of Aboriginal Australia. *Nature* 538:207–213.

Malik I, PK Seth, CH Southwick 1985. Group fission in free-ranging rhesus monkeys of Tughlaqabad, northern India. *Int J Primatol* 6:411–22.

Malpass MA. 2009. *Daily Life in the Incan Empire.* 2nd ed. Westport, CT: Greenwood.

Mann M. 1986. *The Sources of Social Power: A History of Power from the Beginning to 1760 AD,* vol. 1. Cambridge: Cambridge University Press.

Mantini D, et al. 2012. Interspecies activity correlations reveal functional correspondence between monkey and human brain areas. *Nature Methods* 9:277–282.

Marais E. 1939. *My Friends the Baboons.* New York: Robert M McBride.

Marcus J. 1989. From centralized systems to city-states: Possible models for the Epiclassic. In RA Diehl, JC Berlo, eds. *Mesoamerica after the Decline of Teotihuacan A.D. 700–900.* Washington DC: Dumbarton Oaks. pp. 201–208.

Marean CW. 2010. When the sea saved humanity. *Sci Am* 303:54–61.

———. 2016. The transition to foraging for dense and predictable resources and its impact on the evolution of modern humans. *Philos T Roy Soc B* 371:160–169.

Markin GP. 1970. The seasonal life cycle of the Argentine ant in southern California. *Ann Entomol Soc Am* 63:1238–1242.

Marks J, E Staski. 1988. Individuals and the evolution of biological and cultural systems. *Hum Evol* 3:147–161.

Marlowe FW. 2000. Paternal investment and the human mating system. *Behav Proc* 51: 45–61.

———. 2005. Hunter-gatherers and human evolution. *Evol Anthropol* 14:54–67.

———. 2010. *The Hadza: Hunter-Gatherers of Tanzania.* Berkeley: University of California Press.

Marques JM, VY Yzerbyt, J-P Lyons. 1988. The "black sheep effect": Extremity of judgments towards ingroup members as a function of group identification. *Eur J Soc Psychol* 18:1–16.

Marsh AA, HA Elfenbein, N Ambady. 2003. Nonverbal "accents": Cultural differences in facial expressions of emotion. *Psychol Sci* 14:373–376.

———. 2007. Separated by a common language: Nonverbal accents and cultural stereotypes about Americans and Australians. *J Cross Cult Psychol* 38:284–301.

Marshall AJ, RW Wrangham, AC Arcadi. 1999. Does learning affect the structure of vocalizations in chimpanzees? *Anim Behav* 58:825–830.

Marshall L. 1961. Sharing, talking and giving: Relief of social tensions among !Kung Bushmen. *Africa* 31:231–249.

———. 1976. *The !Kung of Nyae Nyae.* Cambridge, MA: Harvard University Press.

Marshall TH. 1950. *Citizenship and Social Class.* Cambridge: Cambridge University Press.

Martin CL, Parker S. 1995. Folk theories about sex and race differences. *Pers Soc Psychol B* 21:45–57.

Martínez R, R Rodríguez-Bailón, M Moya. 2012. Are they animals or machines? Measuring dehumanization. *Span J Psychol* 15:1110–1122.

Marwick B. 2003. Pleistocene exchange networks as evidence for the evolution of language. *Cambr Archaeol J* 13:67–81.

Marzluff JM, RP Balda. 1992. *The Pinyon Jay.* London: T & AD Poyser.

Massen JJM, SE Koski. 2014. Chimps of a feather sit together: Chimpanzee friendships are based on homophily in personality. *Evol Hum Behav* 35:1–8.

Masters RD, DG Sullivan. 1989. Nonverbal displays and political leadership in France and the United States. *Polit Behav* 11:123–156.

Matthey de l'Etang A, P Bancel, M Ruhlen. 2011. Back to Proto-Sapiens. In D Jones, B Milicic, eds. *Kinship, Language & Prehistory*, Salt Lake City: University of Utah Press. pp. 29–37.

Mattingly DJ. 2014. Identities in the Roman World. In L Brody, GL Hoffman, eds. *Roman in the Provinces: Art in the Periphery of Empire.* Chestnut Hill, MA: McMullen Museum of Art Press. pp. 35–59.

May RJ. 2001. *State and Society in Papua New Guinea.* Hindmarsh, SA: Crawford House.

McAnany PA, N Yoffee, eds. 2010. *Questioning Collapse: Human Resilience, Ecological Vulnerability, and the Aftermath of Empire.* Cambridge: Cambridge University Press.

McBrearty S, AS Brooks. 2000. The revolution that wasn't: A new interpretation of the origin of modern human behavior. *J Hum Evol* 39:453–563.

McComb K, C Packer, A Pusey. 1994. Roaring and numerical assessment in contests between groups of female lions. *Anim Behav* 47:379–387.

McConvell P. 2001. Language shift and language spread among hunter-gatherers. In C Panter-Brick, P Rowley-Conwy, R Layton, eds. *Hunter-Gatherers: Cultural and Biological Perspectives.* Cambridge: Cambridge University Press. pp. 143–169.

McCormick J. 2017. *Understanding the European Union.* London: Palgrave.

McCracken GF, JW Bradbury. 1981. Social organization and kinship in the polygynous bat *Phyllostomus hastatus. Behav Ecol Sociobiol* 8.11–34.

McCreery EK. 2000. Spatial relationships as an indicator of successful pack formation in free-ranging African wild dogs. *Behav* 137:579–590.

McCurry S. 2010. *Confederate Reckoning: Power and Politics in the Civil War South.* Cambridge, MA: Harvard University Press.

McDougall W. 1920. *The Group Mind.* New York: G.P. Putnam's Sons.

McElreath R, R Boyd, PJ Richerson. 2003. Shared norms and the evolution of ethnic markers. *Curr Anthropol* 44:122–130.

McGrew WC, et al. 2001. Intergroup differences in a social custom of wild chimpanzees: The grooming hand-clasp of the Mahale Mountains 1. *Curr Anthropol* 42:148–153.

McIntyre RT, DW Smith. 2000. The death of a queen: Yellowstone mutiny ends tyrannical rule of Druid pack. *International Wolf* 10:8–11, 26.

McKie R. 2010. Chimps with everything: Jane Goodall's 50 years in the jungle. *The Observer*, 31 July.

McLemore SD. 1970. Simmel's 'stranger': A critique of the concept. *Pacific Sociol Rev* 13:86–94.

McNeill WH. 1976. *Plagues and Peoples.* Garden City, NY: Anchor.

———. 1986. *Polyethnicity and National Unity in World History.* Toronto: University of Toronto Press.

———. 1995. *Keeping Together in Time: Dance and Drill in Human History.* Cambridge, MA: Harvard University Press.

McNiven I, et al. 2015. Phased redevelopment of an ancient Gunditjmara fish trap over the past 800 years. *Aust Archaeol* 81:44–58.

Mech LD, L Boitani, eds. 2003. *Wolves: Behavior, Ecology, and Conservation.* Chicago: University of Chicago Press.

Meggitt MJ. 1962. *The Desert People: A Study of the Walbiri Aborigines of Central Australia.* Sydney: Angus, Robertson.

———. 1977. *Blood Is Their Argument: Warfare Among the Mae Enga Tribesmen of the New Guinea Highlands.* Houston: Mayfield Publishing Co.

Mellars P, JC French. 2011. Tenfold population increase in Western Europe at the Neandertal-to-modern human transition. *Science* 333:623–627.

Menand L. 2006. What it is like to like. *New Yorker.* June 20, 73–76.

Mercader J, et al. 2007. 4,300-year-old chimpanzee sites and the origins of percussive stone technology. *Proc Nat Acad Sci* 104:3043–3048.

Michener W. 2012. The individual psychology of group hate. *J Hate Stud* 10:15–48.

Milgram S. 1974. *Obedience to Authority.* New York: HarperCollins.

Milicic B. 2013. Talk is not cheap: Kinship terminologies and the origins of language. *Structure and Dynamics* 6: http://escholarship.org/uc/item/6zw317jh.

Miller D. 1995. *On Nationality.* Oxford: Oxford University Press.

Miller DT, C McFarland. 1987. Pluralistic ignorance: When similarity is interpreted as dissimilarity. *J Pers Soc Psychol* 53:298–305.

Miller R, RH Denniston. 1979. Interband dominance in feral horses. *Zeitschrift für Tierpsychologie* 51:41–47.

Mills DS, SM McDonnell. 2005. *The Domestic Horse.* Cambridge: Cambridge University Press.

Milton K. 1991. Comparative aspects of diet in Amazonian forest-dwellers. *Philos T Roy Soc B:* 334:253–263.

Mitani JC, SJ Amsler. 2003. Social and spatial aspects of male subgrouping in a community of wild chimpanzees. *Behav* 140:869–884.

Mitani JC, J Gros-Louis. 1998. Chorusing and call convergence in chimpanzees: Tests of three hypotheses. *Behav* 135:1041–1064.

Mitani JC, DP Watts, SJ Amsler. 2010. Lethal intergroup aggression leads to territorial expansion in wild chimpanzees. *Curr Biol* 20:R507–R508.

Mitchell D. 1984. Predatory warfare, social status, and the North Pacific slave trade. *Ethnology* 23:39–48.

Mitchell TL. 1839. *Three Expeditions into the Interior of Eastern Australia.* London: T.W. Boone.

Modlmeier AP, JE Liebmann, S Foitzik. 2012. Diverse societies are more productive: a lesson from ants. *Proc Roy Soc B* 279: 2142–2150.

Moffett MW. 1989a. Trap-jaw ants. *Natl Geogr* 175:394–400.

———. 1989b. Life in a nutshell. *Natl Geogr* 6:783–796.

———. 1994. *The High Frontier: Exploring the Tropical Rainforest Canopy.* Cambridge, MA: Harvard University Press.

———. 1995. Leafcutters: Gardeners of the ant world. *Natl Geogr* 188:98–111.

———. 2000. What's "up"? A critical look at the basic terms of canopy biology. *Biotropica* 32:569–596.

———. 2010. *Adventures Among Ants.* Berkeley: University of California Press.

———. 2011. Ants and the art of war. *Sci Am* 305:84–89.

———. 2012. Supercolonies of billions in an invasive ant: What is a society? *Behav Ecol* 23:925–933.

———. 2013. Human identity and the evolution of societies. *Hum Nature* 24:219–267.

Moïse RE. 2014. Do Pygmies have a history? revisited: The autochthonous tradition in the history of Equatorial Africa. In BS Hewlett, ed. *Hunter-Gatherers of the Congo Basin.* New Brunswick NJ: Transaction Publishers. pp. 85–116.

Monteith MJ, CI Voils. 2001. Exerting control over prejudiced responses. In GB Moskowitz, ed. *Cognitive Social Psychology.* Mahwah, NJ: Lawrence Erlbaum. pp. 375–388.

Morgan C, RL Bettinger. 2012. Great Basin foraging strategies. In TR Pauketat, ed. *The Oxford Handbook of North American Archaeology.* New York: Oxford University Press.

Morgan J, W Buckley. 1852. *The Life and Adventures of William Buckley.* Hobart, Tasmania: A MacDougall.

Moss CJ, et al., eds. 2011. *The Amboseli Elephants.* Chicago: University of Chicago Press.

Mueller UG. 2002. Ant versus fungus versus mutualism. *Am Nat* 160:S67–S98.

Mullen B, RM Calogero, TI Leader. 2007. A social psychological study of ethnonyms: Cognitive representation of the in-group and intergroup hostility. *J Pers Soc Psychol* 92:612–630.

Mulvaney DJ. 1976. The chain of connection: The material evidence. In N Peterson, ed. *Tribes and Boundaries in Australia.* Atlantic Highlands: Humanities Press. pp. 72–94.

Mulvaney DJ, JP White. 1987. *Australians to 1788.* Broadway, NSW: Fairfax, Syme & Weldon.

Mummendey A, M Wenzel. 1999. Social discrimination and tolerance in intergroup relations: Reactions to intergroup difference. *Pers Soc Psychol Rev* 3:158–174.

Mummert A, et al. 2011. Stature and robusticity during the agricultural transition: Evidence from the bioarchaeological record. *Econ Hum Biol* 9:284–301.

Murphy MC, JA Richeson, DC Molden. 2011. Leveraging motivational mindsets to foster positive interracial interactions. *Soc Pers Psychol Compass* 5:118–131.

Murphy PL. 1991. *Anadarko Agency Genealogy Record Book of the Kiowa, Comanche-Apache & some 25 Sioux Families, 1902.* Lawton, OK: Privately published.

Murphy RF, Y Murphy. 1960. Shoshone-Bannock subsistence and society. *Anthropol Records* 16:293–338.

Nakamichi M, N Koyama. 1997. Social relationships among ring-tailed lemurs in two free-ranging troops at Berenty Reserve, Madagascar. *Int J Primatol* 18:73–93.

Nazzi T, PW Jusczyk, EK Johnson. 2000. Language discrimination by English-learning 5-month-olds: Effects of rhythm and familiarity. *J Mem Lang* 43:1–19.

Nelson E. 1899. The Eskimo about Bering Strait. Washington, DC: Government Printing Office.

Nettle D. 1999. Language variation and the evolution of societies. In RIM Dunbar, C Knight, C Power, eds. *The Evolution of Culture.* Piscataway: Rutgers University Press. pp. 214–227.

Newell RR, et al. 1990. *An Inquiry into the Ethnic Resolution of Mesolithic Regional Groups: The Study of their Decorative Ornaments in Time and Space.* Leiden, Netherlands: Brill.

Nishida T. 1968. The social group of wild chimpanzees in the Mahali mountains. *Primates* 9:167–224.

Nosek BA, MR Banaji, JT Jost. 2009. The politics of intergroup attitudes. In JT Jost, AC Kay, H Thorisdottir, eds. *Social and Psychological Bases of Ideology and System Justification.* New York: Oxford University Press. pp. 480–506.

Nowak MA. 2006. Five rules for the evolution of cooperation. *Science* 314:1560–1563.

Nowicki S. 1983. Flock-specific recognition of chickadee calls. *Behav Ecol Sociobiol* 12:317–320.

Noy D. 2000. *Foreigners at Rome: Citizens and Strangers.* London: Duckworth.

O'Brien GV. 2003. Indigestible food, conquering hordes, and waste materials: Metaphors of immigrants and the early immigration restriction debate in the United States. *Metaphor Symb* 18:33–47.

O'Connell RL. 1995. *The Ride of the Second Horseman: The Birth and Death of War.* Oxford: Oxford University Press.

O'Gorman HJ. 1975. Pluralistic ignorance and white estimates of white support for racial segregation. *Public Opin Quart* 39:313–330.

Oldmeadow J, ST Fiske. 2007. System-justifying ideologies moderate status = competence stereotypes: Roles for belief in a just world and social dominance orientation. *Eur J Soc Psychol* 37:1135–1148.

Olsen CL. 1987. The demography of colony fission from 1878–1970 among the Hutterites of North America. *Am Anthropol* 89:823–837.

Olzak S. 1992. *The Dynamics of Ethnic Competition and Conflict.* Stanford, CA: Stanford University Press.

Opotow S. 1990. Moral exclusion and injustice: An introduction. *J Soc Issues* 46:1–20.

Orgad L. 2011. Creating new Americans: The essence of Americanism under the citizenship test. *Houston Law Rev* 47:1–46.

———. 2015. *The Cultural Defense of Nations.* Oxford: Oxford University Press.

O'Riain MJ, JUM Jarvis, CG Faulkes. 1996. A dispersive morph in the naked mole-rat. *Nature* 380:619–621.

Ortman SG, et al. 2014. The pre-history of urban scaling. *PloS ONE* 9:e87902.

Orton J, et al. 2013. An early date for cattle from Namaqualand, South Africa: Implications for the origins of herding in southern Africa. *Antiquity* 87:108–120.

Orwell G. 1946. *Animal Farm: A Fairy Story.* London: Harcourt Brace.

———. 1971. Notes on nationalism. In S Orwell, I Angus, eds. *Collected Essays,* vol. 3. New York: Harcourt, Brace, Jovanovich. pp. 361–380.

Oswald FL, et al. 2015. Using the IAT to predict ethnic and racial discrimination: Small effect sizes of unknown societal significance. *J Pers Soc Psychol* 108:562–571.

Otterbein KF. 2004. *How War Began.* College Station: Texas A&M University Press.

Pabst MA, et al. 2009. The tattoos of the Tyrolean Iceman: A light microscopical, ultra-structural and element analytical study. *J Archaeol Sci* 36:2335–2341.

Packer DJ. 2008. On being both with us and against us: A normative conflict model of dissent in social groups. *Pers Soc Psychol Rev* 12:50–72.

Pagel M. 2000. The history, rate and pattern of world linguistic evolution. In C Knight et al., eds. *Evolutionary Emergence of Language.* Cambridge: Cambridge University Press. pp. 391–416.

———. 2009. Human language as culturally transmitted replicator. *Nature Rev Genet* 10:405–415.

Pagel M, R Mace. 2004. The cultural wealth of nations. *Nature* 428:275–278.

Painter NI. 2010. *The History of White People.* New York: W.W. Norton.

Palagi E, G Cordoni. 2009. Postconflict third-party affiliation in *Canis lupus*: Do wolves share similarities with the great apes? *Anim Behav* 78:979–986.

Paluck EL. 2009. Reducing intergroup prejudice and conflict using the media: A field experiment in Rwanda. *J Pers Soc Psychol* 96:574–587.

Paranjpe AC. 1998. *Self and Identity in Modern Psychology and Indian Thought.* New York: Plenum Press.

Park RE. 1928. Human migration and the marginal man. *Am J Sociol* 33:881–893.

Parker BJ. 2003. Archaeological manifestations of empire: Assyria's imprint on southeastern Anatolia. *Am J Archaeol* 107:525–557.

Parr LA. 2001. Cognitive and physiological markers of emotional awareness in chimpanzees. *Anim Cogn* 4:223–229.

———. 2011. The evolution of face processing in primates. *Philos T Roy Soc B* 366:1764–1777.

Parr LA, FBM de Waal. 1999. Visual kin recognition in chimpanzees. *Nature* 399:647–648.

Parr LA, WD Hopkins. 2000. Brain temperature asymmetries and emotional perception in chimpanzees. *Physiol Behav* 71:363–371.

Parr LA, BM Waller. 2006. Understanding chimpanzee facial expression: Insights into the evolution of communication. *Soc Cogn Affect Neurosci* 1:221–228.

Pascalis O, et al. 2005. Plasticity of face processing in infancy. *Proc Nat Acad Sci* 102:5297–5300.

Pascalis O, DJ Kelly. 2009. The origins of face processing in humans: Phylogeny and ontogeny. *Persp Psychol Sci* 4:200–209.

Passarge S. 1907. *Die Buschmänner der Kalahari.* Berlin: D Reimer (E Vohsen).

Patterson O. 1982. *Slavery and Social Death.* Cambridge, MA: Harvard University Press.

Paukner A, SJ Suomi, E Visalberghi, PF Ferrari. 2009. Capuchin monkeys display affiliation toward humans who imitate them. *Science* 325:880-883.

Paxton P, A Mughan. 2006. What's to fear from immigrants? Creating an assimilationist threat scale. *Polit Psychol* 27:549–568.

Payne BK. 2001. Prejudice and perception: The role of automatic and controlled processes in misperceiving a weapon. *J Pers Soc Psychol* 81:181–192.

Paz-y-Miño G, et al. 2004. Pinyon jays use transitive inference to predict social dominance. *Nature* 430:778–781.

Peasley WJ. 2010. *The Last of the Nomads.* Fremantle: Fremantle Art Centre Press.

Pelto PJ. 1968. The difference between "tight" and "loose" societies. *Transaction* 5:37–40.

Perdue T. 1979. *Slavery and the Evolution of Cherokee Society, 1540–1866.* Knoxville: University of Tennessee Press.

Perry G. 2018. *The Lost Boys: Inside Muzafer Sherif's Robbers Cave Experiments.* Brunswick, Australia: Scribe Publications.

Peterson N. 1993. Demand sharing: Reciprocity and the pressure for generosity among foragers. *Am Anthropol* 95:860–874.

Peterson RO, et al. 2002. Leadership behavior in relation to dominance and reproductive status in gray wolves. *Canadian J Zool* 80:1405–1412.

Pettigrew TF. 2009. Secondary transfer effect of contact: Do intergroup contact effects spread to noncontacted outgroups? *Soc Psychol* 40:55–65.

Phelan JE, LA Rudman. 2010. Reactions to ethnic deviance. The role of backlash in racial stereotype maintenance. *J Pers Soc Psychol* 99:265–281.

Phelps EA, et al. 2000. Performance on indirect measures of race evaluation predicts amygdala activation. *J Cogn Neurosci* 12:729-738.

Piaget J, AM Weil. 1951. The development in children of the idea of the homeland and of relations to other countries. *Int Soc Sci J* 3:561–578.

Pietraszewski D, L Cosmides, J Tooby. 2014. The content of our cooperation, not the color of our skin: An alliance detection system regulates categorization by coalition and race, but not sex. *PloS ONE* 9:e88534.

Pimlott DH, JA Shannon, GB Kolenosky. 1969. *The Ecology of the Timber Wolf in Algonquin Provincial Park.* Ontario: Department of Lands and Forests.

Pinker S. 2011. *Better Angels of Our Nature: Why Violence Has Declined.* New York: Penguin.

Plous S. 2003. Is there such a thing as prejudice toward animals. In S Plous, ed. *Understanding Prejudice and Discrimination.* New York: McGraw Hill. pp. 509–528.

Poggi I. 2002. Symbolic gestures: The case of the Italian gestionary. *Gesture* 2:71–98.

Pokorny JJ, FBM de Waal. 2009. Monkeys recognize the faces of group mates in photographs. *Proc Nat Acad Sci* 106:21539–21543.

Poole R. 1999. *Nation and Identity.* London: Routledge.

Portes A, RG Rumbaut. 2014. *Immigrant America: A Portrait.* 4th ed. Berkeley: University of California Press.

Portugal SJ, et al. 2014. Upwash exploitation and downwash avoidance by flap phasing in ibis formation flight. *Nature* 505:399–402.

Postmes T, et al. 2005. Individuality and social influence in groups: Inductive and deductive routes to group identity. *J Pers Soc Psychol* 89:747–763.

Potts L. 1990. *The World Labour Market: A History of Migration.* London: Zed Books.

Pounder DJ. 1983. Ritual mutilation: Subincision of the penis among Australian Aborigines. *Am J Forensic Med Pathol* 4:227–229.

Powell A, S Shennan, MG Thomas. 2009. Late Pleistocene demography and the appearance of modern human behavior. *Science* 324:1298–1301.

Prentice DA, et al. 1994. Asymmetries in attachments to groups and to their members: Distinguishing between common-identity and common-bond groups. *Pers Soc Psychol Bull* 20:484–493.

Preston SD, FBM de Waal. 2002. The communication of emotions and the possibility of empathy in animals. In SG Post et al., eds. *Altruism and Altruistic Love.* New York: Oxford University Press. pp. 284–308.

Price R. 1996. *Maroon Societies.* Baltimore: Johns Hopkins University Press.

Price TD, O Bar-Yosef. 2010. Traces of inequality at the origins of agriculture in the ancient Near East. In TD Price, G Feinman, eds. *Pathways to Power.* New York: Springer. pp. 147–168.

Prud'Homme J. 1991. Group fission in a semifree-ranging population of Barbary macaques. *Primates* 32:9–22.

Pruetz JD. 2007. Evidence of cave use by savanna chimpanzees at Fongoli, Senegal. *Primates* 48:316–319.

Pruetz JD, et al. 2015. New evidence on the tool-assisted hunting exhibited by chimpanzees in a savannah habitat at Fongoli, Sénégal. *Roy Soc Open Sci* 2:e140507.

Pusey AE, C Packer. 1987. The evolution of sex-biased dispersal in lions. *Behav* 101:275–310.

Queller DC, JE Strassmann. 1998. Kin selection and social insects. *Bio Science* 48:165–175.

Raijman R, et al. 2008. What does a nation owe non-citizens? *Int J Comp Sociol* 49:195–220.

Rakić T, et al. 2011. Blinded by the accent! Minor role of looks in ethnic categorization. *J Pers Soc Psych* 100:16–29.

Rantala MJ. 2007. Evolution of nakedness in *Homo sapiens. J Zool* 273:1–7.

Rasa OAE. 1973. Marking behavior and its social significance in the African dwarf mongoose. *Z Tierpsychol* 32:293–318.

Ratner KG, et al. 2013. Is race erased? Decoding race from patterns of neural activity when skin color is not diagnostic of group boundaries. *Soc Cogn Affect Neurosci* 8:750–755.

Ratnieks FLW, KR Foster, T Wenseleers. 2006. Conflict resolution in social insect societies. *Annu Rev Etomol* 51:581–608.

Ratnieks FLW, T Wenseleers. 2005. Policing insect societies. *Science* 307:54–56.

Rayor LS. 1988. Social organization and space-use in Gunnison's prairie dog. *Behav Ecol Sociobiol* 22:69–78.

Read DW. 2011. *How Culture Makes Us Human.* Walnut Creek, CA: Left Coast Press.

Redmond EM. 1994. *Tribal and Chiefly Warfare in South America.* Ann Arbor: University of Michigan Press.

Reese G, O Lauenstein. 2014. The eurozone crisis: Psychological mechanisms undermining and supporting European solidarity. *Soc Sci* 3:160–171.

Reese HE, et al. 2010. Attention training for reducing spider fear in spider-fearful individuals. *J Anxiety Disord* 24:657–662.

Reicher SD. 2001. The psychology of crowd dynamics. In MA Hogg, RS Tindale, eds. *Blackwell Handbook of Social Psychology: Group Processes.* Oxford, England: Blackwell. pp. 182–207.

Renan E. 1990 (1882). What is a nation? In HK Bhabah, ed. *Nation and Narration.* London: Routledge. pp. 8–22.

Rendell LE, H Whitehead. 2001. Culture in whales and dolphins. *Behav Brain Sci* 24:309–324.

Reynolds H. 1981. *The Other Side of the Frontier: Aboriginal Resistance to the European Invasion of Australia.* Townsville, Australia: James Cook University Press.

Reynolds S. 1997. *Kingdoms and Communities in Western Europe, 900–1300.* Oxford: Oxford University Press.

Rheingold H. 2002. *Smart Mobs: The Next Social Revolution.* New York: Basic Books.

Richerson PJ, R Boyd. 1998. The evolution of human ultra-sociality. In I Eibl-Eibesfeldt, FK Salter, eds. *Indoctrinability, Ideology, and Warfare.* Oxford: Berghahn. pp. 71–95.

Riolo RL, et al. 2001. Evolution of cooperation without reciprocity. *Nature* 414:441–443.

Rivaya-Martínez J. 2012. Becoming Comanches. In DW Adams, C DeLuzio, eds. *On the Borders of Love and Power: Families and Kinship in the Intercultural American Southwest.* Berkeley: University of California Press. pp. 47–70.

Riveros AJ, MA Seid, WT Wcislo. 2012. Evolution of brain size in class-based societies of fungus-growing ants. *Anim Behav* 83:1043–1049.

Rizzolatti G, L Craighero. 2004. The mirror-neuron system. *Annu Rev Neurosci* 27:169–192.

Roberts SGB. 2010. Constraints on social networks. In RIM Dunbar, C Gamble, J Gowlett, eds. *Social Brain, Distributed Mind.* Oxford: Oxford University Press. pp. 115–134.

Robinson WP, H Tajfel. 1996. *Social Groups and Identities: Developing the Legacy of Henri Tajfel.* Oxford: Routledge.

Rodseth L, et al. 1991. The human community as a primate society. *Curr Anthropol* 32:221–241.

Roe FG. 1974. *The Indian and the Horse.* Norman: University of Oklahoma Press.

Rogers AR, D Iltis, S Wooding. 2004. Genetic variation at the MC1R locus and the time since loss of human body hair. *Curr Anthropol* 45:105–108.

Rogers EM. 2003. *Diffusion of Innovations*. 5th ed. New York: Free Press.

Ron T. 1996. Who is responsible for fission in a free-ranging troop of baboons? *Ethology* 102:128–133.

Roosevelt AC. 1999. Archaeology of South American hunters and gatherers. In RB Lee, R Daly, eds. *The Cambridge Encyclopedia of Hunters and Gatherers*. New York: Cambridge University Press. pp. 86–91.

———. 2013. The Amazon and the Anthropocene: 13,000 years of human influence in a tropical rainforest. *Anthropocene* 4:69–87.

Roscoe P. 2006. Fish, game, and the foundations of complexity in forager society. *Cross Cult Res* 40:29–46.

———. 2007. Intelligence, coalitional killing, and the antecedents of war. *Am Anthropol* 109:485–495.

Roth WE, R Etheridge. 1897. *Ethnological Studies Among the North-West-Central Queensland Aborigines*. Brisbane: Edmund Gregory.

Rothì DM, E Lyons, X Chryssochoou. 2005. National attachment and patriotism in a European nation: A British study. *Polit Psychol* 26:135–155.

Rowell TE. 1975. Growing up in a monkey group. *Ethos* 3:113–128.

Royce AP. 1982. *Ethnic Identity: Strategies of Diversity*. Bloomington: Indiana University Press.

Rudolph KP, JP McEntee. 2016. Spoils of war and peace: Enemy adoption and queen-right colony fusion follow costly intraspecific conflict in *acacia* ants. *Behav Ecol* 27:793–802.

De Ruiter J, G Weston, SM Lyon. 2011. Dunbar's number: Group size and brain physiology in humans reexamined. *Am Anthropol* 113:557–568.

Rushdie S. 2002. *Step Across This Line: Collected Nonfiction 1992–2002*. London: Vintage.

Russell AF, et al. 2003. Breeding success in cooperative meerkats. Effects of helper number and maternal state. *Behav Ecol* 14:486–492.

Russell RJ. 1993. *The Lemurs' Legacy*. New York: Tarcher/Putnam.

Rutherford A, et al. 2014. Good fences: The importance of setting boundaries for peaceful coexistence. *PloS ONE* 9: e95660.

Rutledge JP. 2000. They all look alike: The inaccuracy of cross-racial identifications. *Am J Crim L* 28:207–228.

Sahlins M. 1968. Notes on the original affluent society. In RB Lee, I DeVore, eds. *Man the Hunter*. Chicago: Aldine. pp. 85–89.

Sai FZ. 2005. The role of the mother's voice in developing mother's face preference. *Infant Child Dev* 14:29–50.

Salamone FA, CH Swanson. 1979. Identity and ethnicity: Ethnic groups and interactions in a multi-ethnic society. *Ethnic Groups* 2:167–183.

Salmon CA. 1998. The evocative nature of kin terminology in political rhetoric. *Polit Life Sci* 17:51–57.

Saminaden A, S Loughnan, N Haslam. 2010. Afterimages of savages: Implicit associations between primitives, animals and children. *Brit J Soc Psychol* 49:91–105.

Sampson CG. 1988. *Stylistic Boundaries Among Mobile Hunter-Gatherers*. Washington, DC: Smithsonian Institution.

Samuels A, JB Silk, J Altmann. 1987. Continuity and change in dominance relations among female baboons. *Anim Behav* 35:785–793.

Sanderson SK. 1999. *Social Transformations*. New York: Rowman & Littlefield.

Sangrigoli S, S De Schonen. 2004. Recognition of own-race and other-race faces by three-month-old infants. *J Child Psychol Psych* 45:1219–1227.

Sangrigoli S, et al. 2005. Reversibility of the other-race effect in face recognition during childhood. *Psychol Sci* 16:440–444.

Sani F. 2009. Why groups fall apart: A social psychological model of the schismatic process. In F Butera, JM Levine, eds. *Coping with Minority Status.* New York: Cambridge University Press. pp. 243–266.

Sani F, et al. 2007. Perceived collective continuity: Seeing groups as entities that move through time. *Eur J Soc Psychol* 37:1118–1134.

Santorelli CJ, et al. 2013. Individual variation of whinnies reflects differences in membership between spider monkey communities. *Int J Primatol* 34:1172–1189.

Santos-Granero F. 2009. *Vital Enemies: Slavery, Predation, and the Amerindian Political Economy of Life.* Austin: University of Texas Press.

Saperstein A, AM Penner. 2012. Racial fluidity and inequality in the United States. *Am J Sociol* 118:676–727.

Sapolsky RM. 2007. *A Primate's Memoir: A Neuroscientist's Unconventional Life Among the Baboons.* New York: Simon & Schuster.

Sarna JD. 1978. From immigrants to ethnics: Toward a new theory of "ethnicization." *Ethnicity* 5:370–378.

Sassaman KE. 2004. Complex hunter–gatherers in evolution and history: A North American perspective. *J Archaeol Res* 12:227–280.

Sayers K, CO Lovejoy. 2008. The chimpanzee has no clothes: A critical examination of Pan troglodytes in models of human evolution. *Curr Anthropol* 49:87–117.

Scarre C, ed. 2013. *Human Past.* 3rd ed. London: Thames & Hudson.

Schaal B, L Marlier, R Soussignan. 2000. Human foetuses learn odours from their pregnant mother's diet. *Chem Senses* 25:729–737.

Schaller GB. 1972. *The Serengeti Lion.* Chicago: University of Chicago Press.

Schaller M, SL Neuberg. 2012. Danger, disease, and the nature of prejudice. *Adv Exp Soc Psychol* 46:1–54.

Schaller M, JH Park. 2011. The behavioral immune system (and why it matters). *Curr Dir Psychol Sci* 20:99–103.

Schamberg I, et al. 2017. Bonobos use call combinations to facilitate inter-party travel recruitment. *Behav Ecol Sociobiol* 71:75.

Schapera I. 1930. *The Khoisan Peoples of South Africa.* London: Routledge.

Schatz RT, E Staub, H Lavine. 1999. On the varieties of national attachment: Blind versus constructive patriotism. *Polit Psychol* 20:151–174.

Schelling TC. 1978. *Micromotives and Macrobehavior.* New York: W.W. Norton.

Schladt M, ed. 1998. *Language, Identity, and Conceptualization Among the Khoisan.* Cologne: Rüdiger Köppe.

Schmidt JO. 2016. *The Sting of the Wild.* Baltimore: John Hopkins University Press.

Schmitt DP, et al. 2008. Why can't a man be more like a woman? Sex differences in big five personality traits across 55 cultures. *J Pers Soc Psychol* 94:168–182.

Schradin C, J Lamprecht. 2000. Female-biased immigration and male peace-keeping in groups of the shell-dwelling cichlid fish. *Behav Ecol Sociobiol* 48:236–242.

———. 2002. Causes of female emigration in the group-living cichlid fish. *Ethology* 108:237–248.

Schultz TR, et al. 2005. Reciprocal illumination: A comparison of agriculture in humans and in fungus-growing ants. In F Vega, M Blackwell, eds. *Insect-Fungal Associations.* Oxford: Oxford University Press. pp. 149–190.

Schultz TR, SG Brady. 2008. Major evolutionary transitions in ant agriculture. *Proc Nat Acad Sci* 105:5435–5440.

Schwartz B. 1985. *The World of Thought in Ancient China.* Cambridge, MA: Harvard University Press.

Scott JC. 2009. *The Art of Not Being Governed: An Anarchist History of Upland Southeast Asia.* New Haven, CT; Yale University Press.

———. 2017. *Against the Grain: A Deep History of the Earliest States.* New Haven, CT: Yale University Press.

Scott LS, A Monesson. 2009. The origin of biases in face perception. *Psychol Sci* 20:676–680.

Seeley TD. 1995. *The Wisdom of the Hive*. Cambridge, MA: Harvard University Press.

———. 2010. *Honeybee Democracy*. Princeton, NJ: Princeton University Press.

Seger CR, ER Smith, DM Mackie. 2009. Subtle activation of a social categorization triggers group-level emotions. *J Exp Soc Psychol* 45:460–467.

Sekulic D, G Massey, R Hodson. 1994. Who were the Yugoslavs? Failed sources of a common identity in the former Yugoslavia. *Am Sociol Rev* 59:83–97.

Sellas AB, RS Wells, PE Rosel. 2005. Mitochondrial and nuclear DNA analyses reveal fine scale geographic structure in bottlenose dolphins in the Gulf of Mexico. *Conserv Genet* 6:715–728.

Sen A. 2006. *Identity and Violence: The Illusion of Destiny*. New York: W.W. Norton.

Seneca, LA. 1970. *Moral and Political Essays*. JM Cooper, JF Procopé, eds. Cambridge: Cambridge University Press.

Seth PK, S Seth. 1983. Population dynamics of free-ranging rhesus monkeys in different ecological conditions in India. *Am J Primatol* 5:61–67.

Seto MC. 2008. *Pedophilia and Sexual Offending Against Children*. Washington, DC: American Psychological Association.

Seyfarth RM, DL Cheney. 2017. Precursors to language: Social cognition and pragmatic inference in primates. *Psychon Bull Rev* 24:79–84.

———. 2012. The evolutionary origins of friendship. *Annu Rev Psychol* 63:153–177.

Shah JY, PC Brazy, ET Higgins. 2004. Promoting us or preventing them: Regulatory focus and manifestations of intergroup bias. *Pers Soc Psychol Bull* 30:433–446.

Shannon TJ. 2008. *Iroquois Diplomacy on the Early American Frontier*. New York: Penguin.

Sharot T, CW Korn, RJ Dolan. 2011. How unrealistic optimism is maintained in the face of reality. *Nature Neurosci* 14:1475–1479.

Sharpe LL. 2005. Frequency of social play does not affect dispersal partnerships in wild meerkats. *Anim Behav* 70:559–569.

Shennan S. 2001. Demography and cultural innovation: A model and its implications for the emergence of modern human culture. *Cambr Archaeol J* 11:5–16.

Shepher J. 1971. Mate selection among second-generation kibbutz adolescents and adults: Incest avoidance and negative imprinting. *Arch Sexual Behav* 1:293–307.

Sherif M. 1966. *In Common Predicament: Social Psychology of Intergroup Conflict and Cooperation*. Boston: Houghton Mifflin.

Sherif M, et al. 1961. *Intergroup Conflict and Cooperation: The Robbers Cave Experiment*. Norman: University of Oklahoma Book Exchange.

Sherman PW, JUM Jarvis, RD Alexander, eds. 1991. *The Biology of the Naked Mole-rat*. Princeton, NJ: Princeton University Press.

Shirky C. 2008. *Here Comes Everybody! The Power of Organizing Without Organizations*. New York: Penguin.

Sidanius J, et al. 1997. The interface between ethnic and national attachment: Ethnic pluralism or ethnic dominance? *Public Opin Quart* 61:102–133.

Sidanius J, et al. 1999. Peering into the jaws of the beast: The integrative dynamics of social identity, symbolic racism, and social dominance. In DA Prentice, DT Miller, eds. *Cultural Divides: Understanding and Overcoming Group Conflicts*. New York: Russell Sage Foundation. pp. 80–132.

Sidanius J, JR Petrocik. 2001. Communal and national identity in a multiethnic state. In RD Ashmore, L Jussim, D Wilder, eds. *Social Identity, Intergroup Conflict, and Conflict Resolution*. Oxford: Oxford University Press. pp. 101–129.

Silberbauer GB. 1965. *Report to the Government of Bechuanaland on the Bushman Survey*. Gaberones: Bechuanaland Government.

———. 1981. *Hunter and Habitat in the Central Kalahari Desert*. Cambridge: Cambridge University Press.

————. 1996. Neither are your ways my ways. In S Kent, ed. *Cultural Diversity Among Twentieth-Century Foragers.* New York: Cambridge University Press. pp. 21–64.

Silk JB. 1999. Why are infants so attractive to others? The form and function of infant handling in bonnet macaques. *Anim Behav* 57:1021–1032.

————. 2002. Kin selection in primate groups. *Int J Primatol* 23:849–875.

Silk JB, et al. 2013. Chimpanzees share food for many reasons: The role of kinship, reciprocity, social bonds and harassment on food transfers. *Science Direct* 85:941–947.

Silver S, WR Miller. 1997. *American Indian Languages.* Tucson: University of Arizona Press.

Simmel G. 1950. *The Sociology of Georg Simmel.* KH Wolff, ed. Glencoe, IL: Free Press.

Simoons FJ. 1994. *Eat Not This Flesh: Food Avoidances from Prehistory to the Present.* Madison: University of Wisconsin Press.

Slobodchikoff CN, et al. 2012. Size and shape information serve as labels in the alarm calls of Gunnison's prairie dogs. *Curr Zool* 58:741–748.

Slobodchikoff CN, BS Perla, JL Verdolin. 2009. *Prairie Dogs: Communication and Community in an Animal Society.* Cambridge, MA: Harvard University Press.

Slovic P. 2000. *The Perception of Risk.* New York: Earthscan.

Smedley A, BD Smedley. 2005. Race as biology is fiction, racism as a social problem is real. *Am Psychol* 60:16–26.

Smith AD. 1986. *The Ethnic Origins of Nations.* Oxford: Blackwell.

Smith DL. 2009. *The Most Dangerous Animal: Human Nature and the Origins of War.* New York: Macmillan.

————. 2011. *Less Than Human: Why we Demean, Enslave, and Exterminate Others.* New York: St. Martin's Press.

Smith DL, I Panaitiu. 2015. Aping the human essence. In WD Hund, CW Mills, S Sebastiani, eds. *Simianization: Apes, Gender, Class, and Race.* Zurich: Verlag & Wein. pp. 77–104.

Smith DW, et al. 2015. Infanticide in wolves: Seasonality of mortalities and attacks at dens support evolution of territoriality. *J Mammal* 96:1174–1183.

Smith ER, CR Seger, DM Mackie. 2007. Can emotions be truly group-level? Evidence regarding four conceptual criteria. *J Pers Soc Psychol* 93:431–446.

Smith KB, et al. 2011. Linking genetics and political attitudes: Reconceptualizing political ideology. *Polit Psychol* 32:369–397.

Smith ME. 2010. The archaeological study of neighborhoods and districts in ancient cities. *J Anthropol Archaeol* 29:137–154.

Smith PC. 2009. *Everything you know about Indians is Wrong.* Minneapolis Press: University of Minnesota.

Smith RM. 1997. *Civic Ideals: Conflicting Visions of Citizenship in U.S. History.* New Haven, CT: Yale University Press.

Smith RM, PA Klinkner 1999. *The Unsteady March.* Chicago: University of Chicago Press.

Smouse PE, et al. 1981. The impact of random and lineal fission on the genetic divergence of small human groups: A case study among the Yanomama. *Genetics* 98:179–197.

Sorabji R. 2005. *The Philosophy of the Commentators, 200–600 AD: A Sourcebook, Volume 1: Psychology (with Ethics and Religion).* Ithaca, NY: Cornell University Press.

Sorensen AA, TM Busch, SB Vinson. 1985. Control of food influx by temporal subcastes in the fire ant. *Behav Ecol Sociobiol* 17:191–198.

Sorger DM, W Booth, A Wassie Eshete, M Lowman, MW Moffett. 2017. Outnumbered: A new dominant ant species with genetically diverse supercolonies. *Insectes Sociaux* 64:141–147.

Southwick CH, et al. 1974. Xenophobia among free-ranging rhesus groups in India. In RL Holloway, ed. *Primate Aggression, Territoriality, and Xenophobia.* New York: Academic. pp. 185–209.

Spencer C. 2010. Territorial expansion and primary state formation. *Proc Nat Acad Sci* 107:7119–7126.

Spencer WB, FJ Gillen. 1899. *The Native Tribes of Central Australia.* London: MacMillan & Co.

Sperber D. 1974. *Rethinking Symbolism.* Cambridge: Cambridge University Press.

Spicer EH. 1971. Persistent cultural systems. *Science* 174:795–800.

Spickard P. 2005. Race and nation, identity and power: Thinking comparatively about ethnic systems. In P Spickard, ed. *Race and Nation: Ethnic Systems in the Modern World.* New York: Taylor & Francis. pp. 1–29.

Spoor JR, JR Kelly. 2004. The evolutionary significance of affect in groups. *Group Proc Intergr Rel* 7:398–412.

Spoor JR, KD Williams. 2007. The evolution of an ostracism detection system. In JP Forgas et al., eds. *Evolution and the Social Mind.* New York: Psychology Press. pp. 279–292.

Sridhar H, G Beauchamp, K Shankar. 2009. Why do birds participate in mixed-species foraging flocks? *Science Direct* 78:337–347.

Stahler DR, DW Smith, R Landis. 2002. The acceptance of a new breeding male into a wild wolf pack. *Can J Zool* 80:360–365.

Stanner WEH. 1965. Aboriginal territorial organization. *Oceania* 36:1–26.

———. 1979. *White Man Got No Dreaming: Essays 1938–78.* Canberra: Australian National University Press.

Staub E. 1989. *The Roots of Evil.* Cambridge: Cambridge University Press.

———. 1997. Blind versus constructive patriotism. In D Bar-Tal, E Staub, eds. *Patriotism in the Lives of Individuals and Nations.* Chicago: Nelson-Hall. pp. 213–228.

Steele C, J Gamble. 1999. Hominid ranging patterns and dietary strategies. In H Ullrich, ed. *Hominid Evolution: Lifestyles and Survival Strategies.* Gelsenkirchen: Edition Archaea. pp. 369–409.

Steele CM, SJ Spencer, J Aronson. 2002. Contending with group image: The psychology of stereotype and social identity threat. *Adv Exp Soc Psychol* 34:379–440.

Steffian AF, PG Saltonstall. 2001. Markers of identity. Labrets and social organization in the Kodiak Archipelago. *Alaskan J Anthropol* 1:1–27.

Stiner MC, SL Kuhn. 2006. Changes in the "connectedness" and resilience of Paleolithic societies in Mediterranean ecosystems. *Hum Ecol* 34:693–712.

Stirling MW. 1938. *Historical and Ethnographical Material on the Jivaro Indians.* Washington, DC: Smithsonian Institution.

Stoneking M. 2009. Widespread prehistoric human cannibalism: Easier to swallow? *Trends Ecol Evol* 10,489–490.

Strandburg-Peshkin A, et al. 2015. Shared decision-making drives collective movement in wild baboons. *Science* 348:1358–1361.

Struch N, SH Schwartz. 1989. Intergroup aggression: Its predictors and distinctness from in-group bias. *J Pers Soc Psychol* 56:364–373.

Struhsaker TT. 2010. *The Red Colobus Monkeys.* New York: Oxford University Press.

Sturgis J, DM Gordon. 2012. Nestmate recognition in ants. *Myrmecol News* 16:101–110.

Suarez AV, et al. 2002. Spatiotemporal patterns of intraspecific aggression in the invasive Argentine ant. *Anim Behav* 64:697–708.

Suetonius 1979 (written AD 121). *The Twelve Caesars.* M. Graves, trans. London: Penguin.

Sueur C, et al. 2011. Group size, grooming and fission in primates: A modeling approach based on group structure. *J Theor Biol* 273:156–166.

Sugita Y. 2008. Face perception in monkeys reared with no exposure to faces. *Proc Nat Acad Sci* 105:394–398.

Sugiyama Y. 1999. Socioecological factors of male chimpanzee migration at Bossou, Guinea. *Primates* 40:61–68.

Sumner WG. 1906. *Folkways: The Study of the Sociological Importance of Usages, Manners, Customs, Mores, and Morals.* Boston: Ginn & Co.

Surbeck M, G Hohmann. 2008. Primate hunting by bonobos at LuiKotale, Salonga National Park. *Curr Biol* 18:R906–R907.

Swann WB Jr, et al. 2012. When group membership gets personal: A theory of identity fusion. *Psychol Rev* 119:441–456.

Taglialatela JP, et al. 2009. Visualizing vocal perception in the chimpanzee brain. *Cereb Cortex* 19:1151–1157.

Tainter JA. 1988. *The Collapse of Complex Societies*. Cambridge: Cambridge University Press.

Tajfel H, et al. 1970. The development of children's preference for their own country: A cross national study. *Int J Psychol* 5:245–253.

Tajfel H, JC Turner. 1979. An integrative theory of intergroup conflict. In W Austin, S Worchel, eds. *The Social Psychology of Intergroup Relations*. Monterey, CA: Brooks/Cole. pp. 33–47.

Takahashi H, et al. 2014. Different impressions of other agents obtained through social interaction uniquely modulate dorsal and ventral pathway activities in the social human brain. *Science Direct* 58:289–300.

Tan J, B Hare. 2013. Bonobos share with strangers. *PLoS ONE* 8:e51922.

Tanaka J. 1980. *The San, Hunter-Gatherers of the Kalahari*. Tokyo: University of Tokyo Press.

Tarr B, J Launay, RIM Dunbar. 2016. Silent disco: Dancing in synchrony leads to elevated pain thresholds and social closeness. *Evol Hum Behav* 37:343–349.

Taylor AM. 2005. *The Divided Family in Civil War America*. Chapel Hill: University of North Carolina Press.

Taylor CB, JM Ferguson, BM Wermuth. 1977. Simple techniques to treat medical phobias. *Postgrad Med J* 53:28–32.

Taylor R. 2008. The polemics of making fire in Tasmania. *Aboriginal Hist* 32:1–26.

Tebbich S, R Bshary. 2004. Cognitive abilities related to tool use in the woodpecker finch. *Anim Behav* 67:689–697.

Tennie C, J Call, M Tomasello. 2009. Ratcheting up the ratchet: On the evolution of cumulative culture. *Philos T Roy Soc B* 364:2405–2415.

Terry DJ, CJ Carey, VJ Callan. 2001. Employee adjustment to an organizational merger: An intergroup perspective. *Pers Soc Psychol Bull* 27:267–280.

Testart A. 1982. Significance of food storage among hunter-gatherers. *Curr Anthropol* 23:523–530.

Testi A. 2005. You Americans aren't the only people obsessed with your flag. *Zócalo*. http://www.zocalopublicsquare.org/2015/06/12/you-americans-arent-the-only-people-obsessed-with-your-flag/ideas/nexus/.

———. 2010. *Capture the Flag*. NG Mazhar, trans. New York: New York University Press.

Thaler RH, CR Sunstein. 2009. *Nudge: Improving Decisions about Health, Wealth, and Happiness*. New York: Penguin.

Theberge J, M Theberge. 1998. *Wolf Country: Eleven Years Tracking the Algonquin Wolves*. Toronto: McClelland & Stewart.

Thierry B. 2005. Hair grows to be cut. *Evol Anthropol: Issues, News, and Reviews* 14:5.

Thomas EM. 1959. *The Harmless People*. New York: Alfred A. Knopf.

Thomas-Symonds N. 2010. *Attlee: A Life in Politics*. New York: IB Tauris.

Thompson B. 1983. Social ties and ethnic settlement patterns. In WC McCready, ed. *Culture, Ethnicity, and Identity*. New York: Academic Press. pp. 341–360.

Thompson PR. 1975. A cross-species analysis of carnivore, primate, and hominid behavior. *J Human Evol* 4:113–124.

Thornton A, J Samson, T Clutton-Brock. 2010. Multi-generational persistence of traditions in neighbouring meerkat groups. *Proc Roy Soc B* 277:3623–3629.

Tibbetts EA, J Dale. 2007. Individual recognition: It is good to be different. *Trends Ecol Evol* 22:529–537.

Tilly C. 1975. Reflections on the history of European state-making. In C Tilly, ed. *The Formation of National States in Western Europe*. Princeton, NJ: Princeton University Press. pp. 3–83.

Tincoff R, PW Jusczyk. 1999. Some beginnings of word comprehension in 6-month-olds. *Psychol Sci* 10:172–175.

Tindale RS, S Sheffey. 2002. Shared information, cognitive load, and group memory. *Group Process Intergr Relat* 5:5–18.

Tishkoff SA, et al. 2007. Convergent adaptation of human lactase persistence in Africa and Europe. *Nat Genet* 39:31–40.

Tobler R, et al. 2017. Aboriginal mitogenomes reveal 50,000 years of regionalism in Australia. *Nature* 544:180–184.

Todorov A. 2017. *Face Value: The Irresitable Influence of First Impressions.* Princeton, NJ: Princeton University Press.

Toft MD. 2003. *The Geography of Ethnic Violence.* Princeton, NJ: Princeton University Press.

Tomasello M. 2011. Human culture in evolutionary perspective. In MJ Gelfand et al., eds. *Advances in Culture and Psychology*, vol 1. New York: Oxford University Press. pp. 5–51.

———. 2014. *A Natural History of Human Thinking.* Cambridge, MA: Harvard University Press.

Tomasello M, et al. 2005. Understanding and sharing intentions: The origins of cultural cognition. *Behav Brain* Sci 28:675–673.

Tonkinson R. 1987. Mardujarra kinship. In DJ Mulvaney, JP White, eds. *Australia to 1788.* Broadway, NSW: Fairfax, Syme, Weldon. pp. 197–220.

——— 2002. *The Mardu Aborigines: Living the Dream in Australia's Desert.* 2nd ed. Belmont, CA: Wadsworth.

———. 2011. Landscape, transformations, and immutability in an Aboriginal Australian culture. *Cult Memories* 4:329–345.

Torres CW, ND Tsutsui. 2010. The effect of social parasitism by *Polyergus breviceps* on the nestmate recognition system of its host, *Formica altipetens. PloS ONE* 11:e0147498.

Townsend JB. 1983. Pre-contact political organization and slavery in Aleut society. In E Tooker, ed. *The Development of Political Organization in Native North America.* Philadelphia: Proceedings of the American Ethnological Society. pp. 120–132.

Townsend SW, LI Hollén, MB Manser. 2010. Meerkat close calls encode group-specific signatures, but receivers fail to discriminate. *Anim Behav* 80:133–138.

Trinkaus E, et al. 2014. *The people of Sunghir.* Oxford: Oxford University Press.

Tschinkel WR. 2006. *The Fire Ants.* Cambridge, MA, Harvard University Press.

Tsutsui ND. 2004. Scents of self: The expression component of self/non-self recognition systems. In *Ann Zool Fenn.* Finnish Zoological and Botanical Publishing Board. pp. 713–727.

Turchin P. 2015. *Ultrasociety: How 10,000 Years of War Made Humans the Greatest Cooperators on Earth.* Chaplin, CT: Beresta Books.

Turchin P, et al. 2013. War, space and the evolution of Old World complex societies. *Proc Nat Acad Sci* 110:16384–16389.

Turchin P, S Gavrilets. 2009. Evolution of complex hierarchical societies. *Soc Evol Hist* 8:167–198.

Turnbull CM. 1965. *Wayward Servants.* London: Eyre and Spottiswoode.

———. 1972. *The Mountain People.* London: Cape.

Turner JC. 1981. The experimental social psychology of intergroup behavior. In JC Turner, H Giles, eds. *Intergroup Behavior.* Oxford: Blackwell. pp. 66–101.

Turner JH, RS Machalek. 2018. *The New Evolutionary Sociology.* New York: Routledge.

Turner TS. 2012. The social skin. *J Ethnog Theory* 2:486–504.

Tuzin D. 2001. *Social Complexity in the Making: A Case Study Among the Arapesh of New Guinea.* London: Routledge.

Tyler TR. 2006. Psychological perspectives on legitimacy and legitimation. *Annu Rev Psychol* 57:375–400.

Uehara E. 1990. Dual exchange theory, social networks, and informal social support. *Am J Sociol* 96:521–57.

Vaes J, MP Paladino. 2010. The uniquely human content of stereotypes. *Process Intergr Relat* 13:23–30.

Vaes J, et al. 2012. We are human, they are not: Driving forces behind outgroup dehumanisation and the humanisation of the ingroup. *Eur Rev Soc Psychol* 23:64–106.

Vaesen K, et al. 2016. Population size does not explain past changes in cultural complexity. *Proc Nat Acad Sci* 113:E2241–E2247.

Valdesolo P, J Ouyang, D DeSteno. 2010. The rhythm of joint action: Synchrony promotes cooperative ability. *J Exp Soc Psychol* 46:693–695.

Van den Berghe PL. 1981. *The Ethnic Phenomenon*. New York: Elsevier.

van der Dennen JMG. 1987. Ethnocentrism and in-group/out-group differentiation. In V Reynolds et al., eds. *The Sociobiology of Ethnocentrism*. London: Croom Helm. pp. 1–47.

———. 1991. Studies of conflict. In M Maxwell, ed. *The Sociobiological Imagination*. Albany: State University of New York Press. pp. 223–241.

———. 1999. Of badges, bonds, and boundaries: In-group/out-group differentiation and ethnocentrism revisited. In K Thienpont, R Cliquet, eds. *In-group/Outgroup Behavior in Modern Societies*. Amsterdam: Vlaamse Gemeeschap/CBGS. pp. 37–74.

———. 2014. Peace and war in nonstate societies: An anatomy of the literature in anthropology and political science. *Common Knowledge* 20:419–489.

Van der Toorn J, et al. 2014. My country, right or wrong: Does activating system justification motivation eliminate the liberal-conservative gap in patriotism? *J Exp Soc Psychol* 54:50–60.

van de Waal E, C Borgeaud, A Whiten. 2013. Potent social learning and conformity shape a wild primate's foraging decisions. *Science* 340:483–485.

van Dijk RE, et al. 2013. The thermoregulatory benefits of the communal nest of sociable weavers *Philetairus socius* are spatially structured within nests. *J Avian Biol* 44:102–110.

van Dijk RE, et al. 2014. Cooperative investment in public goods is kin directed in communal nests of social birds. *Ecol Lett* 17:1141–1148.

Vanhaeren M, F d'Errico. 2006. Aurignacian ethno-linguistic geography of Europe revealed by personal ornaments. *J Archaeol Sci* 33:1105–1128.

Van Horn RC, et al. 2007. Divided destinies: Group choice by female savannah baboons during social group fission. *Behav Ecol Sociobiol* 61:1823–1837.

Van Knippenberg D. 2011. Embodying who we are: Leader group prototypicality and leadership effectiveness. *Leadership Quart* 22:1078–1091.

Van Meter PE. 2009. Hormones, stress and aggression in the spotted hyena. Ph.D. diss. in Zoology. Michigan State University, East Lansing, MI.

Van Vugt M, A Ahuja. 2011. *Naturally Selected: The Evolutionary Science of Leadership*. New York: HarperCollins.

Van Vugt M, CM Hart. 2004. Social identity as social glue: The origins of group loyalty. *J Pers Soc Psychol* 86:585–598.

Van Vugt M, R Hogan, RB Kaiser. 2008. Leadership, followership, and evolution: Some lessons from the past. *Am Psychol* 63:182–196.

Vasquez JA, ed. 2012. *What do We Know About War?* 2nd ed. Lanham, Maryland: Rowman & Littlefield.

Vecoli RJ. 1978. The coming of age of the Italian Americans 1945–1974. *Ethnic Racial Stud* 8:134–158.

Verdolin JL, AL Traud, RR Dunn. 2014. Key players and hierarchical organization of prairie dog social networks. *Ecol Complex* 19:140–147.

Verdu P, et al. 2010. Limited dispersal in mobile hunter–gatherer Baka Pygmies. *Biol Lett* 6:858–861.

Viki GT, R Calitri. 2008. Infrahuman outgroup or suprahuman group: The role of nationalism and patriotism in the infrahumanization of outgroups. *Eur J Soc Psychol* 38:1054–1061.

Viki GT, D Osgood, S Phillips. 2013. Dehumanization and self-reported proclivity to torture prisoners of war. *J Exp Soc Psychol* 49:325–328.

Villa P. 1983. *Terra Amata and the Middle Pleistocene Archaeological Record of Southern France.* Berkeley: University of California Publications in Anthropology 13.

Visscher PK. 2007. Group decision making in nest-site selection among social insects. *Annu Rev Entomol* 52:255–275.

Volpp L. 2001. The citizen and the terrorist. *UCLA Law Rev* 49:1575–1600.

Voltaire. 1901. *A Philosophical Dictionary*, vol. 4. Paris: ER Dumont.

Vonholdt BM, et al. 2008. The genealogy and genetic viability of reintroduced Yellowstone grey wolves. *Mol Ecol* 17:252–274.

Voorpostel M. 2013. Just like family: Fictive kin relationships in the Netherlands. *J Gerontol B Psychol Sci Soc Sci* 68:816–824.

Wadley L. 2001. What is cultural modernity? A general view and a South African perspective from Rose Cottage Cave. *Cambr Archaeol J* 11:201–221.

Walford E, J Gillies. 1853. *The Politics and Economics of Aristotle.* London: HG Bohn.

Walker RH, et al. 2017. Sneeze to leave: African wild dogs use variable quorum thresholds facilitated by sneezes in collective decisions. *Proc R Soc B* 284:20170347.

Walker RS. 2014. Amazonian horticulturalists live in larger, more related groups than hunter–gatherers. *Evol Hum Behav* 35:384–388.

Walker RS, et al. 2011. Evolutionary history of hunter-gatherer marriage practices. *PLoS ONE* 6:e19066.

Walker RS, KR Hill. 2014. Causes, consequences, and kin bias of human group fissions. *Hum Nature* 25.165–175.

Wallensteen P. 2012. Future directions in the scientific study of peace and war. In JA Vasquez, ed. *What Do We Know About War?* 2nd ed. Lanham, MD: Rowman & Littlefield. pp. 257–270.

Walter R, I Smith, C Jacomb. 2006. Sedentism, subsistence and socio-political organization in prehistoric New Zealand. *World Archaeol* 38:274–290.

Wang Y, S Quadflieg. 2015. In our own image? Emotional and neural processing differences when observing human-human vs human-robot interactions. *Soc Cogn Affect Neurosci* 10:1515–1524.

Ward RH. 1972. The genetic structure of a tribal population, the Yanomama Indians V. Comparisons of a series of genetic networks. *Ann Hum Genet* 36:21–43.

Warnecke AM, RD Masters, G Kempter. 1992. The roots of nationalism: Nonverbal behavior and xenophobia. *Ethol Sociobiol* 13:267–282.

Watanabe S, J Sakamoto, M Wakita. 1995. Pigeons discrimination of paintings by Monet and Picasso. *J Exp Anal Behav* 63:165–174.

Watson RD Jr. 2008. *Normans and Saxons: Southern Race Mythology and the Intellectual History of the American Civil War.* Baton Rouge: Louisiana State University Press.

Watson-Jones RE, et al. 2014. Task-specific effects of ostracism on imitative fidelity in early childhood. *Evol Hum Behav* 35:204–210.

Watts DP, et al. 2000. Redirection, consolation, and male policing. In F Aureli, FBM de Waal, eds. *Natural Conflict Resolution.* Berkeley: University of California Press. pp. 281–301.

Watts DP, et al. 2006. Lethal intergroup aggression by chimpanzees in Kibale National Park, Uganda. *Am J Primatol* 68:161–180.

Waytz A, N Epley, JT Cacioppo. 2010. Social cognition unbound: Insights into anthropomorphism and dehumanization. *Curr Dir Psychol Sci* 19:58–62.

Weber E. 1976. *Peasants into Frenchmen: The Modernization of Rural France 1870–1914.* Stanford, CA: Stanford University Press.

Weddle RS. 1985. *Spanish Sea: The Gulf of Mexico in North American Discovery 1500–1685.* College Station: Texas A & M University Press.

Wegner DM. 1994. Ironic processes of mental control. *Psychol Rev* 101:34–52.

Weinstein EA. 1957. Development of the concept of flag and the sense of national identity. *Child Dev* 28:167–174.

Weisler MI. 1995. Henderson Island prehistory: Colonization and extinction on a remote Polynesian island. *Biol J Linn Soc* 56:377–404.

Weitz MA. 2008. *More Damning than Slaughter: Desertion in the Confederate Army*. Lincoln: University of Nebraska Press.

Wells RS. 2003. Dolphin social complexity. In FBM de Waal, PL Tyack, eds. *Animal Social Complexity: Intelligence, Culture and Individualized Societies*. Cambridge, MA: Harvard University Press. pp. 32–56.

Wendorf F. 1968. Site 117: A Nubian final paleolithic graveyard near Jebel Sahaba, Sudan. In F Wendorf, ed. *The Prehistory of Nubia*. Dallas: Southern Methodist University Press. pp. 954–1040.

West SA, I Pen, AS Griffin. 2002. Cooperation and competition between relatives. *Science* 296:72–75.

Weston K. 1991. *Families We Choose: Lesbians, Gays, Kinship*. New York: Columbia University Press.

Whallon R. 2006. Social networks and information: non-"utilitarian" mobility among hunter-gatherers. *J Anthropol Archaeol* 25:259–270.

Wheeler A, et al. 2011. Caucasian infants scan own- and other-race faces differently. *PloS ONE* 6: e18621.

White HT. 2011. *Burma*. Cambridge: Cambridge University Press.

Whitehead NL. 1992. Tribes make states and states make tribes: Warfare and the creation of colonial tribes and states in northeastern South America. In RB Ferguson, NL Whitehead, eds. *War in the Tribal Zone*. Santa Fe, NM: SAR Press. pp. 127–150.

Whitehouse H, et al. 2014a. The ties that bind us: Ritual, fusion, and identification. *Curr Anthropol* 55:674–695.

Whitehouse H, et al. 2014b. Brothers in arms: Libyan revolutionaries bond like family. *Proc Nat Acad Sci* 111:17783–17785.

Whitehouse H, RN McCauley. 2005. *Mind and Religion: Psychological and Cognitive Foundations of Religiosity*. Walnut Creek, CA: Altamira.

Whiten A. 2011. The scope of culture in chimpanzees, humans and ancestral apes. *Philos T Roy Soc B* 366:997–1007.

Widdig A, et al. 2006. Consequences of group fission for the patterns of relatedness among rhesus macaques. *Mol Ecol* 15:3825–3832.

Wiessner PW. 1977. Hxaro: A regional system of reciprocity for reducing risk among the !Kung San. Ph.D. diss. University of Michigan, Ann Arbor, MI.

———. 1982. Risk, reciprocity and social influences on !Kung San economics. In E Leacock, R Lee, eds. *Politics and History in Band Societies*. Cambridge: Cambridge University Press. pp. 61–84.

———. 1983. Style and social information in Kalahari San projectile points. *Am Antiquity* 48:253–276.

———. 1984. Reconsidering the behavioral basis for style: A case study among the Kalahari San. *J Anthropol Arch* 3:190–234.

———. 2002. Hunting, healing, and *hxaro* exchange: A long-term perspective on !Kung (Ju/'hoansi) large-game hunting. *Evol Hum Behav* 23:407–436.

———. 2014. Embers of society: Firelight talk among the Ju/'hoansi Bushmen. *Proc Nat Acad Sci* 111:14027–14035.

Wiessner P, A Tumu. 1998. *Historical Vines: Enga Networks of Exchange, Ritual and Warfare in Papua New Guinea*. Washington, DC: Smithsonian Institution Press.

Wildschut T, et al. 2003. Beyond the group mind: A quantitative review of the interindividual-intergroup discontinuity effect. *Psychol Bull* 129:698–722.

Wilkinson GS, JW Boughman. 1998. Social calls coordinate foraging in greater spearnosed bats. *Anim Behav* 55:337–350.

Willer R, K Kuwabara, MW Macy. 2009. The false enforcement of unpopular norms. *Am J Sociol* 115:451–490.

Williams J. 2006. *Clam Gardens: Aboriginal Mariculture on Canada's West Coast.* Point Roberts, WA: New Star Books.

Williams JM, et al. 2004. Why do male chimpanzees defend a group range? *Anim Behav* 68:523–532.

Williams JM, et al. 2008. Causes of death in the Kasakela chimpanzees of Gombe National Park, Tanzania. *Am J Primatol* 70:766–777.

Wilshusen RH, JM Potter. 2010. The emergence of early villages in the American Southwest: Cultural issues and historical perspectives. In MS Bandy, JR Fox, eds. *Becoming Villagers: Comparing Early Village Societies.* Tucson: University of Arizona Press. pp. 165–183.

Wilson DS, EO Wilson. 2008. Evolution "for the Good of the Group." *Am Sci* 96:380–389.

Wilson EO. 1975. *Sociobiology: The New Synthesis.* Cambridge, MA: Harvard University Press.

———. 1978. *On Human Nature.* Cambridge, MA: Harvard University Press.

———. 1980. Caste and division of labor in leaf-cutter ants. I. The overall pattern in *A. sexdens. Behav Ecol Sociobiol* 7:143–156.

———. 2012. *Social Conquest of Earth.* New York: W.W. Norton.

Wilson ML, et al. 2014. Lethal aggression in *Pan* is better explained by adaptive strategies than human impacts. *Nature* 513:414–417.

Wilson ML, M Hauser, R Wrangham. 2001. Does participation in intergroup conflict depend on numerical assessment, range location or rank in wild chimps? *Anim Behav* 61:1203–1216.

Wilson ML, WR Wallauer, AE Pusey. 2004. New cases of intergroup violence among chimpanzees in Gombe National Park, Tanzania. *Int J Primatol* 25:523–549.

Wilson ML, RW Wrangham. 2003. Intergroup relations in chimpanzees. *Annu Rev Anthropol* 32:363–392.

Wilson TD. 2002. *Strangers to Ourselves: Discovering the Adaptive Unconscious.* Cambridge, MA: Harvard University Press.

Wimmer A, B Min. 2006. From empire to nation-state: Explaining wars in the modern world, 1816–2001. *Am Sociol Rev* 71:867–897.

Winston ML, GW Otis. 1978. Ages of bees in swarms and afterswarms of the Africanized honeybee. *J Apic Res* 17:123–129.

Witkin HA, JW Berry. 1975. Psychological differentiation in cross-cultural perspective. *J Cross Cult Psychol* 6:4–87.

Wittemyer G, et al. 2007. Social dominance, seasonal movements, and spatial segregation in African elephants. *Behav Ecol Sociobiol* 61:1919–1931.

Wittemyer G, et al. 2009. Where sociality and relatedness diverge: The genetic basis for hierarchical social organization in African elephants. *Proc Roy Soc B* 276:3513–3521.

Wobst HM. 1974. Boundary conditions for Paleolithic social systems. *Am Antiquity* 39:147–178.

———. 1977. Stylistic behavior and information exchange. In CE Cleland, ed. *Research Essays in Honor of James B. Griffin.* Ann Arbor, MI: Museum of Anthropology. pp. 317–342.

Wohl MJA, MJ Hornsey, CR Philpot. 2011. A critical review of official public apologies: Aims, pitfalls, and a staircase model of effectiveness. *Soc Issues Policy Rev* 5:70–100.

Wohl MJA, et al. 2012. Why group apologies succeed and fail: Intergroup forgiveness and the role of primary and secondary emotions. *J Pers Soc Psychol* 102:306–322.

Wolsko C, B Park, CM Judd. 2006. Considering the tower of Babel. *Soc Justice Res* 19:277–306.

Womack M. 2005. *Symbols and Meaning: A Concise Introduction.* Walnut Creek CA: Altamira.

Wood G. 2015. What ISIS really wants. *The Atlantic* 315:78–94.

Woodburn J. 1982. Social dimensions of death in four African hunting and gathering societies. In M Bloch, J Parry, eds. *Death and the Regeneration of Life*. Cambridge: Cambridge University Press. pp. 187–210.

Word CO, MP Zanna, J Cooper. 1974. The nonverbal mediation of self-fulfilling prophecies in interracial interaction. *J Exp Soc Psychol* 10:109–120.

Wrangham RW. 1977. Feeding behaviour of chimpanzees in Gombe National Park, Tanzania. In TH Clutton-Brock, ed. *Primate Ecology*. London: Academic Press. pp. 504–538.

———. 1999. Evolution of coalitionary killing. *Am J Phys Anthropol* 110:1–30.

———. 2009. *Catching Fire: How Cooking Made Us Human*. New York: Basic Books.

———. 2014. Ecology and social relationships in two species of chimpanzee. In DI Rubenstein, RW Wrangham, eds. *Ecological Aspects of Social Evolution: Birds and Mammals*. Princeton, NJ: Princeton University Press. pp. 352–378.

———. 2019. *The Goodness Paradox: The Strange Relationship between Virtue and Violence in Human Evoution*. New York: Pantheon Books.

Wrangham RW, L Glowacki. 2012. Intergroup aggression in chimpanzees and war in nomadic hunter-gatherers. *Hum Nature* 23:5–29.

Wrangham RW, D Peterson. 1996. *Demonic Males: Apes and the Origins of Human Violence*. New York: Houghton Mifflin Harcourt.

Wrangham RW, et al. 2016. Distribution of a chimpanzee social custom is explained by matrilineal relationship rather than conformity. *Curr Biol* 26:3033–3037.

Wrangham RW, ML Wilson, MN Muller. 2006. Comparative rates of aggression in chimpanzees and humans. *Primates* 47:14–26.

Wray MK, et al. 2011. Collective personalities in honeybee colonies are linked to fitness. *Anim Behav* 81:559–568.

Wright R. 2004. *A Short History of Progress*. New York: Carroll & Graf.

Wright SC, et al. 1997. The extended contact effect: Knowledge of cross-group friendsips and prejudice. *J Pers Soc Psychol* 73:73–90.

Wu F. 2002. *Yellow: Race in America Beyond Black and White*. New York: Basic Books.

Wurgaft BA. 2006. Incensed: Food smells and ethnic tension. *Gastronomica* 6:57–60.

Yates FA. 1966. *The Art of Memory*. London: Routledge & Kegan Paul.

Yellen J, H Harpending. 1972. Hunter-gatherer populations and archaeological inference. *World Archaeol* 4:244–253.

Yoffee N. 1995. Collapse of ancient Mesopotamian states and civilization. In N Yoffee, G Cowgill, eds. *Collapse of Ancient States and Civilizations*. Tucson: University of Arizona. pp 44–68.

Yogeeswaran K, N Dasgupta. 2010. Will the "real" American please stand up? The effect of implicit national prototypes on discriminatory behavior and judgments. *Pers Soc Psychol Bull* 36:1332–1345.

Yonezawa M. 2005. Memories of Japanese identity and racial hierarchy. In P Spickard, ed. *Race and Nation*. New York: Routledge. pp. 115–132.

Zajonc R, et al. 1987. Convergence in the physical appearance of spouses. *Motiv Emot* 11:335–346.

Zayan R, J Vauclair. 1998. Categories as paradigms for comparative cognition. *Behav Proc* 42:87–99.

Zeder MA, et al., eds. 2006. *Documenting Domestication: New Genetic and Archaeological Paradigms*. Berkeley: University of California Press.

Zentall TR. 2015. Intelligence in non-primates. In S Goldstein, D Princiotta, JA Naglieri, eds. *Handbook of Intelligence*. New York: Springer. pp. 11–25.

Zerubavel E. 2003. *Time Maps: Collective Memory and the Social Shape of the Past*. Chicago: University of Chicago Press.

Zick A, et al. 2008. The syndrome of group-focused enmity. *J Soc Issues* 64:363–383.

Zinn H. 2005. *A People's History of the United States*. New York: HarperCollins.

Zvelebil M, M Lillie. 2000. The transition to agriculture in eastern Europe. In TD Price, ed. *Europe's First Farmers*. Cambridge: Cambridge University Press. pp. 57–92.

Index

ants (*continued*)
  breeding and society lifecycle, 75–77,
    245, 246, 247, 268
  conflict among, 67–68, 78, 223–224
  collective decision-making of, 119
  division of labor for, 60, 62–64, 65, 133
  flexibility and adaptability of, 70–71
  housing and infrastructure of, 58,
    60–61
  humans contrasted/compared with, 6,
    53, 57–58, 60–65, 80, 93, 101, 142,
    193, 356
  individual recognition absent in,
    68–69, 372
  resource provision for, 61–62, 126
  in Sateré-Mawé rituals, 199
  as sisterhoods, 62
  slavemaking, 70–71, 284, 287
  societal markers for, 66–67, 69–71, 250,
    344–345, 374
  societies as superorganisms, 89
  society size for, 57–58, 60–65, 78, 247,
    372
  spiders insertion into colony of, 71
  See also Argentine ants; army ants;
    harvester ants; leafcutting ants;
    trap-jaw ants
apartheid, 318
Arawak tribe, 23–24, 189, 229–230
archaeological evidence for societies,
  141–147, 153–156
Argentine ants, 6, 64–71, 75–78, 212, 218,
  247, 268, 276, 380
  societal markers for, 66–67, 69–70, 85,
    87, 250, 344–345
  species other than with supercolonies,
    372
army ants, 119, 224, 245, 394
Ash conformity, 386
assimilation, 148, 314–318, 322–323, 324,
  326, 328, 329, 395, 397
  as never carried to completion, 316,
    317, 326, 334

baboons, savanna, 32, 33, 39–40, 43, 46,
  50, 124, 191, 206–207, 226, 245, 289,
  320–321, 369, 376, 387
band (a society of horses). *See* horses
band (an encampment of hunter-
    gatherers), 97–98, 100–103, 105,
    111, 112–114, 116, 120, 195, 207,
    220, 248, 279, 280, 324, 374, 375
  being part of a larger society,103–104
  chimp or bonobo parties compared
    to, 102

evolutionary origins of, 102
size and dynamics of, 100, 101,
  120–121, 128, 220, 248
social organization of, 112–121
villages of some tribes comparable to,
  279–280, 394
band societies, hunter-gatherer
  adaptability of, 101, 252
  as an ancient condition for human
    ancestors, 136, 353
  as anonymous societies, 108
  as affluent, 376
  agricultural practices impacts on,
    98–99, 137–138
  alliances among, 104, 105–107,
    229–232, 234, 236, 237, 283, 350
  alternative names for, 374
  archaeological evidence of, 143–146
  boundaries and delineations of,
    103–106
  communication across, 42
  conflict and violence between,
    130–131, 220–222
  cosmologies of, 166
  decision-making in, 119–120
  definition of, 97–98, 374
  division of labor in, 101, 112–117, 132
  egalitarianism in, 117–119, 121
  ethnicity and race for, 106–108, 168
  European contact with and impact on,
    99–100, 106, 355
  families/kinship in, 207, 209–210
  freedom in, 88, 115–116
  gathering together of the bands of,
    124–125
  Göbekli Tepe, 143–144
  "imagined communities" of, 18
  individualism in, 115–116, 117
  jack-of-all-trades strategy of, 114–117,
    132
  lack of materialism in, 113, 135
  leisure time for, 114, 135, 286, 376
  minimal goods and infrastructure in,
    113
  novelties and fads adoption by,
    254–255
  oral narrative function for, 180–181
  peace within and between, 105–106,
    229–237
  power and leadership in, 117–119,
    121, 123, 129, 130–131, 278,
    321–322, 378
  protection and security for, 101,
    120–121
  raids between, 221–222, 296

LARA HEIMERT

Mark W. Moffett is a tropical biologist and research associate at the Smithsonian and a visiting scholar in the Department of Human Evolutionary Biology at Harvard University. Called a "daring eco-adventurer" by Margaret Atwood, and recipient of the Lowell Thomas Medal for exploration, Dr. Moffett's writing has appeared in *The Best American Science and Nature Writing* and he has been a guest on *The Colbert Report, Conan,* NPR's *Fresh Air,* and CBS *Sunday Morning.* This is his fourth book.